AF598250

VOLUME FIVE HUNDRED AND THIRTY EIGHT

METHODS IN ENZYMOLOGY

Methods of Adipose Tissue Biology, Part B

METHODS IN ENZYMOLOGY

Editors-in-Chief

JOHN N. ABELSON and MELVIN I. SIMON
Division of Biology
California Institute of Technology
Pasadena, California

ANNA MARIE PYLE
Departments of Molecular, Cellular and Developmental Biology and Department of Chemistry
Investigator, Howard Hughes Medical Institute
Yale University

Founding Editors

SIDNEY P. COLOWICK and NATHAN O. KAPLAN

VOLUME FIVE HUNDRED AND THIRTY EIGHT

METHODS IN ENZYMOLOGY

Methods of Adipose Tissue Biology, Part B

Edited by

ORMOND A. MACDOUGALD

Department of Molecular and Integrative Physiology; Division of Metabolism, Endocrinology and Diabetes, Department of Internal Medicine, School of Medicine, University of Michigan, Ann Arbor, Michigan, USA

ELSEVIER

AMSTERDAM • BOSTON • HEIDELBERG • LONDON
NEW YORK • OXFORD • PARIS • SAN DIEGO
SAN FRANCISCO • SINGAPORE • SYDNEY • TOKYO

Academic Press is an imprint of Elsevier

AP

Academic Press is an imprint of Elsevier
525 B Street, Suite 1800, San Diego, CA 92101-4495, USA
225 Wyman Street, Waltham, MA 02451, USA
Radarweg 29, PO Box 211, 1000 AE Amsterdam, The Netherlands
The Boulevard, Langford Lane, Kidlington, Oxford, OX5 1GB, UK
32 Jamestown Road, London NW1 7BY, UK

First edition 2014

For information on all Academic Press publications
visit our website at store.elsevier.com

ISBN: 978-0-12-800280-3
ISSN: 0076-6879

Printed and bound in United States of America
14 15 16 17 11 10 9 8 7 6 5 4 3 2 1

CONTENTS

CONTRIBUTORS

Rodney C. Baker
Department of Pharmacology and Toxicology, University of Mississippi Medical Center, Jackson, Mississippi, USA

David A. Bernlohr
Department of Biochemistry, Molecular Biology and Biophysics, The University of Minnesota, Minneapolis, Minnesota, USA

Anne Bugge
Metabolic Signaling and Disease Program, Diabetes and Obesity Research Center, Sanford-Burnham Medical Research Institute, Orlando, Florida, USA

Cong Chen
Department of Biological and Agricultural Engineering, Louisiana State University and Agricultural Center, Baton Rouge, Louisiana, USA

Rostislav Chudnovskiy
Department of Nutritional Science and Toxicology, UC Berkeley, Berkeley, California, USA

Tae-Hwa Chun
Department of Internal Medicine, University of Michigan Medical School, and Biointerfaces Institute, the University of Michigan, Ann Arbor, Michigan, USA

Sheila Collins
Metabolic Signaling and Disease Program, Diabetes and Obesity Research Center, Sanford-Burnham Medical Research Institute, Orlando, Florida, USA

Chad A. Cowan
Department of Stem Cell and Regenerative Biology, Harvard University, Cambridge; Center for Regenerative Medicine, Massachusetts General Hospital, Boston, and Harvard Stem Cell Institute, Cambridge, Massachusetts, USA

Benjamin F. Cravatt
Department of Chemical Physiology, The Skaggs Institute for Chemical Biology, The Scripps Research Institute, La Jolla, California, USA

Lea Dib
Metabolic Signaling and Disease Program, Diabetes and Obesity Research Center, Sanford-Burnham Medical Research Institute, Orlando, Florida, USA

Eduardo Dominguez
Department of Chemical Physiology, The Skaggs Institute for Chemical Biology, The Scripps Research Institute, La Jolla, California, USA

Elena Dubikovskaya
Department of Chemistry, École Polytechnique Fédérale de Lausanne, Lausanne, Switzerland

Colleen E. Dugan
Department of Chemistry, University of Michigan, Ann Arbor, Michigan, USA

Thomas O. Eichmann
Institute of Molecular Biosciences, University of Graz, Graz, Austria

Susan K. Fried
Division of Endocrinology, Diabetes and Nutrition, Boston University School of Medicine, Boston, Massachusetts, USA

Andrea Galmozzi
Department of Chemical Physiology, The Skaggs Institute for Chemical Biology, The Scripps Research Institute, La Jolla, California, USA

Barbara Gawronska-Kozak
Institute of Animal Reproduction and Food Research of Polish Academy of Sciences, Olsztyn, Poland

Jeffrey M. Gimble
LaCell LLC, New Orleans, Louisiana, USA

James G. Granneman
Center for Integrative Metabolic and Endocrine Research, Wayne State University School of Medicine, and John D. Dingell Veterans Affairs Medical Center, Detroit, Michigan, USA

Julian L. Griffin
MRC Human Nutrition Research, The Elsie Widdowson Laboratory, and Department of Biochemistry and Cambridge Systems Biology Centre, University of Cambridge, Cambridge, United Kingdom

Wendy S. Hahn
Department of Biochemistry, Molecular Biology and Biophysics, The University of Minnesota, Minneapolis, Minnesota, USA

Jaeseok Han
Center for Neuroscience, Aging, and Stem Cell Research, Sanford Burnham Medical Research Institute, La Jolla, California, USA

Daniel J. Hayes
Department of Biological and Agricultural Engineering, Louisiana State University and Agricultural Center, Baton Rouge, Louisiana, USA

Mayumi Inoue
Department of Internal Medicine, University of Michigan Medical School, and Biointerfaces Institute, The University of Michigan, Ann Arbor, Michigan, USA

Grigory Karateev
Department of Chemistry, École Polytechnique Fédérale de Lausanne, Lausanne, Switzerland

Randal J. Kaufman
Center for Neuroscience, Aging, and Stem Cell Research, Sanford Burnham Medical Research Institute, La Jolla, California, USA

Robert T. Kennedy
Department of Chemistry, University of Michigan, Ann Arbor, Michigan, USA

Achim Lass
Institute of Molecular Biosciences, University of Graz, Graz, Austria

Mi-Jeong Lee
Division of Endocrinology, Diabetes and Nutrition, Boston University School of Medicine, Boston, Massachusetts, USA

Youn-Kyoung Lee
Department of Stem Cell and Regenerative Biology, Harvard University, Cambridge, and Center for Regenerative Medicine, Massachusetts General Hospital, Boston, Massachusetts, USA

Xi Li
Key Laboratory of Molecular Medicine, The Ministry of Education, Department of Biochemistry and Molecular Biology, Fudan University Shanghai Medical College, Shanghai, PR China

Hsiao-Ping H. Moore
College of Arts and Sciences, Lawrence Technological University, Southfield, Michigan, USA

Emilio P. Mottillo
Center for Integrative Metabolic and Endocrine Research, Wayne State University School of Medicine, Detroit, Michigan, USA

Yana Nikitina
Division of Endocrinology, Department of Medicine, University of Mississippi Medical Center, Jackson, Mississippi, USA

Hyo Min Park
Department of Nutritional Science and Toxicology, UC Berkeley, Berkeley, California, USA

George M. Paul
Center for Integrative Metabolic and Endocrine Research, Wayne State University School of Medicine, Detroit, Michigan, USA

Ammar T. Qureshi
Department of Biological and Agricultural Engineering, Louisiana State University and Agricultural Center, Baton Rouge, Louisiana, USA

Lee D. Roberts
MRC Human Nutrition Research, The Elsie Widdowson Laboratory, and Department of Biochemistry and Cambridge Systems Biology Centre, University of Cambridge, Cambridge, United Kingdom

Enrique Saez
Department of Chemical Physiology, The Skaggs Institute for Chemical Biology, The Scripps Research Institute, La Jolla, California, USA

Martina Schweiger
Institute of Molecular Biosciences, University of Graz, Graz, Austria

Forum Shah
LaCell LLC, New Orleans, Louisiana, USA

Andreas Stahl
Department of Nutritional Science and Toxicology, UC Berkeley, Berkeley, California, USA

Angela R. Subauste
Division of Endocrinology, Department of Medicine, University of Mississippi Medical Center, Jackson, Mississippi, USA

Qi-Qun Tang
Key Laboratory of Molecular Medicine, The Ministry of Education, Department of Biochemistry and Molecular Biology, Fudan University Shanghai Medical College, Shanghai, PR China

Ulrike Taschler
Institute of Molecular Biosciences, University of Graz, Graz, Austria

Caasy Thomas-Porch
Tulane University Biomedical Science Department, New Orleans, Louisiana, USA

Antonio Vidal-Puig
Metabolic Research Laboratories, Level 4, Institute of Metabolic Science, Addenbrooke's Hospital, Cambridge, United Kingdom

James A. West
MRC Human Nutrition Research, The Elsie Widdowson Laboratory, and Department of Biochemistry and Cambridge Systems Biology Centre, University of Cambridge, Cambridge, United Kingdom

Qinghui Xu
Department of Biochemistry, Molecular Biology and Biophysics, The University of Minnesota, Minneapolis, Minnesota, USA

Rudolf Zechner
Institute of Molecular Biosciences, University of Graz, Graz, Austria

Rong Zeng
Key Laboratory of Systems Biology, Shanghai Institutes for Biological, Sciences, Chinese Academy of Sciences, Shanghai, PR China

Robert Zimmermann
Institute of Molecular Biosciences, University of Graz, Graz, Austria

PREFACE

This book will be informative to those interested in obesity, stem cells, or the development and physiology of adipose tissues.

Although white adipose tissue is often maligned due to societal pressures and the diseases associated with obesity, in reality these dynamic tissues have important but incompletely understood roles in regulation of whole body metabolism and maintenance of health. Thus, these volumes present a wide array of state-of-the-art methods to facilitate further study of development, physiology, and pathophysiology of adipocytes in cultured cells, animal models, and humans. In addition, research on energy-consuming brown adipocytes has exploded over the past few years because of the potential for activation or expansion of brown adipose tissues in humans to help alleviate incidence of obesity. Consequently, a number of methods for visualization and investigation of brown adipose tissue are also detailed. Finally, white adipose depots are a source of readily available stem cells, whose multipotency and other properties have considerable potential to treat human diseases. Accordingly, methods for purification and study of these important precursors are also described.

The study of adipose tissues goes hand in hand with global efforts to understand and reverse the epidemic of obesity and associated medical complications. Thus, tremendous progress has recently been made in our ability to investigate aspects of white and brown adipose tissue biology. Contributors include those researchers that have made substantive contributions to our ability to explore adipose tissue biology at the biochemical, cellular, tissue, and/or organismal levels. Authors were recruited not only based on their contributions to the field but also on their ability to communicate their cutting-edge methodological advances in a cogent and unambiguous style. These investigators have documented their "lab protocol," including small but critical details for which there is often not space within a standard journal article. Where possible, they have also included general suggestions on how to optimize or modify protocols for the specific application of the reader.

The editor wants to express appreciation to the contributors for providing their contributions in a timely fashion, to the senior editors for guidance, and to the staff at Elsevier for helpful input.

Ormond A. MacDougald
Cambridge, UK
October, 2013

CHAPTER ONE

Preparation and Differentiation of Mesenchymal Stem Cells from Ears of Adult Mice

Barbara Gawronska-Kozak[1]
Institute of Animal Reproduction and Food Research of Polish Academy of Sciences, Olsztyn, Poland
[1]Corresponding author: e-mail address: b.kozak@pan.olsztyn.pl

Contents

Abstract

External murine ears collected postmortem, as well as ear punches obtained during standard marking of live animals, are the source of mesenchymal stem cells, termed ear mesenchymal stem cells (EMSC). These cells provide an easily obtainable, primary culture model system for the study of lineage commitment and differentiation. EMSC are capable of differentiating into adipocytes, osteocytes, chondrocytes, and contractile myocytes. Facile adipogenic differentiation of EMSC provides an excellent model for the study of adipogenesis. In this chapter, methods for isolation, culture, and differentiation of EMSC are described.

1. INTRODUCTION

The obesity epidemic causing the increase in metabolic disorders has spurred scientific approaches to understand the origin of fat tissue and its expansion during ontogenetic development. To dissect the molecular signals and metabolic pathways that govern adipogenesis, different cell culture

Methods in Enzymology, Volume 538
ISSN 0076-6879
http://dx.doi.org/10.1016/B978-0-12-800280-3.00001-3

systems have been developed. Immortalized murine cell lines (3T3-L1, 3T3-F442A, ob1771, and OP9), multipotent murine embryonic cell line C3H10T1/2, and primary culture model (stromal-vascular fraction of fat depots) have brought advances in our understanding of the cascade of molecular events during the adipogenic differentiation (Green & Kehinde, 1975; Lee et al., 2013; Park et al., 2008; Wolins et al., 2006). Major progress in *in vitro* techniques has facilitated the isolation and culture of stem cells/progenitor cells from adult tissues including adipose tissues (Gimble & Guilak, 2003; Pittenger et al., 1999; Toma et al., 2001; Zuk et al., 2001). However, it is well established that different adipose depots from the same subject vary in adipogenic capacities/function. Several studies have demonstrated that such differences are linked to depot-derived preadipocytes and their capacity for adipogenesis (Caserta et al., 2001; Gesta et al., 2006; Tchkonia et al., 2005; Tchkonia et al., 2007). It has also been observed that the expression of adipocyte genes, lipid synthesis, lipolysis, and production of secreted proteins differ between preadipocytes from different fat depots (Bastelica et al., 2002; Berman, Nicklas, Rogus, Dennis, & Goldberg, 1998; Van Harmelen et al., 1998). Additionally, stem cells/progenitor cells that reside in adipose tissues are potentially committed to the adipocyte lineage. Summarizing, it is important to develop easily accessible, primary, unbiased/uncommitted cell culture system that will facilitate the elucidation of the earliest steps in mesenchymal stem cell commitment and adipogenic differentiation.

Ear mesenchymal stem cells (EMSC), a primary culture model of mesenchymal stem cells, are capable of readily differentiating into the four main lineages: adipocytes, osteocytes, chondrocytes, and spontaneously contracting myocytes (Gawronska-Kozak, 2004; Gawronska-Kozak, Manuel, & Prpic, 2007; Rim, Mynatt, & Gawronska-Kozak, 2005). This culture system promises to provide a model for the analysis of the early molecular events controlling stem cell commitment to the adipocyte lineage and its subsequent adipogenic differentiation (Rim et al., 2005).

EMSC were initially isolated from mice that have the capacity for regenerative healing of ear punches (Hsd: Nude-nu; Gawronska-Kozak, 2004). Follow-up studies revealed that EMSC populate external ears of all studied strains of mice: C57BL/6 J, FVB, aP2-agouti, and BAP-agouti transgenic mice (Gawronska-Kozak, 2004), *Sfrp5* mutant (Mori et al., 2012), and rats (Sart, Schneider, & Agathos, 2009, 2010) regardless of their regenerative ability. Although currently we do not know whether EMSC participate in regeneration processes *in vivo*, their potential to differentiate into four lineages *in vitro* makes them a useful tool for the study of cell lineage-specific emergence and lineage differentiation. One of the enormous advantages of the EMSC model is that in addition to accessing the entire external ear collected postmortem, small pieces of ear tissue (ear punches) collected from live adult mice provide sufficient numbers of EMSC to isolate, culture, and differentiate (Gawronska-Kozak et al., 2007). This noninvasive surgical procedure for obtaining EMSC from live animals allows the simultaneous conduct of *in vivo* and *in vitro* studies on the same set of animals. Additionally, EMSC are an excellent alternative to mouse embryonic fibroblasts that can be collected from transgenic and KO mice in order to study commitment/differentiation towards adipogenesis, chondrogenesis, myogenesis, and osteogenesis as well as cellular metabolisms of lineage-differentiated EMSC. Recent data from Eilertsen lab emphasize the profound utility of EMSC as a superb source of cells for pluripotent stem cell (iPS) induction (Gao et al., 2013). The pluripotency of mouse EMSC-derived miPS, reprogrammed by overexpressing the four pluripotency factors Oct4, Klf4, Sox2, and c-Myc, was confirmed in *in vitro* and *in vivo* (teratoma formation in nude mice) studies. The data indicate that mEMSC-derived iPS share functional characteristics with (embryonic stem cell) ES cell clones (Gao et al., 2013).

1.1. EMSC characteristics

Cells isolated from outer ears of mice yield approximately 8×10^6 nucleated cells per gram of tissue. Processing of tissues collected from a pool of six animals routinely yields 1×10^5–1.5×10^5 cells per animal (Gawronska-Kozak, 2004). The size of cells (analyzed at passage 1) ranges from 8.1 to 26.6 μm in diameter with most of the cell population (≈72%) between 12 and 20 μm. Assessment of EMSC growth characteristics from the primary culture up to passage four showed a doubling time that ranged between 2.7 ± 0.6 and 3.3 ± 1.0 days. A significant decrease in doubling time was observed when cells reached passages five and six (2.1 ± 0.2) (Staszkiewicz et al., 2010). Phenotypic characterization of freshly isolated, passaged, and cryopreserved EMSC revealed that the expression of stem cell markers (CD117 and Sca-1), stromal cell markers (CD44, CD73, and CD90), and hematopoietic markers (CD45 and CD4) was retained (Gawronska-Kozak et al., 2007; Staszkiewicz et al., 2010). The number of ear cells expressing particular markers changed among passages primarily between freshly isolated ($p=0$) and cultured cells ($p=1$) (Staszkiewicz et al., 2010). The stem cell marker Sca-1 was initially ($p=0$) expressed in approximately 60% of the cells and increased at passage 1 up to 80%, whereas the number of cells with hematopoietic markers CD45 decreased from 12% at $p=0$ to 3% at further passages (Staszkiewicz et al., 2010). No differences were observed in cell expansion/number, morphology, or differentiation capacity among strains of mice. The differentiation ability of EMSC was confirmed on freshly isolated, passaged, and clonally expanded cultures (Gawronska-Kozak, 2004; Gawronska-Kozak et al., 2007; Rim et al., 2005).

The *in vitro* characterization of EMSC indicates very facile adipogenic differentiation. Adipogenic potential of EMSC is maintained up to the fifth passage. A comparison of EMSC to the stromal-vascular (S-V) fraction of fat depots, under identical culture conditions, showed more robust and consistent adipogenesis in EMSC than S-V fraction. The adipogenic differentiation of EMSC was confirmed by morphological, histochemical, molecular, and physiological methods. EMSC developed morphological features characteristic of mature adipocytes after 7–9 days in adipogenic culture media. The morphological changes in differentiated EMSC were characterized by the abundance of lipid droplets stained by Oil Red O (Rim et al., 2005; Staszkiewicz, Gimble, Manuel, & Gawronska-Kozak, 2008; Staszkiewicz et al., 2010). The EMSC adipogenic differentiation is associated with the expression of adipocyte-specific genes: PPARγ, aP2, LPL, C/EBPα, Pref-1, C/EBPβ, C/EBPδ, and Wnt-10b (Rim et al., 2005; Staszkiewicz et al., 2008). Furthermore, differentiated EMSC secrete leptin and show increased glucose uptake and lipolysis in response to insulin and β-adrenergic stimulation, respectively (Rim et al., 2005; Staszkiewicz et al., 2008). The basic protocol of EMSC differentiation into adipolineage was slightly modified in MacDougald laboratory (Mori et al., 2012). To increase EMSC commitment and frequency of EMSC to undergo adipogenesis, cells were pretreated prior to differentiation with bFGF (Mori et al., 2012).

Chondrogenic activity of EMSC was examined in monolayered and micromass culture models. After 21 days in chondrogenic medium, differentiated EMSC showed typical features of chondrocytes: low number of cells per unit, accumulation of extracellular matrix, and expression of procollagen IIα, a chondrocyte-associated gene (Gawronska-Kozak, 2004).

Osteogenic differentiation of EMSC was confirmed after 14 days in culture by increased expression of the osteocalcin gene and by histochemical staining with alizarin red solution, which stains mineralized matrix, the hallmark of a differentiated osteoblast. Osteogenic differentiation of EMSC was accompanied by low adipogenic differentiation (Gawronska-Kozak, 2004).

Myogenic differentiation stimulated by treatment of EMSC with epidermal growth factor (EGF) changed their morphology from a fibroblastic morphology into sticklike structures that showed spontaneous contractions.

Under conditions that promote myogenic differentiation, EMSC expressed mRNA for myoD and ventricule-specific myosin light chain (MLC-2v) and protein for connexin 43, sarcomeric α-actinin, myocyte enhancer factor 2c (MEF2c), myosin heavy chain (MyHC), myogenin, and SERCA 1 (Gawronska-Kozak et al., 2007). Differentiated EMSC displayed features of spontaneous, nonstimulated contractile activity, which was observed after only 3 days of differentiation. The frequency of calcium transients can be visualized by fluorescence intensity of Fluo-3-AM-loaded cells measured in a spontaneously beating region of differentiated EMSC (Gawronska-Kozak et al., 2007).

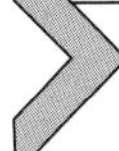

2. EMSC ISOLATION, CULTURE, CRYOPRESERVATION, AND DIFFERENTIATION

2.1. Materials

- Two sets of sterile dissecting scissors and forceps
- 15 and 50 ml polypropylene conical tubes (BD Falcon)
- Sterile 70 μm cell strainers (Becton Dickinson Labware)
- 35, 60, and 100 mm culture petri dishes (BD Falcon)
- Tissue culture plates, 6-well and 12-well (BD Falcon)
- Nalgene filter units disposable, sterile (Fisher Scientific)
- Nalgene "Mr. Frosty" freezing container (Fisher Scientific)
- Corning cryogenic vials (Fisher Scientific)
- Hanks' balanced salt solution (HBSS; Invitrogen/Life Technologies)
- Primocin (InvivoGen)
- DMEM/F12 medium (Invitrogen/Life Technologies)
- DMEM high-glucose (Invitrogen/Life Technologies)
- Fetal bovine serum (FBS; Invitrogen/Life Technologies)
- Collagenase I (Worthington Biochemical Corporation)
- Penicillin–streptomycin (Invitrogen/Life Technologies)
- Red Blood Cell Lysis Buffer (Sigma)
- 0.05% trypsin/0.53 n*M* EDTA (Invitrogen/Life Technologies)
- Dimethylsulfoxide DMSO (Sigma)
- Insulin, transferrin, and selenium (ITS; Invitrogen/Life Technologies)
- Sodium pyruvate (Invitrogen/Life Technologies)

Preparation of media components

- Isobutyl-methylxanthine (IBMX; Sigma)

  50 mM IBMX: dissolve initially in 1 *M* NaOH to have a 0.5 *M* solution. Dilute in water to have 50 m*M* solution, and store in aliquots at −20 °C.

- Insulin (Sigma)
  170 μM insulin: dissolve in 0.01 N HCl, and store in aliquots at −20 °C.
- Dexamethasone (Sigma)
  1 mM dexamethasone: dissolve in water, and store in aliquots at −20 °C.
- Troglitazone (TZD; Sigma)
  10 mM TZD: dissolve in DMSO, and store in aliquots at −20 °C.
- *β*-Glycerophosphate (Sigma)
  7×10^{-1} M β-glycerophosphate: dissolve in water, and store in aliquots at −20 °C.
- Ascorbic acid (Sigma)
  2×10^{-4} M ascorbic acid: dissolve in water, and store in aliquots at −20 °C.
- Dihydroxyvitamin D3 (Biomol)
  10 μM Vit D3: dissolve vit D3 in absolute ethanol, and store in aliquots at −20 °C. Protect from light.
- Transforming growth factor-β (Invitrogen/Life Technologies)
  1 μg/ml TGFβ: dissolve TGFβ in 0.1% BSA, and store in aliquots at −20 °C.
- L-Proline (Sigma)
  40 mg/ml L-proline: dissolve L-proline in water, and store in aliquots at −20 °C.
- Epidermal growth factor (EGF-murine; Invitrogen/Life Technologies)
  20 μg/ml EGF: dissolve EGF in 0.1% BSA, and store in aliquots at −20 °C. Do not store in glass.

2.2. Solution preparation

- HBSS with Primocin: 100 μl of Primocin per 50 ml of HBSS. Filter sterilize. Alternatively, HBSS (100 ml) with penicillin–streptomycin (1 ml). Filter sterilize.
- Growth medium—DMEM/F12 (100 ml) with FBS (15 ml) and Primocin (200 μl). Filter sterilize. Alternatively, DMEM/F12 (100 ml) with FBS (15 ml) with penicillin–streptomycin (1 ml). Filter sterilize.
- Collagenase I solution, 2 mg/ml, in HBSS (with Primocin). Filter sterilize.
- Freezing medium: 9 ml of medium (DMEM/F12 with 20% FBS) mixed with 1 ml of DMSO. Filter sterilize.

- Adipogenic medium I: 100 ml of DMEM/F12 (with 5% FBS), 1 ml of 170 μ*M* insulin, 100 μl of 1 m*M* dexamethasone, 1 ml of 50 m*M* IBMX.

  Final concentration: insulin 1.7 μ*M*, dexamethasone 1 μ*M*, IBMX 0.5 m*M*.
- Adipogenic medium II: 100 ml of DMEM/F12 (with 5% FBS), 10 μl of 170 μ*M* insulin, 20 μl of 10 m*M* TZD.

  Final concentration: insulin 17 n*M*, TZD 2 μ*M*.
- Osteogenic medium: 100 ml of DMEM/F12 (with 10% FBS), 1 ml of 7×10^{-1} *M* β-glycerophosphate, 10 μl of 1 m*M* dexamethasone. Just before use, add 1 ml of 20 m*M* ascorbic acid and 100 μl of 10 μ*M* vit D3.

  Final concentration: β-glycerophosphate 7×10^{-3} *M*, dexamethasone 100 n*M*, vit D3 10 n*M*, ascorbic acid 20 μ*M*.
- Chondrogenic medium:

  Micromass culture: 100 ml of DMEM high-glucose (with 10% FBS), 10 μl of 1 m*M* dexamethasone, 100 μl of 40 mg/ml proline, 1 ml of sodium pyruvate, 1 ml of ITS. Just before use, supplement with 1 ml of 20 m*M* ascorbic acid and 1 ml of 1 μg/ml TGFβ.

  Final concentration: dexamethasone 100 n*M*, proline 40 μg/ml, sodium pyruvate 1 m*M*, ascorbic acid 200 μ*M*, TGFβ 10 ng/ml.

  Monolayer culture: 100 ml of DMEM/F12 (with 10% FBS), 10 μl of 1 m*M* dexamethasone. Just before use, supplement with 1 ml of 20 m*M* ascorbic acid and 100 μl of 1 μg/ml TGFβ.

  Final concentration: dexamethasone 100 n*M*, ascorbic acid 200 μ*M*, TGFβ 1 ng/ml.
- Myogenic medium: 100 ml of DMEM/F12 (with 5% FBS), 1 ml ITS and 1 ml of 20 μg/ml EGF.

  Final concentration: EGF 0.2 μg/ml.

2.3. Methods

2.3.1 Isolation of cells from external mouse ears

1. Euthanize adult mice by CO_2 asphyxiation followed by cervical dislocation.
2. Wet the ears of euthanized mice with 70% ethanol. Cut off external ears, and wash them thoroughly in 70% ethanol to reduce the chance of microbial contamination.
3. Use forceps to transfer tissue samples to the HBSS solution (with Primocin) in a sterile 50 ml tube.

   From this point, work under a sterile hood.

4. Wearing gloves, wipe the outside of the tubes with ear samples with 70% ethanol and place under the hood.
5. Using a second set of sterile forceps, transfer ear tissue into a sterile 60 mm culture petri dish.
6. Add a small volume of collagenase solution (1–2 ml) and using sharp sterile scissors, mince ear samples thoroughly into very small pieces.
7. Transfer the ear tissues into a 50 ml sterile tube and add prewarmed (37 °C) collagenase solution.
8. Incubate for 1 h at 37 °C with vigorous motion to allow collagenase solution to thoroughly penetrate minced tissues.
9. Filter the resulting cell suspension through a 70 μm cell strainer to remove undigested tissue fragments.
10. Collect the cells by centrifugation (360 × *g*, 5–8 min). Aspirate supernatant without disturbing the cell pellet.
11. Resuspend pelleted cells for 1 min in 1 ml red blood lysis buffer with vigorously pipetting. Add 10 ml of growth medium (DMEM/F12 with 15 ml FBS) and mix well.
12. Collect cells by centrifugation (360 × *g*, 5–8 min) and carefully aspirate the supernatant.
13. Resuspend pelleted cells in 5 ml of growth medium (DMEM/F12 with 15 ml FBS) and mix well.
14. Collect cells by centrifugation (360 × *g*, 5–8 min). Aspirate the supernatant.
15. Resuspend pelleted cells in growth medium (DMEM/F12 with 15 ml FBS) and plate in 60 mm petri dishes (passage 0; $P=0$).
16. Keep the culture in a humidified 5% CO_2 incubator at 37 °C.

2.3.2 Culture of EMSC under undifferentiated conditions

1. After 24 h of $p=0$ culture, remove nonadherent cells by changing medium. Change growth medium every 2–3 days.
2. To split the cells, aspirate media from nonconfluent $p=0$ cultures (80–90% confluency). Wash cells with DMEM/F12 (without FBS) and detach cells by incubating for 2–5 min in 0.05% trypsin/0.53 n*M* EDTA.
3. Quench trypsin by adding growth medium (DMEM/F12 with 15 ml FBS), mix well to receive single cell suspension, collect into 15 ml tube, and spin for 5–8 min at 360 × *g* to pellet cells.
4. Aspirate supernatant, and resuspend pellet in 1 ml of growth media, mixing well to receive single cell suspension.

5. Use the hemocytometer and trypan blue to count the number of viable cells in the suspension. Alternatively, an automatic cell counter can be used to determine the vitality, size, and the total number of nucleated cells.
6. Plate 5×10^4 viable cells per milliliter of growth media (per single well in 24-well plate).

2.3.3 Cryopreservation of EMSC

Pelleted cells from $p=0$ (see Section 2.3.2.4) or from other passages can be cryopreserved for further experiments.

1. Count the cells and suspend them in DMEM/F12 medium (with 20% FBS) at the concentration 2×10^6 cells per 1 ml of medium.
2. Take equal volumes of cell suspension and of freezing media (i.e., 0.5 ml of cell suspension + 0.5 ml freezing media).
3. Freeze 1×10^6 of EMSC per vial in freezing container at −80 °C for 24 h.
4. Transfer to liquid nitrogen for long storage.

2.3.4 EMSC differentiation

2.3.4.1 Adipogenic differentiation

1. Plate EMSC in 6- or 12-well culture plates at a density of 5×10^4 cells per milliliter and maintain in growth medium until confluent (considered day 0). Confluent cells are exposed for 2 days (48 h) to adipogenic medium I. For the next 7 days, expose cells to adipogenic medium II changing medium every 2 days.
2. Harvest cells at day 0, 3rd, 5th, 7th, and 9th for RNA and protein purification and media for ELISA assays. Fix some cultures for staining with Oil Red O (Fig. 1.1A).

2.3.4.2 Chondrogenic differentiation

1. Micromass culture: suspend 250,000 EMSC in chondrogenic differentiation medium, centrifuge at 360 × *g* for 8–10 min, and leave them in the bottom of the tube as a pellet. Using pipette tip carefully, free the pellet from the bottom of the tube to allow it to float in the medium. Change medium carefully by aspiration every 3–4 days. After 21 days in culture, the pellet will increase up to 1 mm in diameter. Collect pellets, fix, embed in paraffin, and section to stain with hematoxylin–eosin and for the presence of proteoglycans with alcian blue (Fig. 1.1B). Since micromass cultures form hard pellets that are difficult to homogenize for RNA and/or protein isolation, monolayer cultures have been developed.

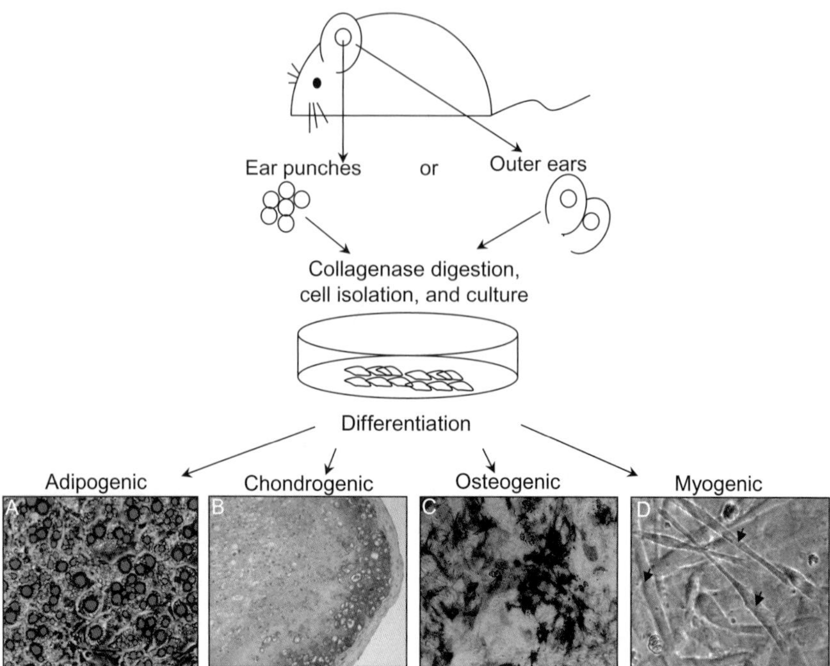

Figure 1.1 EMSC isolation, culture, and adipogenic (A), chondrogenic (B), osteogenic (C), and myogenic (D) differentiation. (A) Confluent EMSC were stimulated with adipogenic differentiation medium. Adipogenesis is indicated by accumulation of lipid vacuoles that stain with Oil Red O; (B) EMSC in high-glucose micromass culture promoted the chondrogenic phenotype, hematoxylin, and eosin staining of paraffin-embedded pelleted cell sections; (C) EMSC differentiation along the osteoblast lineage was monitored with alkaline phosphatase staining; (D) phase-contrast photomicrograph of EMSC differentiated for 7 days in myogenic medium; arrows indicate spontaneously contracting myocytes. (See color plate.)

2. Monolayer culture: plate EMSC in 6-well culture plates at a density of 5×10^4 cells per milliliter and maintain in growth medium until confluency (considered day 0). Confluent monolayer cell cultures are exposed for 14 days to chondrogenic medium, which is changed every 2–3 days. Aqueous safranin O solution (0.1%) can be used for detection of proteoglycans characteristic of a cartilaginous matrix. Harvest cells for RNA/protein isolation.

2.3.4.3 Osteogenic differentiation

1. Confluent cells are stimulated for 20 days by culturing in osteogenic induction medium, changing medium every 3–4 days. At the end of

mineralization period, rinse cells with PBS, fix in 70% ethanol for 1 h, and stain for mineralized matrix with alizarin red solution (Fig. 1.1C).

2.3.4.4 Myogenic differentiation

1. Nonconfluent cultures (70–80% confluency) are stimulated with myogenic medium for 7 days, changing medium every 2–3 days. Monitor cultures microscopically for the appearance of spontaneous contracting cells (Fig. 1.1D).

3. CONCLUSIONS

EMSC has provided a new primary culture stem cell model to study lineage commitment/differentiation. Readily accessible populations of EMSC can be isolated postmortem and from ear punches of live animals. Moreover, recently, it has been demonstrated that mouse EMSC express the imprinted gene cluster Dlk1-Dio3 and produced high-quality iPS clones at a high frequency (Gao et al., 2013). It is possible that corresponding human cells could be used for similar *in vitro* studies for the development of clinical applications.

ACKNOWLEDGMENTS

The research in the Gawronska-Kozak laboratory is partially supported by the European Community's Seventh Framework Programme REGPOT-2010-1; Grant No. 264103 REFRESH and grant from Narodowe Centrum Nauki (NCN, Poland) DEC-2012/05/B/NZ5/01537. I thank Leslie P. Kozak for reviewing this chapter.

REFERENCES

Bastelica, D., Morange, P., Berthet, B., Borghi, H., Lacroix, O., Grino, M., et al. (2002). Stromal cells are the main plasminogen activator inhibitor-1-producing cells in human fat: Evidence of differences between visceral and subcutaneous deposits. *Arteriosclerosis, Thrombosis, and Vascular Biology*, *22*(1), 173–178.

Berman, D. M., Nicklas, B. J., Rogus, E. M., Dennis, K. E., & Goldberg, A. P. (1998). Regional differences in adrenoceptor binding and fat cell lipolysis in obese, postmenopausal women. *Metabolism*, *47*(4), 467–473.

Caserta, F., Tchkonia, T., Civelek, V. N., Prentki, M., Brown, N. F., McGarry, J. D., et al. (2001). Fat depot origin affects fatty acid handling in cultured rat and human preadipocytes. *American Journal of Physiology. Endocrinology and Metabolism*, *280*(2), E238–E247.

Gao, R., Rim, J. S., Strickler, K. L., Barnes, C. W., Harkins, L. L., Staszkiewicz, J., et al. (2013). Reprogramming mouse ear mesenchymal stem cells (EMSC) expressing the Dlk1-Dio3 imprinted gene cluster. *Stem Cell Discovery*, *3*(1), 64–71.

Gawronska-Kozak, B. (2004). Regeneration in the ears of immunodeficient mice: Identification and lineage analysis of mesenchymal stem cells. *Tissue Engineering*, *10*(7–8), 1251–1265.

Gawronska-Kozak, B., Manuel, J. A., & Prpic, V. (2007). Ear mesenchymal stem cells (EMSC) can differentiate into spontaneously contracting muscle cells. *Journal of Cellular Biochemistry*, *102*(1), 122–135.

Gesta, S., Bluher, M., Yamamoto, Y., Norris, A. W., Berndt, J., Kralisch, S., et al. (2006). Evidence for a role of developmental genes in the origin of obesity and body fat distribution. *Proceedings of the National Academy of Sciences of the United States of America*, *103*(17), 6676–6681.

Gimble, J., & Guilak, F. (2003). Adipose-derived adult stem cells: Isolation, characterization, and differentiation potential. *Cytotherapy*, *5*(5), 362–369.

Green, H., & Kehinde, O. (1975). An established preadipose cell line and its differentiation in culture. II. Factors affecting the adipose conversion. *Cell*, *5*(1), 19–27.

Lee, K. Y., Yamamoto, Y., Boucher, J., Winnay, J. N., Gesta, S., Cobb, J., et al. (2013). Shox2 is a molecular determinant of depot-specific adipocyte function. *Proceedings of the National Academy of Sciences of the United States of America*, *110*(28), 11409–11414.

Mori, H., Prestwich, T. C., Reid, M. A., Longo, K. A., Gerin, I., Cawthorn, W. P., et al. (2012). Secreted frizzled-related protein 5 suppresses adipocyte mitochondrial metabolism through WNT inhibition. *The Journal of Clinical Investigation*, *122*(7), 2405–2416.

Park, K. W., Waki, H., Villanueva, C. J., Monticelli, L. A., Hong, C., Kang, S., et al. (2008). Inhibitor of DNA binding 2 is a small molecule-inducible modulator of peroxisome proliferator-activated receptor-gamma expression and adipocyte differentiation. *Molecular Endocrinology*, *22*(9), 2038–2048.

Pittenger, M. F., Mackay, A. M., Beck, S. C., Jaiswal, R. K., Douglas, R., Mosca, J. D., et al. (1999). Multilineage potential of adult human mesenchymal stem cells. *Science*, *284*(5411), 143–147.

Rim, J. S., Mynatt, R. L., & Gawronska-Kozak, B. (2005). Mesenchymal stem cells from the outer ear: A novel adult stem cell model system for the study of adipogenesis. *FASEB Journal*, *19*(9), 1205–1207.

Sart, S., Schneider, Y.-J., & Agathos, S. N. (2009). Ear mesenchymal stem cells: An efficient adult multipotent cell population fit for rapid and scalable expansion. *Journal of Biotechnology*, *139*(4), 291–299.

Sart, S., Schneider, Y.-J., & Agathos, S. N. (2010). Influence of culture parameters on ear mesenchymal stem cells expanded on microcarriers. *Journal of Biotechnology*, *150*(1), 149–160.

Staszkiewicz, J., Frazier, T. P., Rowan, B. G., Bunnell, B. A., Chiu, E. S., Gimble, J. M., et al. (2010). Cell growth characteristics, differentiation frequency, and immunophenotype of adult ear mesenchymal stem cells. *Stem Cells and Development*, *19*(1), 83–92.

Staszkiewicz, J., Gimble, J., Manuel, J. A., & Gawronska-Kozak, B. (2008). IFATS series: Stem cell antigen-1 positive ear mesenchymal stem cells (EMSC) display enhanced adipogenic potential. *Stem Cells*, *26*(10), 2666–2673.

Tchkonia, T., Lenburg, M., Thomou, T., Giorgadze, N., Frampton, G., Pirtskhalava, T., et al. (2007). Identification of depot-specific human fat cell progenitors through distinct expression profiles and developmental gene patterns. *American Journal of Physiology. Endocrinology and Metabolism*, *292*(1), E298–E307.

Tchkonia, T., Tchoukalova, Y. D., Giorgadze, N., Pirtskhalava, T., Karagiannides, I., Forse, R. A., et al. (2005). Abundance of two human preadipocyte subtypes with distinct capacities for replication, adipogenesis, and apoptosis varies among fat depots. *American Journal of Physiology. Endocrinology and Metabolism*, *288*(1), E267–E277.

Toma, J. G., Akhavan, M., Fernandes, K. J., Barnabe-Heider, F., Sadikot, A., Kaplan, D. R., et al. (2001). Isolation of multipotent adult stem cells from the dermis of mammalian skin. *Nature Cell Biology*, *3*(9), 778–784.

Van Harmelen, V., Reynisdottir, S., Eriksson, P., Thorne, A., Hoffstedt, J., Lonnqvist, F., et al. (1998). Leptin secretion from subcutaneous and visceral adipose tissue in women. *Diabetes*, 47(6), 913–917.

Wolins, N. E., Quaynor, B. K., Skinner, J. R., Tzekov, A., Park, C., Choi, K., et al. (2006). OP9 mouse stromal cells rapidly differentiate into adipocytes: Characterization of a useful new model of adipogenesis. *Journal of Lipid Research*, 47(2), 450–460.

Zuk, P. A., Zhu, M., Mizuno, H., Huang, J., Futrell, J. W., Katz, A. J., et al. (2001). Multilineage cells from human adipose tissue: Implications for cell-based therapies. *Tissue Engineering*, 7(2), 211–228.

CHAPTER TWO

3-D Adipocyte Differentiation and Peri-adipocyte Collagen Turnover

Tae-Hwa Chun[*,†,1], **Mayumi Inoue**[*,†]
[*]Department of Internal Medicine, University of Michigan Medical School, Ann Arbor, Michigan, USA
[†]Biointerfaces Institute, the University of Michigan, Ann Arbor, Michigan, USA
[1]Corresponding author: e-mail address: taehwa@umich.edu

Contents

Abstract

Peri-adipocyte extracellular matrix (ECM) remodeling is a key biological process observed during adipose tissue development and expansion. The genetic loss of a pericellular collagenase, MMP14 (also known as MT1-MMP), renders mice lipodystrophic with the accumulation of undigested collagen fibers in adipose tissues. *MMP14* is not necessary for adipocyte differentiation (adipogenesis) per se under a conventional two-dimensional (2-D) culture condition; however, *MMP14* plays a critical role in adipogenesis *in vivo*. The role of MMP14 in adipogenesis and adipocyte gene expression was uncovered *in vitro* only when tested within a three-dimensional (3-D) collagen gel, which recapitulated the *in vivo* ECM-rich environment. Studying adipogenesis in 3-D may serve as an effective experimental approach to bridge gaps in our understanding of *in vivo* adipocyte biology. Moreover, by assessing the content of collagen family members and their rate of degradation in adipose tissues, we should be able to better define the role of dynamic ECM remodeling in the pathogenesis of obesity and diabetes.

Methods in Enzymology, Volume 538
ISSN 0076-6879
http://dx.doi.org/10.1016/B978-0-12-800280-3.00002-5

1. INTRODUCTION

1.1. Adipose tissue development *in vivo*

In mammals, white adipose tissues develop from the clustered islands of mesenchymal cells that are rich in adipocyte precursor cells (Cawthorn, Scheller, & MacDougald, 2012; Hausman & Richardson, 2004; Napolitano, 1963). Adipocyte precursor cells are fibroblast-like cells with a stretched cell shape within the three-dimensional (3-D) adipose tissue environment. The 3-D architecture in the body is defined by a composite of extracellular matrix (ECM) proteins, which constitute solid fibrillar networks (e.g., type I, II, III, and IV collagens, elastin, and fibrin), and loosely connected ground substances (e.g., laminin, hyaluronate, aggrecan, and chondroitin sulfate). During postnatal development, type I collagen is the most abundant fibrillar ECM protein, and it plays a key role in the regulation of cell shape, proliferation, and migration of mesenchymal cells (Ricard-Blum, 2011).

Although the transcriptional regulation of adipocyte differentiation (adipogenesis) has been extensively defined to this date, the role of the ECM microenvironment in the regulation of adipogenesis and adipocyte function has not been fully studied (Cristancho & Lazar, 2011; Rosen & MacDougald, 2006). Indeed, the gene expression profile of adipocytes differentiated *in vitro* and that of adipocytes isolated from adipose tissues *in vivo* displays marked discrepancies (Soukas, Socci, Saatkamp, Novelli, & Friedman, 2001), which may be at least partially resolved through dissection of cellular behavior in *in vivo*-like settings, as represented by 3-D tissue culture technique. By reconstituting transparent 3-D ECM gels, cell morphology, proliferation, migration, and differentiation can be probed in real time under an inverted microscope. In addition, this 3-D cell culture approach is amenable to molecular analysis of gene expression, protein expression, and subcellular protein localization (Fig. 2.1).

1.2. 3-D cell biology

Fibroblasts behave differently in 3-D ECM-rich environment (Tomasek, Hay, & Fujiwara, 1982). The biological significance of 3-D cell culture in understanding cellular behavior and function is highlighted by a series of studies with various cell types, such as Madin–Darby canine kidney epithelial cells (Montesano, Schaller, & Orci, 1991), mammary epithelial cells

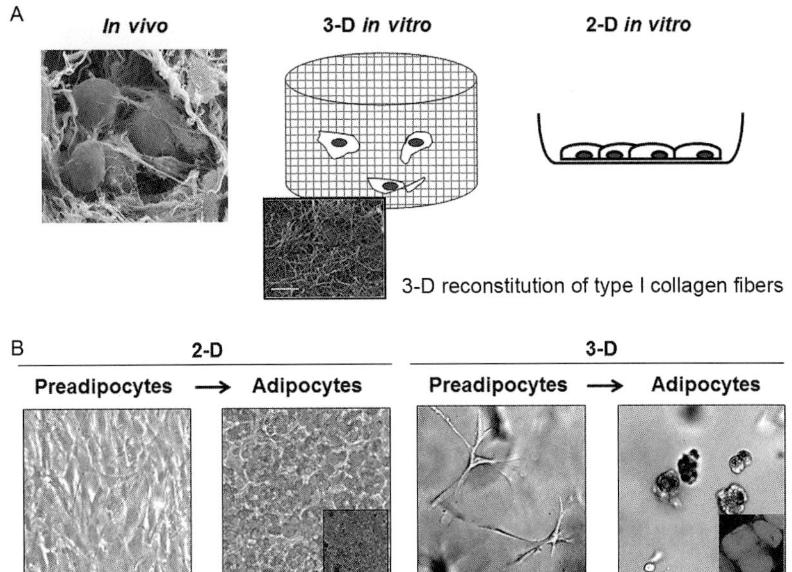

Figure 2.1 3-D *in vitro* adipogenesis. (A) 3-D but not 2-D *in vitro* adipogenesis recapitulates 3-D ECM environment found *in vivo*. *In vivo* adipocytes and preadipocytes are enwrapped by the network of collagen fibers (see scanning electron micrograph—SEM, left). By reconstituting rat tail-derived type I collagen as a gel (SEM, middle), we can recreate three-dimensional collagen-rich environment *in vitro*, which is as easily observable under a microscope as 2-D cultured cells. (B) Microscopic observation of 3-D adipogenesis. By adding adipogenic cocktail (insulin, IBMX, dexamethasone), vascular stromal cells (preadipocytes) isolated from white adipose tissue differentiate into adipocytes 2-D *in vitro*. Lipid accumulation is stained with Oil Red O (inset). Likewise, 3-D embedded preadipocytes differentiate into lipid-laden adipocytes 3-D *in vitro*. Lipid stained with Nile red (inset). (See color plate.)

(Wang et al., 1998), and vascular endothelial cells (Chun et al., 2004; Hiraoka, Allen, Apel, Gyetko, & Weiss, 1998). Under 3-D culture conditions, mesenchymal cells employ an array of adhesion molecules in a manner distinct from that employed under two-dimensional (2-D) culture conditions (Cukierman, Pankov, Stevens, & Yamada, 2001). For example, 2-D cell migration is mainly directed by lamellipodia (Pollard & Borisy, 2003), whereas 3-D cell migration employs the formation of filopodia, pseudopodia/lobopodia (Greenburg & Hay, 1988; Petrie, Gavara, Chadwick, & Yamada, 2012), and blebbing or amoeboid-like movement (Charras & Paluch, 2008; Wolf et al., 2003) (Fig. 2.2). As such, molecular machinery employed under 3-D conditions may substantially differ from that under

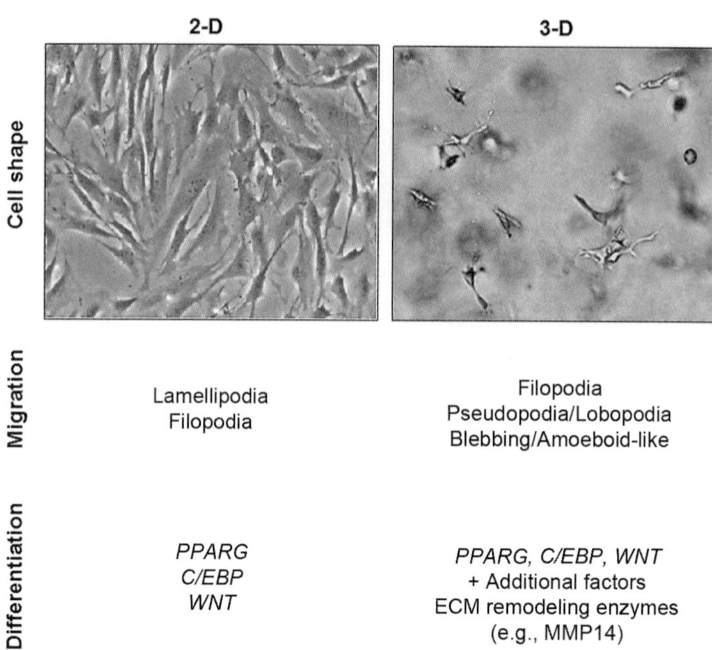

Figure 2.2 Cell shape, migration, and differentiation under 2-D versus 3-D culture condition. Vascular stromal cells isolated from inguinal adipose tissues are shown under 2-D and 3-D culture conditions. 3-D adipogenesis may require nonclassical regulators, for example, ECM remodeling enzymes, in addition to classical adipogenesis regulators, that is, *PPARG*, *C/EBPs*, and *WNTs*.

2-D conditions, not only in migration but also in differentiation. Whereas the principal roles in adipogenesis played by major adipogenesis regulators such as *PPARG*, *C/EBPs*, and *WNTs* will remain important, heretofore undefined regulators may play additional roles in the process of *in vivo* adipogenesis (Fig. 2.2).

1.3. Matrix metalloproteinase and collagen remodeling

The ECM environment maintains a very dynamic equilibrium between the synthesis and degradation of ECM proteins. The family of matrix metalloproteinases (MMPs) consists of 24 zinc-dependent endopeptidases that degrade ECM proteins and thus allow cells to negotiate ECM environment (Sternlicht & Werb, 2001). Most MMPs are expressed as latent zymogens and are activated in pathological states; however, a few exceptions such as MMP3, MMP14 (MT1-MMP), and MMP28 are expressed in active forms by undergoing intracellular activation (Pei & Weiss, 1995; Sternlicht & Werb, 2001; Yana & Weiss, 2000). Under physiological conditions, tissue inhibitors of matrix metalloproteinase (TIMP) family members act as molecular brakes against MMP-dependent ECM degradation (Murphy, 2011). MMPs and TIMPs are differentially regulated during the progression of obesity (Chavey et al., 2003), and the roles of TIMPs in the regulation of obesity via either peripheral or central mechanism have been reported by others (Gerin et al., 2009; Jaworski et al., 2011; Lijnen, Demeulemeester, Van Hoef, Collen, & Maquoi, 2003).

Among MMP family members, MMP14 (MT1-MMP) is a membrane-type MMP that acts as the major pericellular collagenase during postnatal adipose tissue development and obesity progression (Chun et al., 2006, 2010). *MMP14*-null mice are lipodystrophic with their rudimental white adipose tissues entwined within a dense network of fibrillar collagens (Chun et al., 2006). Unexpectedly, *MMP14*-null cells are fully capable of differentiating into adipocytes *in vitro*. We hypothesized that providing a wild-type host environment to *MMP14*-null cells might rescue the lipodystrophic phenotype of *MMP14*-null adipocytes; however, when transplanted onto wild-type background, *MMP14*-null cells continued to display impaired adipogenesis (Fig. 2.3) (Chun et al., 2006). These data suggest that the loss of *MMP14* (*MT1-MMP*) causes cell-autonomous defects in adipogenesis selectively *in vivo* but not *in vitro* (2-D). Thus, MMP14 (MT1-MMP) is a cell-autonomous *in vivo*-specific adipogenic factor, whose function cannot be unmasked when tested solely using conventional 2-D culture conditions (Fig. 2.3).

1.4. ECM remodeling and adipogenesis

MMP14 (MT1-MMP) and MMP15 (MT2-MMP) play major roles in pericellular collagenolysis *in vitro* and *in vivo* (Chun et al., 2004; Hotary, Allen, Punturieri, Yana, & Weiss, 2000). In rodents, white adipose tissues express MMP14 but not MMP15, and vascular stromal cells isolated from *MMP14*-null mice are unable to degrade and remodel type I collagen fibrils (Chun et al., 2006). We initially speculated that type I collagen was the direct target of MMP14 during adipogenesis; however, adipogenic potential of *MMP14*-null cells was not impaired even when cultured atop 2-D fibrillar type I collagen gels (Chun et al., 2006). When preadipocytes were embedded within a 3-D collagen gel, however, we recapitulated the severe impairment of adipogenesis found *in vivo* (Fig. 2.4). We also found that the types of ECM protein that constitute 3-D scaffold exert modifying effects on cellular function and gene expression. Indeed, we observed that the expression of a select set of genes is altered even atop 2-D type I collagen gels in an MMP-dependent manner, albeit to a degree that does not interfere with

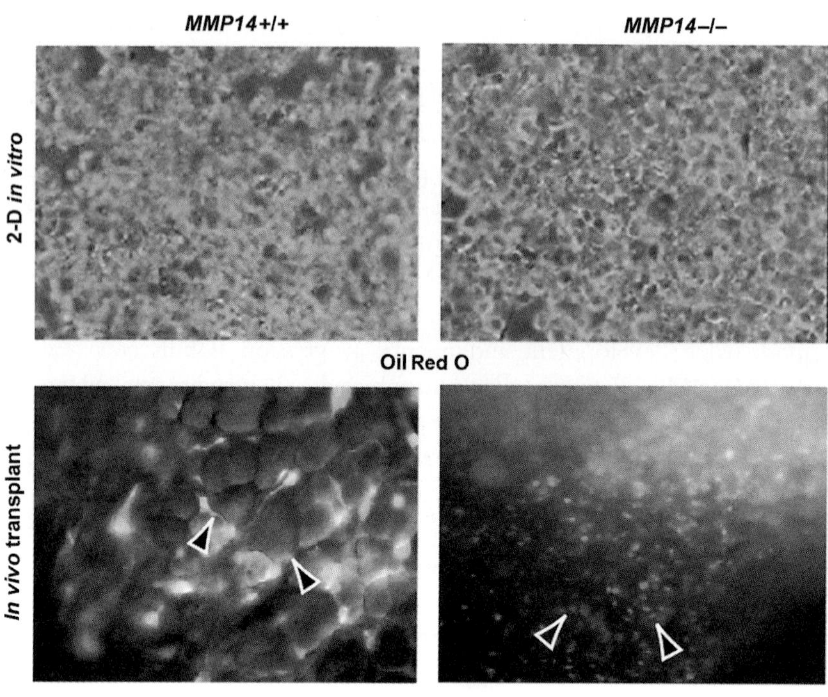

Figure 2.3 MMP14 as *in vivo*-specific cell-autonomous adipogenesis regulator. *MMP14* deficiency does not interfere with 2-D *in vitro* adipogenesis; however, the loss of *MMP14 in vivo* leads to aborted adipogenesis. Primary vascular stromal cells are differentiated 2-D *in vitro*, and lipids are stained with Oil Red O; GFP-labeled wild-type and *MMP14−/−* cells are transplanted subcutaneously into wild-type host, and lipid droplets are detected with Nile red. *Reprinted from Chun et al., 2006 with permission from Elsevier.* (See color plate.)

adipogenesis per se (e.g., lipid accumulation). The MMP-dependent transcriptome modification of adipocytes cultured atop type I collagen was mediated, at least in part, by altered histone modifications (Sato-Kusubata, Jiang, Ueno, & Chun, 2011). As such, two *in vivo* parameters—the type of ECM protein and 3-D geometry—play an additive or synergistic role in determining the genetic and epigenetic landscape of adipogenesis *in vivo*.

1.5. ECM remodeling during obesity progression

When fed a high-fat diet, adipose tissues undergo a rapid ECM remodeling to rebuild adipose tissue mass and function in adaptation to an increased caloric intake. MMP14 plays a critical role in the resulting cleavage and

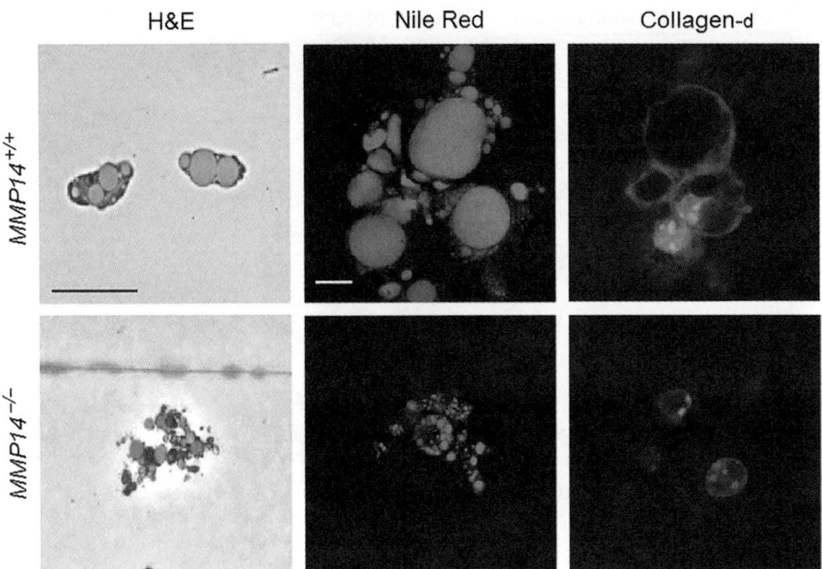

Figure 2.4 *MMP14*-dependent adipogenesis in 3-D collagen gels. Loss of *MMP14* impairs adipogenesis *in vitro* within 3-D collagen gels. MMP14 is essential for hypertrophic maturation of adipocytes, which is coupled with collagen degradation. Lipid droplets shown with Nile red (red), collagen degradation (green), and nucleus (blue). *Reprinted from Chun et al., 2006 with permission from Elsevier.* (See color plate.)

degradation of type I collagen fibers (Fig. 2.5) (Chun et al., 2010). Within a week of high-fat diet, adipose tissues start to express selective sets of genes that belong to two biological pathways—lipogenesis and ECM remodeling. While the mechanism is still unclear, the biological processes of lipid synthesis and ECM remodeling appear to be tightly coupled in expanding adipose tissues.

Compared to subcutaneous adipose tissues, visceral adipose tissues, for example, epididymal adipose tissues, are enwrapped by a loose network of type I collagen fibers. Interestingly, however, the synthesis of collagens is similar between visceral and subcutaneous adipose tissues (unpublished data). It is assumed that visceral adipose tissues may stay at a higher level of equilibrium of ECM remodeling, that is, increased synthesis and degradation of ECM proteins. The loss of equilibrium in collagen turnover may lead to an excess collagen deposition—tissue fibrosis. While the pathological role of collagen deposition in adipose tissues has not been as much highlighted as in the lung, liver, and kidney, adipose tissue fibrosis has been particularly observed in obese, insulin-resistant humans (Divoux &

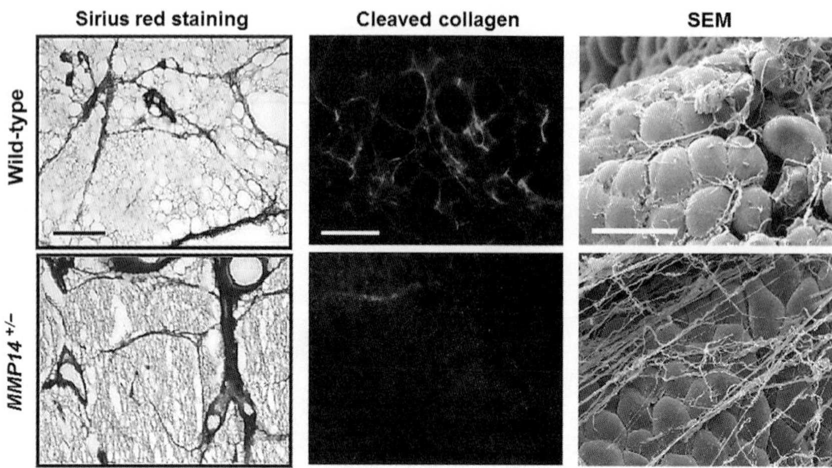

Figure 2.5 *MMP14*-dependent collagen turnover in high-fat diet-induced obesity. Cross-linked fibrillar collagen is detected with Sirius red staining. In wild-type mice, high-fat diet decreases the density of collagen fibers via MMP14-dependent collagen cleavage. Cleaved collagen products are detected with anticleaved collagen antibody. SEM confirms the disruption of collagen fibers upon HFD challenge. This diet-induced collagen turnover is markedly suppressed in MMP14 heterozygous mice. *Reprinted from Chun et al., 2010 with permission.* (See color plate.)

Clément, 2011). It remains to be determined whether the degree of adipose tissue fibrosis may correlate with insulin resistance and other human metabolic disorders (Chun, 2012). Assessing ECM remodeling of adipose tissues or identifying the biological markers that reflect the degree of dysregulation in adipose ECM remodeling may help develop a new diagnostic approach to the dissection of the pathological link between obesity and diabetes.

2. MATERIALS AND METHODS

2.1. 3-D adipogenesis in collagen gels

2.1.1 Type I collagen preparation

2.1.1.1 Materials

- Aged rat tails: Frozen rat tails can be obtained from Harlan Laboratories (Indianapolis, IN).
- Sterile scissors and forceps to isolate tendons from rat tails.
- 20 m*M* acetic acid (AcOH), Sigma-Aldrich (St Louis, MO).
- Labconco Freeze Dry System.
- Dialysis tubing, Fischer Scientific.

2.1.1.2 Method

The method of type I collagen purification is described by others in detail and modified later (Elsdale & Bard, 1972; Rajan, Habermehl, Cote, Doillon, & Mantovani, 2007).

1. Thaw frozen rat tails, rinse in 70% ethanol, and dry.
2. Prepare 1 g of rat tail tendon and cut into small pieces with scissors or chopped in a blender.
   i. When taking out tendons from rat tails, carefully remove blood vessels between tendons to avoid any contaminating MMP activities.
3. Add 1 l of 20 m*M* acetic acid (AcOH) to 1 g tendons.
4. Gently stir using a magnetic stirrer for 48 h in cold room.
5. Centrifuge at 4200 rpm (Beckman J2-HS, $5000 \times g$) for 30 min, and transfer the supernatant to a new container.
6. Add 330 ml of 3 *M* NaCl and 20 m*M* AcOH (pH 3.5) to the 660 ml of supernatant. Incubate at 4 °C for 1 h.
7. Centrifuge at 4200 rpm for 30 min and discard the supernatant.
8. Add 200 ml of 20 m*M* AcOH to collagen precipitate and dissolve overnight.
9. Dialyze newly dissolved collagen solution against 4 l of 20 m*M* AcOH for 3 days. Change the dialysis buffer every day.
10. After dialysis, assess the protein concentration using 1 mg/ml collagen as a standard.
11. Aliquot 25–30 ml of collagen solution into 50 ml Falcon tubes and store at −80 °C.
12. Freeze-dry the aliquots and store collagen powder at −80 °C.

2.1.2 Isolation of primary vascular stromal cells from WAT

2.1.2.1 Materials

- Collagenase type III, Worthington (Lakewood, NJ) LS004182 (1 g)
- Hank's balanced salt solution (HBSS), calcium, magnesium, no phenol red, Life Technologies (Carlsbad, CA), 14025-092
- Primary media: High-glucose DMEM with 10% FBS, antibiotic–antimycotic (Life Technologies, #15240)
- Steriflip-GP filter unit, Millipore (Billerica, MA) SCGP00525
- Cell strainer, 100 μm, yellow, BD Biosciences (San Jose, CA) 352360

2.1.2.2 Methods

1. Prepare 20 ml of type III collagenase at 5 mg/ml dissolved in HBSS (+Ca, +Mg). Adjust pH to 7.4 and sterile filter. Aliquot into four 50 ml Falcon tubes.

2. Harvest subcutaneous or visceral fat pads from mice. Keep tissues in sterile PBS in 6 cm dishes until ready for collagenase digestion.
3. After harvesting tissues, carefully aspirate off PBS and mince the tissues into small pieces before adding them to collagenase.
4. Shake at 37 °C for 30–45 min until tissues are fully digested.
5. Add 20 ml primary media (with antibiotics–antimycotic) to digested tissues and pipette up and down 10 times, and then strain cells using a 100 μm cell strainer and centrifuge for 10 min at 1500 rpm.
6. Carefully remove media by decanting or aspiration and add 5 ml of water to pellet and pipette five times; wait 3 min to lyse red blood cells. Add 25 ml of primary media to recover and strain cells again.
7. Centrifuge for 10 min at 1200 × *g*. Decant media and suspend pellet in 1 ml primary media.
8. Plate 1×10^5 cells onto a well in a 24-well plate.
9. Observe how many cells are attached next day to ensure the effective recovery and adhesion of cells. On second day, change media.
10. Change media every day and pass cells at 1:3 dilution when cells become 80% confluent.[1]

2.1.3 Embedding preadipocytes in type I collagen gels

2.1.3.1 Materials

- Primary vascular stromal cells isolated from WATs
- 3T3-L1 cells are obtained from ATCC (CL-173). 3T3-L1 cells are maintained in DMEM, high-glucose (Life Technologies), supplemented with 10% newborn calf bovine serum (HyClone) and penicillin/streptomycin (Gibco)
- Type I collagen solution
- 10 × MEM, Life Technologies, 11430-030
- 0.34 N NaOH, Fisher Scientific
- 1 *M* HEPES, pH 7.2–7.5, Life Technologies 15630-080

2.1.3.2 Method

We use three different types of ECM proteins for 3-D cell culture, that is, type I collagen, fibrinogen, and matrigel. In this chapter, we explain the use of type I collagen gels, which is a model most relevant to physiological

[1] Ideally, use cells at second passage for experiments. Adipogenic potential of these cells will be lost when cultured beyond three passages. Splitting cells at low cell density impairs proliferation and adipogenic potential.

conditions found in connective tissues. The methods of 3-D fibrin and matrigel culture are described by others (Hiraoka et al., 1998; Mroue & Bissell, 2013).

1. Type I collagen is prepared from rat tail tendons and stored as described earlier (Elsdale & Bard, 1972; Rajan et al., 2007). A few days before experiments, weigh a piece of freeze-dried collagen and dissolve in 0.2% acetic acid to the final concentration of 2.7 mg/ml by continuously rocking the tube overnight in cold room.
   i. Please ensure that freeze-dried collagen fibers are completely dissolved in acetic acid. Keep the tubes of collagen in acetic acid on ice all the time.
2. In a tissue culture hood, all reagents should be kept on ice. Gently mix listed reagents in a new 50 ml Falcon tube in the following order. Try to avoid making bubbles as much as possible during mixing reagents.
   a. Transfer 8 vol. of rat tail collagen (2.7 mg/ml) in 1:500 dilution of glacial acetic acid (33 m*M*) into a new tube on ice.
   b. Add 1 vol. of 10 × MEM.
   c. Add 25 μl of 1 *M* HEPES, pH 7.2–7.5, to every 1 ml of final collagen solution (final concentration 25 m*M*).
   d. Add 1 vol. of 0.34 N NaOH solution; gently swirl the tube to mix.
3. Add 5×10^5 cells in less than 20 μl volume of media at the corner of a well. You can slightly tilt a plate if needed.
4. Using 1 ml pipette, add 0.5 ml of collagen gel (~2.1 mg/ml) to the cells and mix collagen gel with cells by pipetting more than 10 times without making bubbles.
5. Repeat mixing cells with collagen solution in each well.
   i. Do not prepare more than six wells at one time because some cells in gels may sink if you spend too much time for one plate. It is better to use multiple plates with a few samples in each plate than using one plate for many samples.
6. For polymerization, incubate the mixture at 37 °C in 5% CO_2 incubator. Final collagen concentration will be 2.1 mg/ml. The concentration of collagen gel can be modified by changing the concentration of extracted collagen.
   i. The pH of the collagen solution before gelation is 7.5 at 4 °C on ice. During incubation time, pH may change; however, according our assessment, the pH of collagen solution at the end of incubation should stay between 7.4 and 7.5. The pH of final solution may have impact on collagen fiber size and quality.

7. After 40 min to ~1 h of incubation, confirm the polymerization of collagen gels, and add 1 ml of DMEM with 10% FBS to each well in a 24-well plate.

2.1.4 Induction of adipocyte differentiation

2.1.4.1 Materials

- Porcine insulin, Sigma-Aldrich (St Louis, MO) I-5523: Dissolve as 1 mg/ml (1000 ×) in 0.01 N HCl and store at −20 °C; working concentration, 1 μg/ml (170 n*M*).
- 3-Isobutyl-1-methylxanthine (IBMX), Sigma-Aldrich I-5879: Dissolve 55.6 mg in 1 ml (500 ×) of 0.35 *M* KOH and store at −20 °C. This stock concentration is 250 m*M*, and working concentration is 0.5 m*M*.
- Dexamethasone, Sigma-Aldrich D-1756: Dissolve in ethanol as 0.98 mg/ml (10,000 ×) and store at −20 °C. The stock concentration is 2.5 m*M*, and working concentration is 0.25 μ*M*.
- 3,3′,5-Triiodo-L-thyronine (T3), Sigma-Aldrich T6397: To prepare 20 μg/ml (30 μ*M*) stock solution, add 1 ml 1 N NaOH per 1 mg of T3, and gently swirl to dissolve; to this, add 49 ml sterile medium per 1 ml 1 N NaOH added (Sigma-Aldrich). Final working concentration is 10 n*M*.
- Troglitazone (Cayman Chemical 71750): Dissolve 4.4 mg in 1 ml DMSO (10 m*M*) as stock solution. Final working concentration is 1–10 μ*M*.

2.1.4.2 Method

1. 48–72 h after embedding cells in collagen gels, cells should display a stretched, fibroblast-like cell shape within the 3-D collagen gels of the 24-well plate.
2. Adipocyte differentiation is induced with a mixture of 1 μg/ml insulin, 0.5 m*M* IBMX, and 0.25 μ*M* dexamethasone in DMEM with 10% FBS.
3. For the induction of adipogenesis in primary vascular stromal cells, we use a higher insulin concentration (10 μg/ml).
4. 3 days later, change media to DMEM with 10%FBS and 1 μg/ml insulin. Cell shape change and accumulation of lipid droplets can be observed under a phase-contrast microscopy (Olympus).
5. Alternative method of adipocyte differentiation: Add the following combination of the adipogenic reagents (excluding IBMX):

a. Triiodothyronine (T3) 10 n*M*
b. Troglitazone 10 μ*M*
c. Dexamethasone 0.25 μ*M*
d. Insulin 1 μg/ml

2.2. Gene expression analysis of 3-D adipocytes

2.2.1 Harvesting RNA from adipocytes differentiated in 3-D

RNA isolation of cells cultured within 3-D collagen gel will be performed in a manner similar to that of RNA isolation from tissues.

1. After careful aspiration of media, gently detach a gel from the wall and the bottom of a well using a spatula.
2. Transfer the gel to 1.5 ml microtube and centrifuge at 1000 × *g* for 1 min at 4 °C to spin down collagen matrix. Carefully remove the supernatant.
3. Store the samples at −80 °C until you are ready for RNA isolation.
4. After adding Trizol (Invitrogen) or the lysis buffer of RNeasy kit (Qiagen), homogenize gels using a Dounce homogenizer. After spinning down, collect supernatants and proceed with RNA isolation following a manufacturer's protocol.
5. The expression of a specific gene is assessed with real-time PCR; unbiased gene expression survey can be performed using DNA microarray or RNA-seq.

2.3. Assessment of insulin signaling of 3-D adipocytes

2.3.1 Western blot of 3-D samples

1. The media of 3-D collagen culture are changed to fresh serum-free DMEM (1 ml in each well). After 1 h incubation, media are changed to fresh serum-free media again to remove the nutrients diffused out from gels.
2. Next morning, media are replaced by fresh 0.5 ml serum-free media and incubated for an additional 2 h.
3. Add insulin to the media at a specified concentration (e.g., 200 n*M*) and harvest samples at appropriate times for your end points, based on your preliminary studies.
4. After removing media, gels are detached from the walls and bottoms of wells using a spatula. The detached gels are cut by a knife into small pieces (less than 5 mm cubes) and transferred to a 1.5 ml microtube.
5. Add 300 μl of 2 × lysis buffer (diluted from 10 × cell lysis buffer, Cell Signaling, Danvers, MA, #9803), supplemented with proteinase inhibitor cocktail set III (Millipore, Billerica, MA, #59134).

6. After 1 h rocking at 4 °C (cold room), the microtubes are centrifuged (13,000 × g for 3 min), and the supernatant is transferred to a new tube on ice.
7. After measuring the protein concentration with BCA protein assay kit (Pierce, Rockford, IL, #23225), run lysates in SDS-PAGE and perform Western blots.

For the assessment of insulin-dependent Akt activation, the antibodies of Phospho-Akt Ser^{473}, Phospho-Akt Thr^{308}, and total Akt (#9271, #9275, #9272 all rabbit polyclonal antibodies, Cell Signaling) are used.

2.4. Imaging of 3-D adipocytes

2.4.1 Lipid accumulation

1. After removing media, detach collagen gels from the walls. Detached collagen gels are fixed in PBS with 4% paraformaldehyde (Electron Microscopy Sciences, Hatfield, PA) by rocking for 1 h at room temperature.
2. Rinse gels in PBS with rocking for 10 min, three times. Neutral lipid accumulation in adipocytes can then be assessed with Nile red (Invitrogen) and nuclei with Hoechst 33342 (Invitrogen), which is detected under an epifluorescent microscope (using filters with Ex/Em, 350/450, and 560/585, respectively).

2.4.2 Protein localization

1. For immunofluorescent staining, the cell-containing gel is incubated in blocking buffer (5% goat serum in PBS; both from Invitrogen) for 1 h and permeabilized with 0.5% Triton X-100 for 30 min.
2. After washing with PBS, gels are incubated with primary antibodies against your protein of interest. In our case, antibodies against caveolin-2 (anti-mouse caveolin-2 mouse monoclonal antibody D4A6 #8522, BD Biosciences 1:50) or Glut4 (anti-mouse Glut4 rabbit polyclonal antibody Alpha Diagnostic #GT41-A 1:100) are incubated overnight at 4 °C followed by the secondary antibody (Alexa goat anti-mouse antibody 488 and goat anti-rabbit antibody 594 1:500 for both, Molecular Probes).
3. Stained samples are mounted in Vector Shield mounting medium (Vector Laboratories) and protected with a coverslip. The image is captured by laser scanning confocal microscopy (Olympus FluoView 500). The optical sections are stacked as a *Z*-series for 3-D images.

2.5. Collagen remodeling *in vivo*

2.5.1 Sirius red staining to assess fibrillar collagen content

2.5.1.1 Materials

- *Solution A.* Picrosirius red: Dissolve 0.5 g sirius red F3B (C.I. 35782), which is available as Direct Red 80 from Sigma-Aldrich (St Louis, MO) (365548), in 500 ml saturated picric acid solution (Sigma-Aldrich P6744).
- *Solution B.* Acidified water: Add 5 ml glacial acetic acid to 1 l of water (tap or distilled).

2.5.1.2 Method

This method is described by others previously (Junqueira, Bignolas, & Brentani, 1979; Kiernan, 1990, 2002).

1. Dewax and hydrate paraffin sections.
2. (Optional) Stain nuclei with Weigert's hematoxylin. Wash with slides for 10 min in running tap water.
3. Stain in solution A for 1 h.
4. Wash in two changes of solution B.
5. Physically remove most of the water from the slides.
6. Dehydrate in three changes of 100% ethanol.
7. Clear in xylene and mount in a resinous medium.

2.5.2 Staining of collagen family members in vitro

2.5.2.1 Materials

- 3T3-L1 cells (ATCC)
- Primary vascular stromal cells
- Lab-Tek II chamber slide, 4 wells per slide, 154526 (Fisher Scientific, Pittsburgh, PA)
- Paraformaldehyde—16%, Electron Microscopy Sciences (EMS) 15710
- Glycine, Sigma-Aldrich (St Louis, MO) G8898
- Prolong Gold antifade reagent, Invitrogen P36930
- Antibodies: Anti-collagen antibodies (rabbit), from Rockland Immunochemicals (Gilbertsville, PA)
    - Anti-collagen type I antibody 600-401-103
    - Anti-collagen type III antibody 600-401-105
    - Anti-collagen type IV antibody 600-401-106
    - Anti-collagen type V antibody 600-401-107
    - Anti-collagen type VI antibody 600-401-108

2.5.2.2 Method

1. Culture preadipocytes (3T3-L1 or primary vascular stromal cells) on a Lab-Tek Chamber.
2. Fix cells in 4% paraformaldehyde for 20 min, and quench excess paraformaldehyde with 100 m*M* glycine for 15 min and wash with PBS.
3. Incubate samples in a blocking buffer—PBS with 2% goat serum for 30 min.
4. *Do not permeabilize samples if only the extracellular collagens are to be detected.*
5. Incubate samples with primary antibody (1:100) overnight at 4 °C.
6. Next day, wash samples with PBS three times, 5 min each, and incubate with secondary antibody at 1:500 dilutions (goat anti-rabbit IgG, Molecular Probes) for 30 min.
7. Wash samples with PBS for 5 min, and repeat three times.
8. Mount samples in ProLong Gold antifade reagent (Molecular Probes).

2.5.3 Whole-mount tissue staining

2.5.3.1 Method

1. Immediately after harvesting adipose tissues from a mouse, fix the samples in 2% paraformaldehyde for 1 h.
2. After washing with PBS followed by blocking with 5% BSA for 30 min, incubate samples in a buffer with primary antibody (same as for cultured cells) for 1 h at room temperature with rocking.
3. Wash samples with PBS three times, 5 min each.
4. Incubate samples in a buffer with secondary antibody (same as for cultured cells), BODIPY, Hoechst 33342 (Invitrogen) for 30 min at room temperature with rocking.
5. Wash with PBS three times, 5 min each.
6. Keep samples in ProLong Gold antifade reagent and observe under a fluorescent microscope (Fig. 2.6).

2.5.4 MMP-dependent cleavage of collagen fibers

2.5.4.1 Materials

- C1,2C (Col 2 3/4C_{short}) polyclonal rabbit antibody, IBEX Pharmaceuticals Inc. (Montreal, Quebec), 50-1035

  This antibody was initially developed against a cryptic sequence of type II collagen, which is exposed after MMP-dependent cleavage (Billinghurst et al., 1997). This antibody also detects MMP-dependent cleavage of type I collagen (Chun et al., 2010).

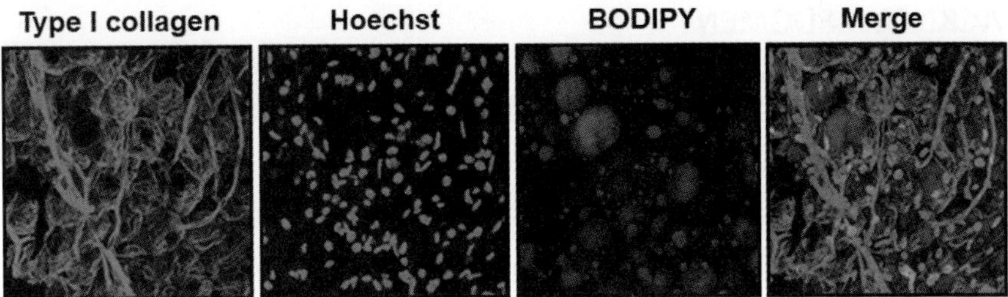

Figure 2.6 Whole-mount adipose tissue staining for type I collagen and lipids. A piece of mouse inguinal adipose tissue was stained with anti-type I collagen antibody (red), Hoechst 33342 (nuclei, blue), and BODIPY (lipid droplets, green). A merged image is shown on the far right. *Modified from Chun, 2012.* (See color plate.)

- Tissue-Tek cryomolds, OCT compound, 10% sucrose in PBS, isopentane, liquid nitrogen, acetone
- Blocking buffer: PBS, 5% goat serum, 0.1% Triton X-100
- Washing buffer: PBS, 0.1% Triton X-100
- Alexa-Fluor anti-rabbit IgG antibody (Fab fragment), Life Technologies

2.5.4.2 Method

1. Fix tissue samples or gels in 4% paraformaldehyde/PBS for 10 min.
2. Immerse samples in 10% sucrose.
3. Mount samples in OCT compound and freeze them in isopentane cooled in liquid nitrogen.
4. Cut 5–20 μm thick cryostat sections and mount on superfrost plus slide. Store samples at −80 °C. Adipose tissue may need to be cut thicker depending on the fragility of tissue samples mounted.
5. When staining, thaw samples at 30 °C and briefly postfix in cold acetone for 15 s.
6. Wash in PBS and immerse in a blocking buffer for 1 h.
7. Incubate with C1, 2C polyclonal rabbit antibody overnight.
8. Wash with washing buffer four times, 5–10 min each.
9. Incubate with Alexa-Fluor anti-rabbit IgG antibody for 1 h.
10. Wash with washing buffer four times, 5 min each. Before the last wash, incubate with Hoechst 33342 to stain nuclei if necessary.
11. Add ProLong Gold antifade reagent, mount a cover glass, and let the samples sit overnight in a dark drawer.
12. Next day, seal a cover glass using a nail polisher and observe the samples under a fluorescent microscope.

ACKNOWLEDGMENT

We thank all the lab members who have developed the experimental protocols. This work is supported by R01DK095137 and R21HL HL106332.

REFERENCES

Billinghurst, R. C., Dahlberg, L., Ionescu, M., Reiner, A., Bourne, R., Rorabeck, C., et al. (1997). Enhanced cleavage of type II collagen by collagenases in osteoarthritic articular cartilage. *The Journal of Clinical Investigation*, *99*(7), 1534–1545. http://dx.doi.org/10.1172/jci119316.

Cawthorn, W. P., Scheller, E. L., & MacDougald, O. A. (2012). Adipose tissue stem cells meet preadipocyte commitment: Going back to the future. *Journal of Lipid Research*, *53*(2), 227–246. http://dx.doi.org/10.1194/jlr.R021089.

Charras, G., & Paluch, E. (2008). Blebs lead the way: How to migrate without lamellipodia. *Nature Reviews. Molecular Cell Biology*, *9*(9), 730–736. http://dx.doi.org/10.1038/nrm2453.

Chavey, C., Mari, B., Monthouel, M. N., Bonnafous, S., Anglard, P., Van Obberghen, E., et al. (2003). Matrix metalloproteinases are differentially expressed in adipose tissue during obesity and modulate adipocyte differentiation. *The Journal of Biological Chemistry*, *278*(14), 11888–11896. http://dx.doi.org/10.1074/jbc.M209196200 [doi] M209196200 [pii].

Chun, T. H. (2012). Peri-adipocyte ECM remodeling in obesity and adipose tissue fibrosis. *Adipocyte*, *1*(2), 89–95.

Chun, T. H., Hotary, K. B., Sabeh, F., Saltiel, A. R., Allen, E. D., & Weiss, S. J. (2006). A pericellular collagenase directs the 3-dimensional development of white adipose tissue. *Cell*, *125*(3), 577–591. http://dx.doi.org/10.1016/j.cell.2006.02.050, S0092-8674(06)00447-8 [pii].

Chun, T.-H., Inoue, M., Morisaki, H., Yamanaka, I., Miyamoto, Y., Okamura, T., et al. (2010). Genetic link between obesity and MMP14-dependent adipogenic collagen turnover. *Diabetes*, *59*(10), 2484–2494. http://dx.doi.org/10.2337/db10-0073.

Chun, T. H., Sabeh, F., Ota, I., Murphy, H., McDonagh, K. T., Holmbeck, K., et al. (2004). MT1-MMP-dependent neovessel formation within the confines of the three-dimensional extracellular matrix. *The Journal of Cell Biology*, *167*(4), 757–767. http://dx.doi.org/10.1083/jcb.200405001, doi: jcb.200405001 [pii].

Cristancho, A. G., & Lazar, M. A. (2011). Forming functional fat: A growing understanding of adipocyte differentiation. *Nature Reviews. Molecular Cell Biology*, *12*(11), 722–734. http://dx.doi.org/10.1038/nrm3198.

Cukierman, E., Pankov, R., Stevens, D. R., & Yamada, K. M. (2001). Taking cell-matrix adhesions to the third dimension. *Science*, *294*(5547), 1708–1712. http://dx.doi.org/10.1126/science.1064829.

Divoux, A., & Clément, K. (2011). Architecture and the extracellular matrix: The still unappreciated components of the adipose tissue. *Obesity Reviews*, *12*(5), e494–e503. http://dx.doi.org/10.1111/j.1467-789X.2010.00811.x.

Elsdale, T., & Bard, J. (1972). Collagen substrate for studies on cell behavior. *The Journal of Cell Biology*, *54*(3), 626–637. http://dx.doi.org/10.1083/jcb.54.3.626.

Gerin, I., Louis, G. W., Zhang, X., Prestwich, T. C., Kumar, T. R., Myers, M. G., et al. (2009). Hyperphagia and obesity in female mice lacking tissue inhibitor of metalloproteinase-1. *Endocrinology*, *150*(4), 1697–1704. http://dx.doi.org/10.1210/en.2008-1409.

Greenburg, G., & Hay, E. D. (1988). Cytoskeleton and thyroglobulin expression change during transformation of thyroid epithelium to mesenchyme-like cells. *Development*, *102*(3), 605–622.

Hausman, G. J., & Richardson, R. L. (2004). Adipose tissue angiogenesis. *Journal of Animal Science*, *82*(3), 925–934.

Hiraoka, N., Allen, E., Apel, I. J., Gyetko, M. R., & Weiss, S. J. (1998). Matrix metalloproteinases regulate neovascularization by acting as pericellular fibrinolysins. *Cell*, *95*(3), 365–377. http://dx.doi.org/10.1016/s0092-8674(00)81768-7.

Hotary, K., Allen, E., Punturieri, A., Yana, I., & Weiss, S. J. (2000). Regulation of cell invasion and morphogenesis in a three-dimensional type i collagen matrix by membrane-type matrix metalloproteinases 1, 2, and 3. *The Journal of Cell Biology*, *149*(6), 1309–1323. http://dx.doi.org/10.1083/jcb.149.6.1309.

Jaworski, D. M., Sideleva, O., Stradecki, H. M., Langlois, G. D., Habibovic, A., Satish, B., et al. (2011). Sexually dimorphic diet-induced insulin resistance in obese tissue inhibitor of metalloproteinase-2 (TIMP-2)-deficient mice. *Endocrinology*, *152*(4), 1300–1313. http://dx.doi.org/10.1210/en.2010-1029.

Junqueira, L. C., Bignolas, G., & Brentani, R. R. (1979). Picrosirius staining plus polarization microscopy, a specific method for collagen detection in tissue sections. *The Histochemical Journal*, *11*(4), 447–455.

Kiernan, J. A. (1990). *Histological and histochemical methods: Theory and practice* (2nd ed.). New York: Pergamon Press.

Kiernan, J. A. (2002). Collagen type I staining. *Biotechnic & Histochemistry*, 77(4), 231.

Lijnen, H. R., Demeulemeester, D., Van Hoef, B., Collen, D., & Maquoi, E. (2003). Deficiency of tissue inhibitor of matrix metalloproteinase-1 (TIMP-1) impairs nutritionally induced obesity in mice. *Thrombosis and Haemostasis*, *89*(2), 249–255. http://dx.doi.org/10.1267/THRO03020249 [doi] 03020249 [pii].

Montesano, R., Schaller, G., & Orci, L. (1991). Induction of epithelial tubular morphogenesis *in vitro* by fibroblast-derived soluble factors. *Cell*, *66*(4), 697–711. http://dx.doi.org/10.1016/0092-8674(91)90115-F.

Mroue, R., & Bissell, M. J. (2013). Three-dimensional cultures of mouse mammary epithelial cells. *Methods in Molecular Biology*, *945*, 221–250. http://dx.doi.org/10.1007/978-1-62703-125-7_14.

Murphy, G. (2011). Tissue inhibitors of metalloproteinases. *Genome Biology*, *12*(11), 233.

Napolitano, L. (1963). The differentiation of white adipose cells: An electron microscopy study. *The Journal of Cell Biology*, *18*(3), 663–679. http://dx.doi.org/10.1083/jcb.18.3.663.

Pei, D., & Weiss, S. J. (1995). Furin-dependent intracellular activation of the human stromelysin-3 zymogen. *Nature*, *375*(6528), 244–247. http://dx.doi.org/10.1038/375244a0.

Petrie, R. J., Gavara, N., Chadwick, R. S., & Yamada, K. M. (2012). Nonpolarized signaling reveals two distinct modes of 3D cell migration. *The Journal of Cell Biology*, *197*(3), 439–455. http://dx.doi.org/10.1083/jcb.201201124.

Pollard, T. D., & Borisy, G. G. (2003). Cellular motility driven by assembly and disassembly of actin filaments. *Cell*, *112*(4), 453–465. http://dx.doi.org/10.1016/S0092-8674(03)00120-X.

Rajan, N., Habermehl, J., Cote, M.-F., Doillon, C. J., & Mantovani, D. (2007). Preparation of ready-to-use, storable and reconstituted type I collagen from rat tail tendon for tissue engineering applications. *Nature Protocols*, *1*(6), 2753–2758. http://dx.doi.org/10.1038/nprot.2006.430.

Ricard-Blum, S. (2011). The collagen family. *Cold Spring Harbor Perspectives in Biology*, *3*(1). http://dx.doi.org/10.1101/cshperspect.a004978.

Rosen, E. D., & MacDougald, O. A. (2006). Adipocyte differentiation from the inside out. *Nature Reviews Molecular Cell Biology*, 7(12), 885–896.

Sato-Kusubata, K., Jiang, Y., Ueno, Y., & Chun, T.-H. (2011). Adipogenic histone mark regulation by matrix metalloproteinase 14 in collagen-rich microenvironments. *Molecular Endocrinology*, *25*(5), 745–753. http://dx.doi.org/10.1210/me.2010-0429.

Soukas, A., Socci, N. D., Saatkamp, B. D., Novelli, S., & Friedman, J. M. (2001). Distinct transcriptional profiles of adipogenesis *in vivo* and *in vitro*. *The Journal of Biological Chemistry, 276*, 34167–34174. http://dx.doi.org/10.1074/jbc.M104421200, M104421200.

Sternlicht, M. D., & Werb, Z. (2001). How matrix metalloproteinases regulate cell behavior. *Annual Review of Cell and Developmental Biology, 17*(1), 463–516. http://dx.doi.org/10.1146/annurev.cellbio.17.1.463.

Tomasek, J. J., Hay, E. D., & Fujiwara, K. (1982). Collagen modulates cell shape and cytoskeleton of embryonic corneal and fibroma fibroblasts: Distribution of actin, [alpha]-actinin, and myosin. *Developmental Biology, 92*(1), 107–122.

Wang, F., Weaver, V. M., Petersen, O. W., Larabell, C. A., Dedhar, S., Briand, P., et al. (1998). Reciprocal interactions between β1-integrin and epidermal growth factor receptor in three-dimensional basement membrane breast cultures: A different perspective in epithelial biology. *Proceedings of the National Academy of Sciences, 95*(25), 14821–14826.

Wolf, K., Mazo, I., Leung, H., Engelke, K., von Andrian, U. H., Deryugina, E. I., et al. (2003). Compensation mechanism in tumor cell migration: Mesenchymal–amoeboid transition after blocking of pericellular proteolysis. *The Journal of Cell Biology, 160*(2), 267–277. http://dx.doi.org/10.1083/jcb.200209006.

Yana, I., & Weiss, S. J. (2000). Regulation of membrane type-1 matrix metalloproteinase activation by proprotein convertases. *Molecular Biology of the Cell, 11*(7), 2387–2401.

CHAPTER THREE

Differentiation of White and Brown Adipocytes from Human Pluripotent Stem Cells

Youn-Kyoung Lee[*,†], **Chad A. Cowan**[*,†,‡,1]
[*]Department of Stem Cell and Regenerative Biology, Harvard University, Cambridge, Massachusetts, USA
[†]Center for Regenerative Medicine, Massachusetts General Hospital, Boston, Massachusetts, USA
[‡]Harvard Stem Cell Institute, Cambridge, Massachusetts, USA
[1]Corresponding author: e-mail address: ccowan@fas.harvard.edu

Contents

Abstract

Given the rapid increase in the prevalence of obesity and related metabolic diseases, research to better understand adipose tissue biology and physiology has garnered considerable attention. Adipose has been studied using both cell culture systems and model organisms. However, the mechanisms of adipocyte regulation are not comprehensively understood, as currently available *in vitro* or *in vivo* systems do not fully recapitulate human metabolic processes. Human primary adipocytes are difficult to culture and expand, and current cell systems have limitations such as cell line-to-cell line variability for adipocyte differentiation, decreased proliferation, and differentiation potential upon continued passaging. To overcome these limitations, we developed and established an efficient and robust adipocyte differentiation protocol using human pluripotent stem cells (hPSCs) and inducible expression of key adipogenic transcriptional regulators. Here, we provide a simple and stepwise protocol for programming hPSCs into mature white or brown adipocytes.

Methods in Enzymology, Volume 538
ISSN 0076-6879
http://dx.doi.org/10.1016/B978-0-12-800280-3.00003-7

1. INTRODUCTION

Adipose tissue plays a central role in energy homeostasis while acting as an integrator of various physiological pathways (Rosen & Spiegelman, 2006). White adipose tissue functions primarily to store energy as triglycerides and releases it as fatty acids and glycerol when it is needed in our body. In contrast, brown adipose tissue dissipates energy via a process termed nonshivering thermogenesis. As an endocrine organ, adipose tissue communicates with other tissues through the release of adipokines such as adiponectin and leptin, which are known to mediate inflammation, lipid metabolism, and glucose metabolism globally (Coppari & Bjorbaek, 2012; Hotamisligil, 2006; Miner, 2004; Ouchi, Parker, Lugus, & Walsh, 2011; Rosen & Spiegelman, 2006). Although studies in mouse models of obesity and related metabolic diseases offer significant insights, their applicability to humans is limited by apparent differences in metabolism and physiology. Human white adipose is easily obtained; however, primary adipocytes are difficult to maintain in culture and are not amenable to expansion. Further, recent studies have discovered that adult humans have functional brown adipose depots in inverse correlation to their body mass index (Cypess & Kahn, 2010; Cypess et al., 2009; van Marken Lichtenbelt et al., 2009). This has fueled considerable interest in the therapeutic potential of brown adipocytes. However, it is difficult to obtain human brown adipocytes for further study.

Recent advances in human stem cell research, particularly in the derivation of human embryonic stem cells and generation of induced pluripotent stem cells (hiPSCs), have made it possible to produce patient-specific *in vitro* cell-based models of human diseases (Cowan et al., 2004; Ebert et al., 2009; Park et al., 2008). The primary challenge in studying human diseases using human pluripotent stem cells (hPSCs) has been the ability to direct these cells to become specialized terminally differentiated cell types. Most disease genotypes in the pluripotent state do not exhibit a disease-specific phenotype, but rather require the tissue-specific protein state and signaling environment found only in specialized, differentiated cells to reveal disease phenotypes.

While the most common type of protocol used to differentiate pluripotent cells to adult cell types attempts to recreate in a dish the key steps that occur during the *in vivo* development of an adult cell, they are less applicable to cell types where there may be little to no information about the developmental origins and signaling cascades that give rise to a specific cell type, as is the case for adipocytes. In contrast, programming one cell fate into another cell fate has been performed in many cases, including iPSC generation and muscle differentiation, by recreating the transcriptional and epigenetic landscape present in a particular cell type by delivering both intrinsic (transcription factors) and extrinsic (growth factors) factors known to be necessary and/or sufficient for the cell type desired. For example, the overexpression of OCT4, SOX2, KLF4, and cMYC in fibroblasts turns their cell fate back to the pluripotent state, and the ectopic expression of MyoD, a transcription factor critical for skeletal muscle specification, can convert many nonmuscle cell types into contracting muscle cells (Choi et al., 1990; Takahashi & Yamanaka, 2006). Adopting this approach, we have recently established a protocol for the differentiation of both white and brown adipocytes from hPSCs (Ahfeldt et al., 2012). Peroxisome proliferator-activated receptor γ2 (PPARG2), widely accepted as a master regulator of adipogenesis, is overexpressed for white adipocyte differentiation. In contrast, a combination of PPARG2 and CCAAT/enhancer-binding protein β (CEBPB), or PPARG2, CEBPB, and PR domain containing 16 (PRDM16), is overexpressed for brown adipocytes differentiation, along with the application of adipogenic factors such as insulin, dexamethasone, and rosiglitazone (Kajimura et al., 2009; Seale et al., 2007; Tontonoz, Hu, Graves, Budavari, & Spiegelman, 1994; Tontonoz, Hu, & Spiegelman, 1994). While there are published data on the differentiation of adipocytes from hPSCs, these protocols are inefficient and the resulting cells have not undergone thorough analysis and rigorous characterization (Dani, 1999; van Harmelen et al., 2007; Xiong et al., 2005). We have shown, however, that multiple human pluripotent cell lines can be differentiated consistently into adipocytes with efficiencies of 85–90% and exhibit the properties of mature, functional cells. By transplanting programmed white or brown adipocytes into immunocompromised mice, we were able to demonstrate that these cells exhibit morphological and functional characteristics of mature adipocytes *in vivo*. In this chapter, we describe the details of (1) general hPSC culture techniques, (2) production of inducible lentivirus for adipogenic transcription factors, and (3) differentiation of white and brown adipocytes.

2. HUMAN PLURIPOTENT STEM CELLS

2.1. Required materials

- hPSCs can be obtained from various sources
- Accutase (STEMCELL™ Technologies, Vancouver, Canada; Cat. No. 07920)
- DMEM (Invitrogen, Grand Island, NY; Cat. No. 11995-065)
- DPBS (Invitrogen, Grand Island, NY; Cat. No. 14190-250)
- Geltrex (Invitrogen, Grand Island, NY; Cat. No. A1413202)
- mTeSR (STEMCELL™ Technologies, Vancouver, Canada; Cat. No. 058050)
- Penicillin/streptomycin 100 × solution (P/S) (Invitrogen, Grand Island, NY; Cat. No. 15140-163)
- Plasmocin (InvivoGen, San Diego, CA; Cat. No. Ant-mpp)
- ROCK inhibitor (Y-27632) (Cayman Chemical, Ann Arbor, MI; Cat. No. 717140)

2.2. Cell culture

Culture conditions described here for hPSCs are feeder-free (Fig. 3.1). It is critical to maintain high-quality hPSCs free of mycoplasma; therefore, all hPSC culture media contain Plasmocin at a prophylactic concentration. All procedures should be carried out using sterile/aseptic technique in an appropriate tissue culture room and under a laminar flow hood. Gloves are worn when handling all reagents and materials that come in contact with cells. All workspaces are thoroughly cleaned with 70% ethanol before and after use.

2.2.1 Thawing and plating hPSC cells

1. Coat the 100 mm culture dish with 100 μl of Geltrex in 10 ml ice-cold DMEM solution for about 1 h at room temperature.
2. Thaw the frozen hPSC vials at 37 °C water bath quickly.
3. Transfer cells into 15 ml conical tube and add 5 ml warmed mTeSR media containing 1% penicillin/streptomycin, 5 μg/ml Plasmocin (Table 3.2, hPSC medium).
4. Spin down cells at 1000 rpm for 5 min.
5. Aspirate the medium, and resuspend cell pellets with 10 ml mTeSR containing 4 μ*M* ROCK inhibitor.

6. Plate cells onto Geltrex-coated plate in step 1.
7. Incubate cells at 37 °C, 5% CO_2 incubator.
8. Feed hPSCs with mTeSR medium every day.

2.2.2 Passaging and maintaining hPSC cells

1. When cells reach the point of being 80–90% confluent, passage the cells.
2. Coat the three to four 100 mm culture dishes with 100 μl Geltrex in 10 ml ice-cold DMEM solution per dish for about 1 h at room temperature.
3. Aspirate the medium and wash with 10 ml DPBS.
4. Add 2 ml Accutase diluted in DPBS (1:1 ratio) and incubate at 37 °C for 1 min or until colonies lose cell–cell attachment.
5. Add 8 ml DPBS and pipette up and down to collect detached cells.
6. Transfer the detached cells to a 15 ml conical tube.
7. Spin down cells at 1000 rpm for 5 min.
8. Aspirate the medium, and resuspend cell pellets with 10 ml mTeSR containing 4 μ*M* ROCK inhibitor.
9. Plate cells onto Geltrex-coated 10 mm culture dishes in step 2 (1:3 or 1:4 split).
10. Incubate cells at 37 °C, 5% CO_2 incubator.
11. Feed hPSCs with mTeSR every day.

3. DIFFERENTIATION INTO MESENCHYMAL PROGENITOR CELLS

3.1. Required materials

- Dispase (STEMCELL™ Technologies, Vancouver, Canada; Cat. No. 07923)
- DMEM (Invitrogen, Grand Island, NY; Cat. No. 11995-065)
- DPBS (Invitrogen, Grand Island, NY; Cat. No. 14190-250)
- FBS (Invitrogen, Grand Island, NY; Cat. No. 16140063)
- bFGF (Aldevron, Fargo, ND; Cat. No. 7001)
- Gelatin (Sigma, St Louis, MO; Cat. No. G8150)
- GlutaMax (Invitrogen, Cat. No. 35050-079)
- Knockout serum replacement (KOSR) (Invitrogen, Grand Island, NY; Cat. No. 10828-028)
- Penicillin/streptomycin 100 × Solution (Invitrogen, Grand Island, NY; Cat. No. 15140-163)

- ROCK inhibitor (Y-27632) (Cayman Chemical, Ann Arbor, MI; Cat. No. 717140)
- Trypsin (0.25%) (Invitrogen, Grand Island, NY; Cat. No. 25200-072)
- Ultralow-attachment 6-well culture dish (Corning, Corning, NY; Cat. No. 3471)

3.2. EB Formation from hPSCs

For EB formation, hPSC cells should not dissociate to a single cell level; rather, they should dissociate to small cell clumps. This process induces differentiation and forms cell aggregates in the suspension culture. Special caution is required when feeding suspended cell aggregates (Fig. 3.1).

1. Culture hPSCs in 100 mm culture dishes until they reach 70% confluency.
2. Aspirate the medium and wash with 10 ml DPBS.
3. Add 1 ml Dispase (1 mg/ml) and incubate at 37 °C until colony edges appear white and folded back.
4. Aspirate Dispase and wash cells with 10 ml DMEM twice.
5. Add 10 ml DMEM and disaggregate cells into small clumps containing 5–10 cells using a cell scraper.
6. Transfer the detached cells to a 15 ml conical tube.
7. Spin down cells at 1000 rpm for 5 min.
8. Aspirate the medium and resuspend cell pellets with 12 ml of DMEM containing 10% KOSR, 1% penicillin/streptomycin, and 4 μ*M* ROCK inhibitor (Table 3.2, EB formation medium).
9. Plate cells into ultralow-attachment 6-well culture dish.
10. After 24 h, change the medium very carefully (alternatively, collect cells in 50 ml conical tubes and let them sit for 10 min, remove supernatant very carefully, and add fresh medium to replate them into ultralow-attachment 6-well plates).
11. Every other day, change with fresh medium until day 7.

3.3. Mesenchymal progenitor cells from EBs

1. Coat two to three 100 mm culture dishes with 0.1% gelatin.
2. At day 7, collect EBs in a conical tube and let them sit for 10 min.
3. Remove the supernatant using pipette.
4. Resuspend EBs with medium containing DMEM, 10% FBS, 1% penicillin/streptomycin, and 1% GlutaMax (Table 3.2, EB plating medium).

5. Plate EBs on gelatin-coated 100 mm culture dishes.
6. Culture cells by changing fresh medium every other day for 5–7 days or until they attach to the culture dishes and reach 90% confluency.
7. At 90% confluency, wash cells with 10 ml PBS.
8. Aspirate PBS and add 1 ml 0.25% trypsin and incubate >1 min in 37 °C incubator.
9. Add 5 ml of medium containing DMEM, 15% FBS, 1% penicillin/streptomycin, 1% GlutaMax, and 2.5 ng/ml bFGF (Table 3.2, MPC medium) and collect detached cells.
10. Replate cells on gelatin-coated culture dishes with a 1:3 split ratio.
11. Culture and passage cells for differentiation experiments before passage.

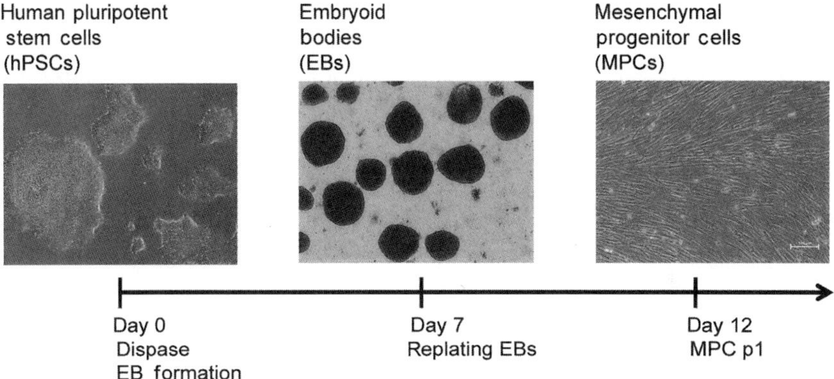

Figure 3.1 Overview of the procedure and timeline used to differentiate hPSCs towards MPCs. Bright-field images showing different stages during the derivation of MPCs.

4. DIFFERENTIATION INTO ADIPOCYTES

4.1. Required materials

- 2× BBS, pH 6.95 (50 m*M* BES, 280 m*M* NaCl, 1.5 m*M* Na_2HPO_4)
- 10× 2.5 *M* $CaCl_2$
- Dexamethasone (Sigma, St Louis, MO; Cat. No. D1756)
- DMEM (Invitrogen, Grand Island, NY; Cat. No. 11995-065)
- Doxycycline hyclate (Sigma, St Louis, MO; Cat. No. D9891)
- DPBS (Invitrogen, Grand Island, NY; Cat. No. 14190-250)

- FBS (Invitrogen, Grand Island, NY; Cat. No. 16140063)
- Stericup 0.22 μm filter unit (Millipore, Billerica, MA; Cat. No. SCGP00525)
- bFGF (Aldevron, Fargo, ND; Cat. No. 7001)
- Gelatin (Sigma, St Louis, MO; Cat. No. G8150)
- HEK293T (ATCC, Manassas, VA; Cat. No. CRL-11268)
- Insulin (Sigma, St Louis, MO; Cat. No. I9278)
- KOSR (Invitrogen, Grand Island, NY; Cat. No. 10828-028)
- Lentivirus plasmids: Lenti-rtTA, lenti-PPARG2, lenti-CEBPB, and lenti-PRDM16 (distributed by Cowan Laboratory upon request)
- Lentivirus packaging plasmids: pMDL, pREV, and V-SVG
- Nonessential amino acids (Invitrogen, Grand Island, NY; Cat. No. 11140-076)
- Penicillin/streptomycin 100 × Solution (Invitrogen, Grand Island, NY; Cat. No. 15140-163)
- Plasmanate (Talecris, Research Triangle Park, NC; Cat. No. 613-25)
- Rosiglitazone (Cayman Chemical, Ann Arbor, MI; Cat. No. 71740)
- Trypsin (0.25%) (Invitrogen, Grand Island, NY; Cat. No. 25200-072)

4.2. Inducible lentivirus production

It is crucial to use high-quality HEK293T cells for getting high virus titer. The plasmids for virus production can be isolated with a commercial kit to ensure purity and quality. Transfection reagents, 2 × BBS, and $CaCl_2$ should be validated by a GFP reporter before use.

1. Culture HEK293T cells in 150 mm culture dish following standard culture technique.
2. At day 1, dissociate HEK293T cells with 0.25% trypsin, split to a 1:4 ratio, and plate them onto 0.1% gelatin-coated culture dish.
3. At day 2, transfect when cells reach 70% confluency.
    1. Prepare plasmid mix: 22.5 μg of lentivirus plasmids to express, 14.7 μg of pMDL, 5.7 μg of pREV, and 7.9 μg of V-SVG in total volume at 250 μl (Table 3.1).
    2. Add 1 ml of 1 × $CaCl_2$ solution.
    3. Add 1 ml of 2×BBS to the mix in a dropwise manner, mix gently by inverting ~3–5 times, and incubate at room temperature for 15 min.
    4. Add the DNA mixture to dish (2.25 ml per 15 mm dish) in a dropwise manner and try to spread the DNA by carefully rocking the plate.
    5. Place cells in 37 °C, 5% CO_2 incubator.

Table 3.1 Plasmid preparation for transfection

	100 mm dish	150 mm dish
Transcription factors	10.0 μg	22.5 μg
pMDLg/p RRE (p*MDL*)	6.5 μg	14.7 μg
pRSV.Rev (pRev)	2.5 μg	5.7 μg
pMD.G (*V-SVG*)	3.5 μg	7.9 μg
1 × $CaCl_2$	500 μl	1 ml
2 × BBS	500 μl	1 ml

4. At day 3, change the medium with DMEM containing 10% FBS and 1% penicillin/streptomycin (Table 3.2, HEK293 medium) and incubate cells in 37 °C, 10% CO_2 incubator.
5. At day 4, collect the medium from the culture dish, which contains the virus particles and store at 4 °C. Feed cells with HEK293 medium.
6. At day 5, collect the medium from the culture dish and combine with day 4 collections to filter through 0.22 μ*M* filter.
7. Virus supernatants can be stored at 4 °C for a few days before experiments or be frozen at −80 °C. Alternatively, virus can be concentrated by centrifugation.
8. Virus titration can be done with various methods (e.g., standard serial infection test and qRT-PCR-based titration).

4.3. White adipocyte differentiation

To transduce mesenchymal progenitor cells (MPCs), it is necessary to test the particular cell lines for seeding and cell density 6 h after splitting cells and before performing actual experiments. The exact number of cells plated and transduced throughout experiments will provide consistency and a reliable differentiation rate for each experiment.

1. Coat the appropriate-sized culture dish with 0.1% gelatin.
2. Seed human PSC-derived MPCs at desired cell density with a medium containing DMEM, 10% FBS, 1% penicillin/streptomycin, and 2.5 ng/ml bFGF.
3. Six hours after splitting cells (about 40–50% confluency), transduce cells with lentiviral supernatants at MOI = 100 with 1:1 ratio of rtTA: PPARG2.
4. Incubate cells at 37 °C overnight.
5. The next morning, remove viral supernatants and wash cells with PBS.

6. Culture cells in growth medium until 100% confluency is reached.
7. Initiate differentiation by feeding cells using adipogenic differentiation medium containing DMEM, 7.5% KOSR, 7.5% human plasmanate, 0.5% nonessential amino acids, 1% penicillin/streptomycin, 1 μ*M* dexamethasone, 10 μg/ml insulin, and 0.5 μ*M* rosiglitazone (Table 3.2) and administration of 700 ng/ml doxycycline to switch on exogenous PPARG2 expression.
8. Feed cells every other day with adipogenic differentiation medium containing doxycycline for 16 days.
9. Culture cells with adipogenic differentiation medium in the absence of doxycycline until day 21 or longer, as experiments require (Fig. 3.2).

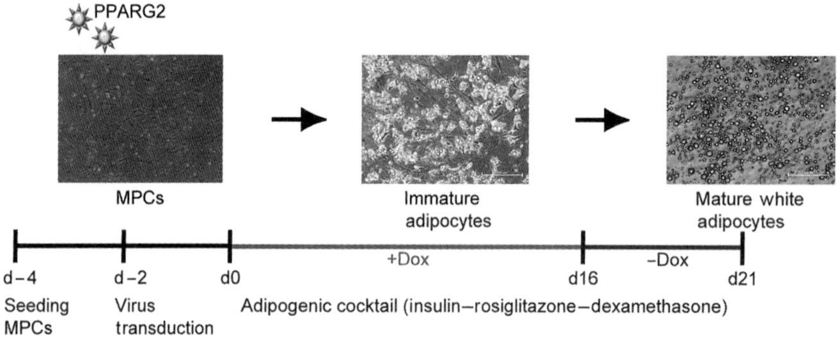

Figure 3.2 Experimental scheme for the differentiation of hPSC into white adipocytes. MPCs were transduced with lentivirus constitutively expressing the lenti-rtTA M2 domain and PPARG2. Cells were cultured in adipogenic differentiation medium containing doxycycline for 16 days and maintained without doxycycline until 21 days. Bright-field images showing different stages during the differentiation of white adipocytes. Pictures were taken at day 10 for immature white adipocytes and at day 21 for mature white adipocytes, which exhibit a single large lipid droplet (×200 magnification). (See color plate.)

4.4. Brown adipocyte differentiation

1. Coat the appropriate-sized culture dish with 0.1% gelatin.
2. Seed human PSC-derived MPCs at desired cell density with a medium containing DMEM, 10% FBS, 1% penicillin/streptomycin, and 2.5 ng/ml bFGF.
3. Six hours after splitting cells (about 40–50% confluency), transduce cells with lentiviral supernatants at MOI = 100 with 2:1:1 ratio of rtTA:PPARG2:CEBPB or with 3:1:1:1 ratio of rtTA:PPARG2:CEBPB:PRDM16.
4. Incubate cells at 37 °C overnight.
5. The next morning, remove viral supernatants and wash cells with PBS.

6. Culture cells in growth medium until 100% confluency is reached.
7. Initiate differentiation by feeding cells using adipogenic differentiation medium and 700 ng/ml doxycycline to switch on exogenous PPARG2 and CEBPB or PPARG2, CEBPB, and PRDM16 expression.
8. Feed cells every other day with adipogenic differentiation medium containing doxycycline for 14 days.
9. Culture cells with adipogenic differentiation medium in the absence of doxycycline until day 21 or longer, as experiments require (Fig. 3.3).

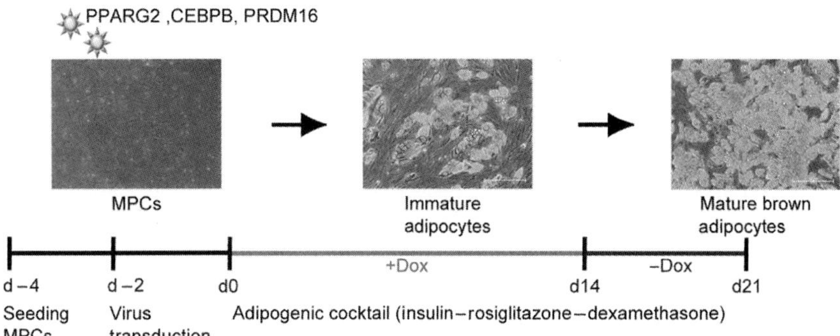

Figure 3.3 Experimental scheme for differentiation of hPSC into brown adipocytes. MPCs were transduced with lentivirus constitutively expressing the lenti-rtTA M2 domain and PPARG2, CEBPB, and PRDM16. Cells were cultured in adipogenic differentiation medium containing doxycycline for 14 days and maintained without doxycycline until 21 days. Bright-field images showing different stages during the differentiation of brown adipocytes. Majority of mature brown adipocytes at day 21 exhibit multiple small lipid droplets (×200 magnification). (See color plate.)

Table 3.2 Media used in this protocol

Ingredient	**Amount**
hPSC medium (500 ml)	
mTeSR™1 basal medium	400 ml
mTeSR™1 5 × supplement	100 ml
Penicillin/streptomycin (100 ×)	5 ml
Plasmocin (2.5 mg/ml)	1 ml
EB formation medium (50 ml)	
DMEM	45 ml
Knockout serum replacement	5 ml
Penicillin/streptomycin (100 ×)	0.5 ml

Continued

Table 3.2 Media used in this protocol—cont'd

Ingredient	**Amount**
ROCK inhibitor (10 m*M*)	20 μl
EB plating medium (500 ml)	
DMEM	450 ml
FBS	50 ml
Penicillin/streptomycin (100 ×)	5 ml
GlutaMax	5 ml
MPC medium (500 ml)	
DMEM	425 ml
FBS	75 ml
Penicillin/streptomycin (100 ×)	5 ml
GlutaMax	5 ml
bFGF (10 μg/ml)	125 μl
HEK293 medium (500 ml)	
DMEM	450 ml
FBS	50 ml
Penicillin/streptomycin (100 ×)	5 ml
Adipogenic differentiation medium (1 L)	
DMEM	850 ml
Knockout serum replacement	75 ml
Plasmanate	75 ml
Nonessential amino acids	5 ml
Penicillin/streptomycin (100 ×)	10 ml
Dexamethasone (1 m*M*)	1 ml
Insulin (10 mg/ml)	1 ml
Rosiglitazone (5 m*M*)	100 μl
Doxycycline (1 mg/ml)	700 μl

Doxycycline is light-sensitive; thus, wrap the medium bottle with foil after adding doxycycline. Adipogenic differentiation medium is supplemented with doxycycline until 16 days for white adipocytes and 14 days for brown adipocytes, and afterward, cells are maintained in culture in the absence of doxycycline.

REFERENCES

Ahfeldt, T., Schinzel, R. T., Lee, Y. K., Hendrickson, D., Kaplan, A., Lum, D. H., et al. (2012). Programming human pluripotent stem cells into white and brown adipocytes. *Nature Cell Biology*, *14*, 209–219.

Choi, J., Costa, M. L., Mermelstein, C. S., Chagas, C., Holtzer, S., & Holtzer, H. (1990). MyoD converts primary dermal fibroblasts, chondroblasts, smooth muscle, and retinal pigmented epithelial cells into striated mononucleated myoblasts and multinucleated myotubes. *Proceedings of the National Academy of Sciences of the United States of America*, *87*, 7988–7992.

Coppari, R., & Bjorbaek, C. (2012). Leptin revisited: Its mechanism of action and potential for treating diabetes. *Nature Reviews. Drug Discovery*, *11*, 692–708.

Cowan, C. A., Klimanskaya, I., McMahon, J., Atienza, J., Witmyer, J., Zucker, J. P., et al. (2004). Derivation of embryonic stem-cell lines from human blastocysts. *The New England Journal of Medicine*, *350*, 1353–1356.

Cypess, A. M., & Kahn, C. R. (2010). Brown fat as a therapy for obesity and diabetes. *Current Opinion in Endocrinology, Diabetes, and Obesity*, *17*, 143–149.

Cypess, A. M., Lehman, S., Williams, G., Tal, I., Rodman, D., Goldfine, A. B., et al. (2009). Identification and importance of brown adipose tissue in adult humans. *The New England Journal of Medicine*, *360*, 1509–1517.

Dani, C. (1999). Embryonic stem cell-derived adipogenesis. *Cells, Tissues, Organs*, *165*, 173–180.

Ebert, A. D., Yu, J., Rose, F. F., Jr., Mattis, V. B., Lorson, C. L., Thomson, J. A., et al. (2009). Induced pluripotent stem cells from a spinal muscular atrophy patient. *Nature*, *457*, 277–280.

Hotamisligil, G. S. (2006). Inflammation and metabolic disorders. *Nature*, *444*, 860–867.

Kajimura, S., Seale, P., Kubota, K., Lunsford, E., Frangioni, J. V., Gygi, S. P., et al. (2009). Initiation of myoblast to brown fat switch by a PRDM16-C/EBP-beta transcriptional complex. *Nature*, *460*, 1154–1158.

Miner, J. L. (2004). The adipocyte as an endocrine cell. *Journal of Animal Science*, *82*, 935–941.

Ouchi, N., Parker, J. L., Lugus, J. J., & Walsh, K. (2011). Adipokines in inflammation and metabolic disease. *Nature Reviews. Immunology*, *11*, 85–97.

Park, I. H., Arora, N., Huo, H., Maherali, N., Ahfeldt, T., Shimamura, A., et al. (2008). Disease-specific induced pluripotent stem cells. *Cell*, *134*, 877–886.

Rosen, E. D., & Spiegelman, B. M. (2006). Adipocytes as regulators of energy balance and glucose homeostasis. *Nature*, *444*, 847–853.

Seale, P., Kajimura, S., Yang, W., Chin, S., Rohas, L. M., Uldry, M., et al. (2007). Transcriptional control of brown fat determination by PRDM16. *Cell Metabolism*, *6*, 38–54.

Takahashi, K., & Yamanaka, S. (2006). Induction of pluripotent stem cells from mouse embryonic and adult fibroblast cultures by defined factors. *Cell*, *126*, 663–676.

Tontonoz, P., Hu, E., Graves, R. A., Budavari, A. I., & Spiegelman, B. M. (1994). mPPAR gamma 2: Tissue-specific regulator of an adipocyte enhancer. *Genes & Development*, *8*, 1224–1234.

Tontonoz, P., Hu, E., & Spiegelman, B. M. (1994). Stimulation of adipogenesis in fibroblasts by PPAR gamma 2, a lipid-activated transcription factor. *Cell*, *79*, 1147–1156.

van Harmelen, V., Astrom, G., Stromberg, A., Sjolin, E., Dicker, A., Hovatta, O., et al. (2007). Differential lipolytic regulation in human embryonic stem cell-derived adipocytes. *Obesity (Silver Spring)*, *15*, 846–852.

van Marken Lichtenbelt, W. D., Vanhommerig, J. W., Smulders, N. M., Drossaerts, J. M., Kemerink, G. J., Bouvy, N. D., et al. (2009). Cold-activated brown adipose tissue in healthy men. *The New England Journal of Medicine*, *360*, 1500–1508.

Xiong, C., Xie, C. Q., Zhang, L., Zhang, J., Xu, K., Fu, M., et al. (2005). Derivation of adipocytes from human embryonic stem cells. *Stem Cells and Development*, *14*, 671–675.

CHAPTER FOUR

Optimal Protocol for the Differentiation and Metabolic Analysis of Human Adipose Stromal Cells

Mi-Jeong Lee[1], Susan K. Fried[1]
Division of Endocrinology, Diabetes and Nutrition, Boston University School of Medicine, Boston, Massachusetts, USA
[1]Corresponding authors: e-mail address: mijlee@bu.edu; skfried@bu.edu

Contents

Abstract

Obesity is reaching epidemic proportions so there is growing interest in the mechanisms that regulates adipose tissue development and function. Although murine adipose cell lines are useful for many mechanistic studies, primary human adipose stromal cells (ASCs), which can be isolated from distinct adipose depots and cultured *in vitro*, have clear translational relevance. We describe the methods to isolate, culture,

Methods in Enzymology, Volume 538
ISSN 0076-6879
http://dx.doi.org/10.1016/B978-0-12-800280-3.00004-9

and differentiate human ASCs to adipocytes that respond to physiologically relevant hormones, such as insulin and β-adrenergic agonists. We also describe methods for assaying hormonal effects on glucose transport and lipolysis.

1. INTRODUCTION

Obesity, defined as accumulation of excess adipose tissue (AT), is increasing worldwide. AT mass expands by increases in the size (hypertrophy) and number of adipocytes (hyperplasia). There is a growing interest in understanding mechanisms of adipocyte differentiation and function as alterations in AT metabolic and endocrine function are thought to play key roles in the pathogenesis of obesity and related metabolic diseases. The life span of subcutaneous (sc) adipocytes in humans is estimated as between ~2–3 (Strawford, Antelo, Christiansen, & Hellerstein, 2004) and 10 years (Spalding et al., 2008). Thus, significant numbers of adipocytes are replaced though the recruitment of adipose precursors and mechanisms that regulate preadipocyte differentiation are important for understanding the maintenance of a healthy AT. Although mouse embryonic fibroblast cell lines, 3T3-L1 and 3T3-F422A, among others, have provided invaluable model systems for mechanistic studies of proliferation and adipogenesis, there is a clear need for cultures of human adipose precursors. Several human adipose cell strains are available including SGBS and hMAD (Bordicchia et al., 2012; Fischer-Posovszky, Newell, Wabitsch, & Tornqvist, 2008), and the use of primary adipose stromal cells (ASCs) that prepared from subjects with fat distribution and/or obesity phenotypes are essential for translational research.

Large number of ASCs, often called AT stem cells or preadipocytes, can be isolated with collagenase digestion of AT. AT specimens can be easily obtained through needle aspiration or during elective surgeries with only minimal risk to the volunteers. ASCs are multipotent cells (adipogenic, chondrogenic, osteogenic, and myogenic; Cawthorn, Scheller, & Macdougald, 2012b), are capable of expansion *in vitro*, and can be cryopreserved for long periods without significant loss in proliferation and differentiation capacity. The differentiated adipocytes display phenotypic characteristics of genuine adipocytes, that is, freshly isolated ones. Specially, they respond to physiologically relevant concentrations of hormones, including insulin and β-adrenergic agonists (Fried et al., 2010; Lee & Fried, 2012;

Lee, Wu, & Fried, 2012; Lystedt et al., 2005; Zierath et al., 1998). Further, ASCs are also useful for assessing donor- and origin-dependent effects (depot, sex, age, obesity, etc.) on cell proliferation and differentiation capacity, which is not possible with cell lines. Visceral adipose depots contain fewer adipogenic precursors and they exhibit lower proliferation and differentiation capacity (Hauner, Wabitsch, & Pfeiffer, 1988; Tchkonia et al., 2002, 2006). Femoral compared to abdominal sc preadipocytes exhibit lower differentiation capacity (Hauner & Entenmann, 1991; Tchoukalova et al., 2010). Obesity- and aging-related declines in ASC proliferation and differentiation capacity have also been reported (Isakson, Hammarstedt, Gustafson, & Smith, 2009; Perez et al., 2013; Sepe, Tchkonia, Thomou, Zamboni, & Kirkland, 2011). Taken together, available data show that cultures of human ASCs reflect the depot-specific biology of human ATs and are therefore valuable models for understanding the underlying mechanisms regulating fat distribution and its influence on metabolic disease.

Conditions to proliferate and differentiate ASCs (e.g., seeding density, culture media and conditions, and subculturing) vary among published studies. Standardization of growth and differentiation conditions is critical as they affect the degree to which the culture will differentiate. We describe methods to proliferate and differentiate ASCs and discuss potential confounders in the processes. We also describe methods to use differentiated human adipocytes for analyses of lipolysis and glucose transport.

2. ISOLATION OF ASCs FROM AT

2.1. Materials

Most chemicals are purchased from Sigma-Aldrich (St Louis, MO), cell culture media and FBS are from Life Technologies (Carlsbad, CA), and cell culture plates and plasticware are from Corning (Corning, NY) or BD Biosciences (Franklin Lakes, NJ). Surgical grade scissors, forceps, and perforated spoons can be purchased from Fine Scientific Tools (Foster City, CA):

- AT, obtained from surgical resection or sc fat aspiration.
- Laminar flow hood.
- Tissue culture incubator, 5% CO_2 atmosphere.
- Sterile scissors, dissecting forceps, and perforated spoons.
- Funnels with 250 μm nylon mesh affixed on top and autoclaved.
- Tissue culture plates and 50 ml tubes.
- Sterile PBS or saline.

- Collagenase (type 1, Worthington Biochemical, Lakewood, NJ) or Liberase (Roche, Indianapolis, IN) solution can be prepared in advance (1 mg/ml HBSS), filtered through 0.2 μm filter, and frozen in aliquots at −20 °C.
- Red blood cell (RBC) lysis buffer (0.154 m*M* NH_4Cl, 10 m*M* K_2HPO_4, and 0.1 m*M* EDTA, pH 7.3).
- Culture media (growth media): alpha-MEM (5 m*M* glucose, 11900-024, Life Technologies, Carlsbad, CA) with 10% FBS, 100 units/ml penicillin, and 100 μg/ml streptomycin (pen/strep) (15140, Life Technologies, Carlsbad, CA).

2.2. Procedures for ASC isolation

All procedures are done under sterile conditions.

1. AT is brought to the lab in room temperature PBS or Medium 199 (12340-030, Life Technologies, Carlsbad, CA). From most fat aspirations and surgical resection, 0.3–10 g AT is obtained. With panniculectomy, a large chunk of AT, often with skin attached, is obtained and thus, it is necessary to dissect AT using scissors and forceps.

   Note: AT can be stored at 4 °C overnight and ASCs can be isolated the next day. The yields, however, may be lower.
2. Mince AT into small pieces, approximately 5–10 mg pieces (1–2 mm^3) using sharp scissors. Tissues from fat aspirations are already fragmented and do not require mincing.
3. Pour the minced tissue through a 250 μm mesh. The mesh is formed into the shape of a funnel, stapled to hold this shape, inserted into a funnel, affixed with tape, and autoclaved. Place this apparatus on top of an empty 500 ml bottle to capture the waste.
4. Rinse AT with saline or PBS. Remove any visible blood clots and obvious connective tissue using sterile forceps.
5. Estimate tissue amount (weigh an empty sterile container such as a petri dish, add AT to the container, and weigh again the container with the tissue, and the difference is tissue weight).
6. Transfer the tissue into 50 ml tubes and add collagenase solution, 1–2 ml/g of AT.
7. Digest AT by incubating in a water bath with shaking (100 rpm) at 37 °C for 2 h until there are few intact pieces of AT and the mixture becomes a homogeneous soupy consistency. Vortexing or rapid swirling by hand will facilitate digestion.

8. Filter the digested AT through a 250 μm mesh affixed to a funnel and capture the flow through into a 50 ml tube. The flow through contains the ASCs.
9. Wash the mesh with culture media. When a large amount of tissue is used, undigested or connective tissue will clog the mesh. Scraping the mesh with a spoon or forceps will facilitate the filtering process and increase yield. Repeat wash as needed.
10. Centrifuge the 50 ml tubes at 500 × *g* for 10 mins at room temperature. After centrifugation, three layers are visible: a fat cake at the top, an intermediate layer of medium, and a cell pellet at the bottom.
11. Carefully aspirate off the upper fat cake and the internatant above the cell pellet.
12. Add 1–5 ml of RBC lysis buffer (depending on the amount of RBC) and pipette up and down to resuspend cells. Incubate in RBC lysis buffer no more than 15 min at room temp or 10 min at 37 °C.

    Note: RBCs are denser than ASCs and are thought to block cell attachment. When a very small amount of tissue is used (e.g., less than 0.5 g), cells can be plated without RBC lysis.
13. Centrifuge the tubes at 500 × *g* for 5 mins.
14. Aspirate off the supernatant and resuspend cells in growth media.
15. Count cell number using a hemocytometer and plate accordingly.
16. After overnight attachment, gently wash the plates with PBS to remove residual RBC and cell debris and refeed.

    Note: Washing and refeeding after just several hours of attachment may enrich preadipocytes/fibroblasts (personal communication—JL Kirkland). Based on cell surface markers (Cawthorn, Scheller, & Macdougald, 2012a), an adipocyte progenitor population can be isolated using FACS or antibody conjugated magnetic beads.

3. PROTOCOLS FOR PROLIFERATION, SUBCULTURE, AND FREEZING DOWN

10% FBS-supplemented DMEM, DMEM/F12, or alpha-MEM is mostly used during ASC culture. FBS, however, is known to decrease the proliferation and differentiation capacity (Hauner, Schmid, & Pfeiffer, 1987; Skurk, Ecklebe, & Hauner, 2007). Human serum or growth factors can substitute for FBS with preserved proliferation and differentiation capacity (Bieback et al., 2012; Chieregato et al., 2011; Im, Chung, Kim, & Kim, 2011; Tunaitis et al., 2011).

3.1. Materials

- Growth media (GM): alpha-MEM with 10% FBS and antibiotics (usually pen/strep).
- Freezing media: 10% dimethyl sulfoxide (DMSO) supplemented GM.
- 0.25% trypsin/EDTA (25200, Life Technologies, Carlsbad, CA).
- Cryogenic tubes.
- Freezing container.
- Culture plates and plasticware.

3.2. Proliferation of cells

1. Grow and proliferate cells in the GM, with refeeding every 2–3 days.
2. When cells reach 70–80% confluence, subculture or freeze down cells. Cells may lift off from the plates if they are kept too long after reaching confluence.

3.3. Subculturing

1. When cells are 70–80% confluent, aspirate off the media and wash once with PBS.
2. Aspirate off the PBS, and add ~1.5 ml of 0.25% trypsin to 10 cm dishes (less volume for smaller dishes).
3. Swirl the plates to distribute the trypsin and then aspirate off most of the liquid, leaving about 0.3 ml per dish.
4. Incubate dishes 2–3 min in the incubator at 37 °C. Check whether cells are lifted off from the plates by examining under a microscope. If cells have not lifted off, incubate longer.
5. Add 5 ml GM to each dish to inactivate the trypsin. Pipette up and down, washing the plate thoroughly, to collect all of the cells off the plate.
6. Transfer the media and cells into 15 or 50 ml tubes and centrifuge at $500 \times g$ for 5 min.
7. Remove the supernatant and resuspend the cells.
8. Count cell numbers and subculture cells accordingly. With 1–4 splitting, it generally takes 3–5 days until they reach 70–80% confluence.

Note: Cells derived from visceral depots grow slower than sc depots. In addition, there is also subject-dependant variation in proliferation rates.

3.4. Freezing and storage

When cells have grown for several days, they survive freezing/thawing cycle better.

1. At 70–80% confluency, trypsinize cells following steps 1–6 described in the previous section.
2. Aspirate off the supernatant, add 1–2 ml freezing media, and resuspend cells.
3. Transfer the cell suspension to prelabeled cryogenic vials. Place the vials into a freezing container and store at −80 °C for several days.
4. Transfer cell vials to a cryostorage system for long-term storage.

Note: In our experience, ASCs can be stored in a cryostorage system without any significant loss in proliferation or differentiation capacity for more than 20 years.

3.5. Protocol for thawing cells

1. Remove vial(s) from cryostorage system and immediately place in 37 °C water bath for 60–90 s. Remove as soon as cells are thawed.
2. Add 1 ml of GM in the cell vial, pipette up and down to mix, and transfer the contents of the vial to centrifuge tubes containing GM.
3. Centrifuge for 5 min at 500 × *g*.
4. Aspirate off the supernatant and resuspend cells in GM.
5. Count the cells for plating or just plate into 10 cm dishes overnight and split on the following day.

4. PROTOCOLS TO DIFFERENTIATE HUMAN ASCs TO ADIPOCYTES

This is a modified protocol from Hauner's (Hauner, Skurk, & Wabitsch, 2001) and Kirkland's (Tchkonia et al., 2005), in which differentiation is carried out in a serum-free, chemically defined media. In standard protocols, adipogenic induction cocktail contains insulin, glucocorticoids, and 1-methyl-3-isobutylxanthine (IBMX) (Hauner et al., 2001; Yu et al., 2011). Thiazolidinediones (TZD), which are PPAR-γ agonists, are often used to improve the degree of differentiation as human ASCs do not differentiate well in the absence of PPAR-γ ligands (Gustafson & Smith, 2012; Nimitphong, Holick, Fried, & Lee, 2012).

4.1. Materials

- Complete differentiation media (CDM): make up stock solutions, prepare the CDM by adding the components and stock solutions, filter through a 0.2 micron filters, and store at 4 °C up to 1 month.

	Final concentrations	Stock concentrations	Amount/l
DMEM/F12, 18.5 m*M* glucose			1 pack
HEPES	15 m*M*		3.57 g
$NaHCO_3$	25 m*M*		2.22 g
Penicillin/streptomycin	100 units/ml (pen) and 100 μg/ml (strep)	10,000 units/ml (pen) and 10,000 μg/ml (strep)	10 ml
d-Biotin	33 μ*M*	3.3 m*M*	10 ml
Pantothenate	17 μ*M*	1.7 m*M*	10 ml
Dexamethasone	100 n*M*	10 m*M*	10 μl
Insulin	100 n*M*	600 μ*M*	167 μl
Rosiglitazone or other TZDs	1 μ*M*	10 m*M*	100 μl
IBMX	0.5 m*M*		110 mg
T_3	2 n*M*	20 μ*M*	100 μl
Transferrin	10 μg/ml	10 mg/ml	1 ml

- Maintenance media (MM): DMEM/F12 with pen/strep, HEPES, $NaHCO_3$, d-biotin, pantothenate, 10 n*M* insulin and 10 n*M* dexamethasone.

4.2. Preparation and storage of stock solutions

- d-Biotin (3.3 m*M*): dissolve in DMEM/F12, sterilize by filtering through a 0.2-micron filter, and store at 4 °C.
- Pantothenate (1.7 m*M*): dissolve in DMEM/F12, sterilize by filtering through a 0.2-micron filter, and store at 4 °C.
- Transferrin (10 mg/ml): reconstitute in water, aliquot, and store at −80 °C.
- T_3 (20 μ*M*): make 2 m*M* stock solution in 1 N NaOH, then dilute into ethanol to make 20 μ*M*, and store at −20 °C.
- IBMX: weigh out the amount needed, add H_2O followed by addition of 1 N KOH dropwise, and vortex until it is dissolved. Once it is reconstituted, immediately transfer the solution into the media.

- Rosiglitazone (10 m*M*, ALX-350-125, Enzo, Farmingdale, NY): make 10 m*M* solution in DMSO, and store at −80 °C in aliquots.
- Insulin (Humulin R, Lilly, Indianapolis, IN): recombinant human insulin from pharmacy.
- Dexamethasone (10 m*M*): make stock solutions in ethanol and store at −20 °C.

4.3. Procedures for differentiating ASCs

1. Count and plate cells depending on the experimental design with a plating density of 5000–15,000 cells/cm^2.
2. Grow cells in the GM until 2 days after confluence. With a plating density of 5000 cells/cm^2, it generally takes 5–7 days until they are ready for differentiation.
3. Once cells are ready for differentiation, aspirate off the GM and refeed with the CDM.

   Note: Both serum-free (Deslex, Negrel, Vannier, Etienne, & Ailhaud, 1987; Entenmann & Hauner, 1996; Shahparaki, Grunder, & Sorisky, 2002; Tchkonia et al., 2002) and supplemented (Bunnell, Flaat, Gagliardi, Patel, & Ripoll, 2008; Yu et al., 2011) conditions are used during differentiation, the two most commonly used conditions being serum-free or 3% FBS-supplemented conditions. Although up to 3% FBS does not significantly affect differentiation, metabolic properties of adipocytes, however, are different between serum-free and 3% FBS protocols (Lee et al., 2012).
4. Replenish the CDM every 2 to 3 days. Lipid droplet accumulation is usually visible after 3 days of differentiation.
5. After 3–7-day induction, remove the CDM and feed cells with the MM.

   Note: In commonly used protocols, IBMX and TZD are present only during the initial 3 days (Hauner et al., 2001; Yu et al., 2011). Longer periods of induction are also used and may actually improve extent of differentiation (Dicker et al., 2004; Fischer-Posovszky et al., 2008; Skurk & Hauner, 2012; Yu et al., 2011), especially for cells with poor differentiation capacity under standard conditions (Lee et al., 2012).
6. Keep cells in the MM and replenish every 2 to 3 days until fully differentiated. We generally consider cells as having matured into adipocytes after at least 12 days of differentiation. Adipocytes can be maintained in the culture > 40 days and the size of lipid droplets grows until they have several enlarged droplets per cell or even become unilocular.

Note: The MM contains insulin (10 n*M*) and dexamethasone (10 n*M*), as the combination of these two hormones best maintains high expression levels of adipocyte genes such as GLUT4 and adiponectin. Extra caution is required during refeeding well-differentiated adipocytes as lipid-filled adipocytes tend to lift off the plates. The use of 1 ml pipetter to remove and refeed media can be considered to minimize disturbing the cells. Unless required, leaving ~20% of the media during refeeding is also helpful.

4.4. Determination of differentiation degree

Substantial changes in cell morphology occur during differentiation (Fig. 4.1). Starting from day 3 of differentiation, a small amount of lipid accumulation is observed. By day 8, cells should be rounding up to a more spherical shape from a more elongated fibroblast shape. Well-differentiated cultures will have over 80% of the cells containing lipid droplets.

1. The formation of lipid droplets is a hallmark of adipogenesis and can be easily observed by phase-contrast microscope. The lipid droplets can be stained with Oil Red O or Nile red. The amount of triacylglycerol in cell lysates can be measured for quantification.
2. Expression levels of adipogenic markers can be determined at the mRNA and protein levels. Typical markers include
   - adipogenic transcription factors: PPAR γ, CEBP β, and CEBP α;
   - adipocyte genes: fatty acid binding protein 4, lipoprotein lipase, adiponectin, leptin, perilipin, adipose triglyceride lipase, etc.

Note: When measuring the expression levels of lipid droplet proteins, especially perilipin, add 5–10% SDS to cell lysis buffer and incubate cell lysates at

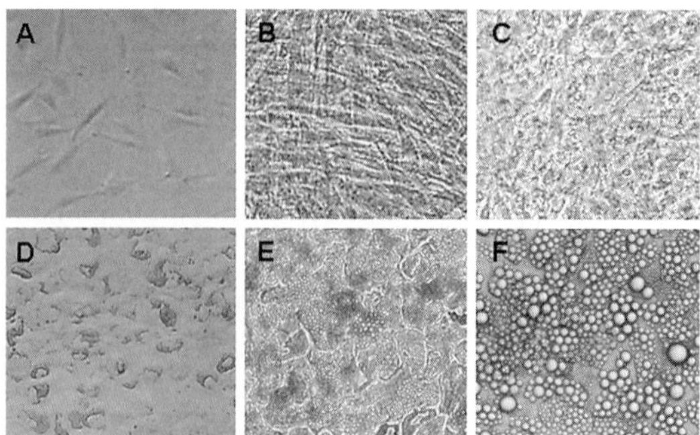

Figure 4.1 Microscopic images of human adipose stromal cells at various stages. Microscopic images are taken at preconfluence (A), 2 days postconfluence (B), d3 (C), d7 (D), d14 (E), and d35 (F) of differentiation. (See color plate.)

37 °C for 1 h with vortexing every 5–10 min. This procedure is required to release the tight association of these hydrophobic proteins with the lipid droplet surface.

Note: Differentiated adipocytes also express significant amount of adiponectin and leptin that can be easily measured in culture media with commercially available ELISA kits.

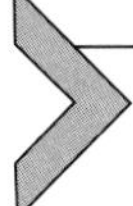

5. METHODS TO USE NEWLY DIFFERENTIATED ADIPOCYTES FOR METABOLIC STUDIES

Newly differentiated adipocytes provide a valuable adipocyte system for metabolic and endocrine functions of adipocytes. In this section, we will describe the protocols for lipolysis and glucose transport assays. The procedures described are based on 12-well plates and adjustment is needed when different formats are used. The results of lipolysis and glucose transport assay performed in sc adipocytes derived from five independent subjects are presented in Fig. 4.2.

5.1. Lipolysis

Newly differentiated adipocytes in culture respond to β-adrenergic agonists or cAMP analogs that stimulate lipolysis and respond to physiological concentrations of insulin that inhibit lipolysis (Figure 4).

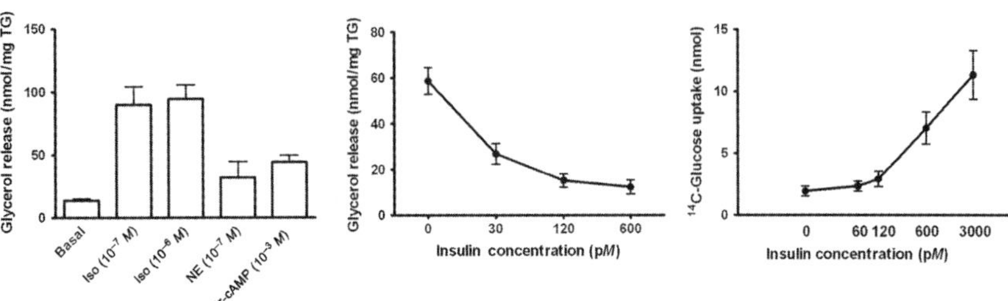

Figure 4.2 Metabolic phenotypes of newly differentiated human adipocytes. Lipolytic rates, basal and in response to isoproterenol (Iso), norepinephrine (NE), or 8-bromo-cAMP (A) and in response to insulin (B), are measured on d14 of differentiation. Basal lipolysis was measured in the conditions of 1 units/ml ADA + 20 n*M* PIA. Inhibition of lipolysis by insulin was measured against 8-bromo-cAMP (1 m*M*)-stimulated conditions. (C) The rates of ^{14}C-2-deoxy-glucose uptake were measured after 3–4 h of starvation at 0, 60, 120, 600, and 3000 p*M* insulin concentrations. Data are obtained from sc ASCs derived from five independent subjects. Some of these data have been published (Lee & Fried, 2012; Lee et al., 2012).

5.1.1 Materials

- DMEM/F12
- Krebs Ringer bicarbonate (KRB) buffer with 5 m*M* glucose and 4% bovine serum albumin (BSA, fatty acid free, EMD Millipore, Darmstadt, Germany).

  10 × mixed salts (MS)

  1. Combine the following salts in approximately 800 ml of deionized and distilled H_2O:
     a. 76.14 g NaCl.
     b. 3.51 g KCl.
     c. 3.06 g $MgSO_4$.
     d. 3.63 g $CaCl_2$.
  2. Bring up to 1 l with H_2O and store at 4 °C.

  10 × bicarbonate/phosphate buffer

  1. Combine:
     a. 20.66 g $NaHCO_3$.
     b. 1.51 g KH_2PO_4.
  2. Dissolve the salts in H_2O, adjust to pH 7.6, bring to 1 l with H_2O, and store at 4 °C.

  KRB with 5 m*M* glucose, 200 n*M* adenosine, and 4% BSA: prepare on the day of the experiment

  1. In a clean flask, combine
     a. 80 parts of H_2O.
     b. 10 parts of 10 × MS solution.
     c. 10 parts of 10 × bicarbonate/phosphate buffer.
  2. Gas for 15 min with 95% O_2/5% CO_2.
  3. Add BSA and glucose and 100 μl of adenosine (200 μ*M* stock, prepared in H_2O and stored in −80 °C) per 100 ml buffer.
  4. Stir with a stir bar until dissolved.
  5. Adjust pH to 7.4.

     Note: There are lot-to-lot variations in BSA that can affect metabolic and secretion properties of adipocytes. Pretesting and buying in bulk are recommended.

- Reagents for lipolysis assay: adenosine, adenosine deaminase (ADA, 10102121001, Roche, Indianapolis, IN), phenylisopropyl adenosine (PIA), β-adrenergic agonists, cAMP analogs, and insulin.

  Note: ADA and PIA are added to standardize any potential variations in adenosine levels (Honnor, Dhillon, & Londos, 1985).

5.1.2 Procedure for lipolysis

1. Plate and differentiate cells according to experimental design. Plate extra wells for cell number and triglyceride quantification.
2. On the day of assay, wash cells with warm basal DMEM/F12 or PBS several times to remove any residual hormones, especially when measuring inhibition of lipolysis by insulin.
3. Add basal DMEM/F12 to the cells and incubate for 3–4 h for hormone starvation.
4. Prepare KRB buffer with 5 m*M* glucose and 4% BSA, pH 7.4.
5. Prepare various conditions for lipolysis assay, hormones, and lipolytic reagents in KRB buffer with 4% BSA, 0.5 ml per well of 12-well cell culture plates.

   Note: Basal lipolytic rates are defined as ADA (1 units/ml) + PIA (20 n*M*) (Fried et al., 2010). Stimulated lipolytic rates are measured with isoproterenol (10^{-8}–10^{-6} *M*), norepinephrine (10^{-7}–10^{-6} *M*), and 8-bromo-cAMP (1 m*M*). Inhibition of lipolysis by insulin can be measured under 8-bromo-cAMP-stimulated condition (1 m*M*). The effective concentration 50 (EC50) of insulin for well-differentiated sc adipocytes is 10–30 p*M* (Lee & Fried, 2012).
6. After preincubation, remove media and add the prepared KRB buffer with different reagents (step 5), 0.5 ml per well.
7. Incubate cells in the incubator at 37 °C for 2 h (rates are linear within this time).
8. After incubation, place cell culture plates on ice directly to stop the reaction. Collect and save incubation media for measurement of glycerol and FFA.
9. Wash cells with ice-cold PBS several times to remove BSA and scrape cells in cell lysis buffer for future use for quantifying cell numbers or triglyceride or immunoblotting. Data are expressed the amount of glycerol released per cell number or triacylglycerol amount.

5.2. Glucose uptake

Uptake of glucose is measured at different insulin concentrations (0–10 n*M*) to establish concentration dependency. The following protocol is optimized for 12-well cell culture plates in total incubation volume of 300 μl per well. Total incubation volume, the amount, and specific activity of labeled glucose can be adjusted.

5.2.1 Materials

- Krebs Ringer HEPES (KRH) buffer with 0.01% BSA, with or without 5 m*M* glucose, 200 n*M* adenosine, pH 7.4.

  10 × *Mixed Salts:* same as previously described.

  10 × HEPES/phosphate buffer:

  a. 23.8 g HEPES.

  b. 3.42 g $NaH_2PO_4H_2O$.

  Dissolve the salts in H_2O, adjust pH 7.6, bring to 1 l with H_2O, and store at 4 °C.

  KRH buffer with 0.01% BSA, with or without 5 m*M* glucose, 200 n*M* adenosine, pH 7.4: prepare on the day of experiment.

  1. In a clean flask, combine:
     80 parts of H_2O
     10 parts of 10 × MS
     10 parts of 10 × HEPES buffer
  2. Add BSA and adenosine to the buffer. Divide the buffer into two parts, one without glucose and one with glucose. Then add glucose to KRH buffer with glucose (5 m*M*).
  3. Stir with a stir bar until dissolved.
  4. Adjust pH to 7.4.
- 2-Deoxyglucose (2-DOG), unlabeled and ^{14}C-labeled
- Insulin
- Lysis buffer: Triton X-100 lysis buffer (1% Triton X-100, 20 m*M* Tris, and 150 m*M* NaCl)
- Cytochalasin B: 0.96 mg/ml (2 m*M*) dissolved in ethanol and stored at −20 °C

5.2.2 Procedure for glucose transport assay

1. Plate and differentiate cells for glucose transport assay. Plate extra wells for cell number counting and nonspecific uptake measurement.
2. On the day of assay, prepare KRH buffer plus 0.01% BSA, with or without 5 m*M* glucose.
3. Wash cells times with warm PBS, being careful not to disturb cells.
4. Add KRH buffer with 5 m*M* glucose into each well and incubate cells for 3–4 h for insulin starvation.
5. Prepare varying incubation conditions, containing different concentrations of insulin in the KRH buffer without glucose (0.27 ml per well of 12-well plates). EC50 for insulin-stimulated glucose transport is 100–300 p*M* (Lee et al., 2012).

6. Wash the preincubated cells with warm KRH buffer without glucose several times to remove any residual glucose.
7. Add the KRH buffer prepared in step 5 into cells and incubate cells for 30 min. For nonspecific uptake, treat separate wells with 1.5 μl cytochalasin B stock.
8. Prepare 10 × 2-DOG mixture (final concentrations 0.25–1 μCi/ml [^{14}C]2-DOG and 0.05–0.2 m*M* unlabeled 2-DOG).

   Note: The amount of 2-DOG (both cold and radioactive) can be adjusted. Be sure to count a known volume of the 2-DOG mixture for specific activity.
9. After preincubation with insulin, add 10 × 2-DOG mixture (30 μl per 300 μl total volume) to each well as quickly as possible, and incubate for a further 10 min in the incubator.
10. Immediately place the cells on ice to stop the reaction after incubation and wash three times with ice-cold PBS to remove unincorporated 2-DOG.
11. Remove PBS, add 0.5 ml of Triton X-100 lysis buffer, and shake the plates on a rocker for 1 h.
12. Transfer the cell lysates to scintillation vials and count radioactivity. Also count the specific activity of the 2-DOG mixture.
13. Subtract cytochalasin B values from each condition to determine the specific uptake and calculate data.

ACKNOWLEDGMENTS

This work was supported by NIH (DK 52398, DK 080448, and P30 DK 046200) to SKF and Evans Biomedical Research Foundation pilot grant to MJLMJL and SKF take full responsibility for this chapter.

REFERENCES

Bieback, K., Hecker, A., Schlechter, T., Hofmann, I., Brousos, N., Redmer, T., et al. (2012). Replicative aging and differentiation potential of human adipose tissue-derived mesenchymal stromal cells expanded in pooled human or fetal bovine serum. *Cytotherapy*, *14*, 570–583.

Bordicchia, M., Liu, D., Amri, E. Z., Ailhaud, G., Dessi-Fulgheri, P., Zhang, C., et al. (2012). Cardiac natriuretic peptides act via p38 MAPK to induce the brown fat thermogenic program in mouse and human adipocytes. *The Journal of Clinical Investigation*, *122*, 1022–1036.

Bunnell, B. A., Flaat, M., Gagliardi, C., Patel, B., & Ripoll, C. (2008). Adipose-derived stem cells: Isolation, expansion and differentiation. *Methods*, *45*, 115–120.

Cawthorn, W. P., Scheller, E. L., & Macdougald, O. A. (2012a). Adipose tissue stem cells meet preadipocyte commitment: Going back to the future. *Journal of Lipid Research*, *53*, 227–246.

Cawthorn, W. P., Scheller, E. L., & Macdougald, O. A. (2012b). Adipose tissue stem cells: The great WAT hope. *Trends in Endocrinology and Metabolism, 23*, 270–277.

Chieregato, K., Castegnaro, S., Madeo, D., Astori, G., Pegoraro, M., & Rodeghiero, F. (2011). Epidermal growth factor, basic fibroblast growth factor and platelet-derived growth factor-bb can substitute for fetal bovine serum and compete with human platelet-rich plasma in the ex vivo expansion of mesenchymal stromal cells derived from adipose tissue. *Cytotherapy, 13*, 933–943.

Deslex, S., Negrel, R., Vannier, C., Etienne, J., & Ailhaud, G. (1987). Differentiation of human adipocyte precursors in a chemically defined serum-free medium. *International Journal of Obesity, 11*, 19–27.

Dicker, A., Ryden, M., Naslund, E., Muehlen, I. E., Wiren, M., Lafontan, M., et al. (2004). Effect of testosterone on lipolysis in human pre-adipocytes from different fat depots. *Diabetologia, 47*, 420–428.

Entenmann, G., & Hauner, H. (1996). Relationship between replication and differentiation in cultured human adipocyte precursor cells. *The American Journal of Physiology, 270*, C1011–C1016.

Fischer-Posovszky, P., Newell, F. S., Wabitsch, M., & Tornqvist, H. E. (2008). Human SGBS cells—A unique tool for studies of human fat cell biology. *Obesity Facts, 1*, 184–189.

Fried, S. K., Tittelbach, T., Blumenthal, J., Sreenivasan, U., Robey, L., Yi, J., et al. (2010). Resistance to the antilipolytic effect of insulin in adipocytes of African-American compared to Caucasian postmenopausal women. *Journal of Lipid Research, 51*, 1193–1200.

Gustafson, B., & Smith, U. (2012). The WNT inhibitor Dickkopf 1 and bone morphogenetic protein 4 rescue adipogenesis in hypertrophic obesity in humans. *Diabetes, 61*, 1217–1224.

Hauner, H., & Entenmann, G. (1991). Regional variation of adipose differentiation in cultured stromal-vascular cells from the abdominal and femoral adipose tissue of obese women. *International Journal of Obesity, 15*, 121–126.

Hauner, H., Schmid, P., & Pfeiffer, E. F. (1987). Glucocorticoids and insulin promote the differentiation of human adipocyte precursor cells into fat cells. *The Journal of Clinical Endocrinology and Metabolism, 64*, 832–835.

Hauner, H., Skurk, T., & Wabitsch, M. (2001). Cultures of human adipose precursor cells. *Methods in Molecular Biology, 155*, 239–247.

Hauner, H., Wabitsch, M., & Pfeiffer, E. F. (1988). Differentiation of adipocyte precursor cells from obese and nonobese adult women and from different adipose tissue sites. *Hormone and Metabolic Research Supplement, 19*, 35–39.

Honnor, R. C., Dhillon, G. S., & Londos, C. (1985). cAMP-dependent protein kinase and lipolysis in rat adipocytes. I. Cell preparation, manipulation, and predictability in behavior. *The Journal of Biological Chemistry, 260*, 15122–15129.

Im, W., Chung, J. Y., Kim, S. H., & Kim, M. (2011). Efficacy of autologous serum in human adipose-derived stem cells; cell markers, growth factors and differentiation. *Cell Mol Biol (Noisy-le-grand), 57*(Suppl.), OL1470–OL1475.

Isakson, P., Hammarstedt, A., Gustafson, B., & Smith, U. (2009). Impaired preadipocyte differentiation in human abdominal obesity: Role of Wnt, tumor necrosis factor-alpha, and inflammation. *Diabetes, 58*, 1550–1557.

Lee, M. J., & Fried, S. K. (2012). Glucocorticoids antagonize tumor necrosis factor-alpha-stimulated lipolysis and resistance to the antilipolytic effect of insulin in human adipocytes. *American Journal of Physiology. Endocrinology and Metabolism, 303*, E1126–E1133.

Lee, M. J., Wu, Y., & Fried, S. K. (2012). A modified protocol to maximize differentiation of human preadipocytes and improve metabolic phenotypes. *Obesity (Silver Spring), 20*, 2334–2340.

Lystedt, E., Westergren, H., Brynhildsen, J., Lindh-Astrand, L., Gustavsson, J., Nystrom, F. H., et al. (2005). Subcutaneous adipocytes from obese hyperinsulinemic women with polycystic ovary syndrome exhibit normal insulin sensitivity but reduced maximal insulin responsiveness. *European Journal of Endocrinology, 153*, 831–835.

Nimitphong, H., Holick, M. F., Fried, S. K., & Lee, M. J. (2012). 25-hydroxyvitamin d(3) and 1,25-dihydroxyvitamin d(3) promote the differentiation of human subcutaneous preadipocytes. *PLoS One, 7*, e52171.

Perez, L. M., Bernal, A., San, M. N., Lorenzo, M., Fernandez-Veledo, S., & Galvez, B. G. (2013). Metabolic rescue of obese adipose-derived stems cells by Lin28/Let7 pathway. *Diabetes, 62*, 2368–2379.

Sepe, A., Tchkonia, T., Thomou, T., Zamboni, M., & Kirkland, J. L. (2011). Aging and regional differences in fat cell progenitors—A mini-review. *Gerontology, 57*, 66–75.

Shahparaki, A., Grunder, L., & Sorisky, A. (2002). Comparison of human abdominal subcutaneous versus omental preadipocyte differentiation in primary culture. *Metabolism, 51*, 1211–1215.

Skurk, T., Ecklebe, S., & Hauner, H. (2007). A novel technique to propagate primary human preadipocytes without loss of differentiation capacity. *Obesity (Silver Spring), 15*, 2925–2931.

Skurk, T., & Hauner, H. (2012). Primary culture of human adipocyte precursor cells: Expansion and differentiation. *Methods in Molecular Biology, 806*, 215–226.

Spalding, K. L., Arner, E., Westermark, P. O., Bernard, S., Buchholz, B. A., Bergmann, O., et al. (2008). Dynamics of fat cell turnover in humans. *Nature, 453*, 783–787.

Strawford, A., Antelo, F., Christiansen, M., & Hellerstein, M. K. (2004). Adipose tissue triglyceride turnover, de novo lipogenesis, and cell proliferation in humans measured with 2H2O. *American Journal of Physiology. Endocrinology and Metabolism, 286*, E577–E588.

Tchkonia, T., Giorgadze, N., Pirtskhalava, T., Tchoukalova, Y., Karagiannides, I., Forse, R. A., et al. (2002). Fat depot origin affects adipogenesis in primary cultured and cloned human preadipocytes. *American Journal of Physiology Regulatory, Integrative and Comparative Physiology, 282*, R1286–R1296.

Tchkonia, T., Giorgadze, N., Pirtskhalava, T., Thomou, T., DePonte, M., Koo, A., et al. (2006). Fat depot-specific characteristics are retained in strains derived from single human preadipocytes. *Diabetes, 55*, 2571–2578.

Tchkonia, T., Tchoukalova, Y. D., Giorgadze, N., Pirtskhalava, T., Karagiannides, I., Forse, R. A., et al. (2005). Abundance of two human preadipocyte subtypes with distinct capacities for replication, adipogenesis, and apoptosis varies among fat depots. *American Journal of Physiology. Endocrinology and Metabolism, 288*, E267–E277.

Tchoukalova, Y. D., Koutsari, C., Votruba, S. B., Tchkonia, T., Giorgadze, N., Thomou, T., et al. (2010). Sex- and depot-dependent differences in adipogenesis in normal-weight humans. *Obesity (Silver Spring), 18*, 1875–1880.

Tunaitis, V., Borutinskaite, V., Navakauskiene, R., Treigyte, G., Unguryte, A., Aldonyte, R., et al. (2011). Effects of different sera on adipose tissue-derived mesenchymal stromal cells. *Journal of Tissue Engineering and Regenerative Medicine, 5*, 733–746.

Yu, G., Floyd, Z. E., Wu, X., Hebert, T., Halvorsen, Y. D., Buehrer, B. M., et al. (2011). Adipogenic differentiation of adipose-derived stem cells. *Methods in Molecular Biology, 702*, 193–200.

Zierath, J. R., Livingston, J. N., Thorne, A., Bolinder, J., Reynisdottir, S., Lonnqvist, F., et al. (1998). Regional difference in insulin inhibition of non-esterified fatty acid release from human adipocytes: Relation to insulin receptor phosphorylation and intracellular signalling through the insulin receptor substrate-1 pathway. *Diabetologia, 41*, 1343–1354.

CHAPTER FIVE

Human Adipose-Derived Stromal/Stem Cell Isolation, Culture, and Osteogenic Differentiation

Ammar T. Qureshi[*], **Cong Chen**[*], **Forum Shah**[†], **Caasy Thomas-Porch**[‡], **Jeffrey M. Gimble**[†], **Daniel J. Hayes**[*,1]

[*]Department of Biological and Agricultural Engineering, Louisiana State University and Agricultural Center, Baton Rouge, Louisiana, USA
[†]LaCell LLC, New Orleans, Louisiana, USA
[‡]Tulane University Biomedical Science Department, New Orleans, Louisiana, USA
[1]Corresponding author: e-mail address: danielhayes@lsu.edu

Contents

Methods in Enzymology, Volume 538
ISSN 0076-6879
http://dx.doi.org/10.1016/B978-0-12-800280-3.00005-0

Abstract

Annually, more than 200,000 elective liposuction procedures are performed in the United States and over a million worldwide. The ease of harvest and abundance make human adipose-derived stromal/stem cells (hASCs) isolated from lipoaspirates an attractive, readily available source of adult stem cells that have become increasingly popular for use in many studies. Here, we describe common methods for hASC culture, preservation, and osteogenic differentiation. We introduce methods of ceramic, polymer, and composite scaffold synthesis with a description of morphological, chemical, and mechanical characterization techniques. Techniques for scaffold loading are compared, and methods for determining cell loading efficiency and proliferation are described. Finally, we provide both qualitative and quantitative techniques for *in vitro* assessment of hASC osteogenic differentiation.

1. INTRODUCTION

1.1. Human adipose-derived stromal/stem cells

To date, elective subcutaneous liposuction procedures remain the gold standard for human adipose tissue collection (Zuk et al., 2002, 2001). The surgical procedure yields anywhere from 300 mL to several liters of lipoaspirate (Katz, Llull, Hedrick, & Futrell, 1999; Zuk et al., 2002). In three consecutive articles published in the *Journal of Biological Chemistry* in 1966, Rodbell described the protocol for isolation of cells from lipoaspirate (or the stromal vascular fraction—SVF; Rodbell, 1966a, 1966b; Rodbell & Jones, 1966). This procedure consisted of homogenization of adipose tissue, consecutive saline washes to remove hematopoietic cells, collagenase digestion, and separation of undigested adipose tissue from the pelleted SVF.

The term processed lipoaspirate (PLA) cells was first used to describe a subpopulation of progenitor cells isolated from human lipoaspirates in 2001 (Zuk et al., 2001). Moreover, the embryological origin of adipose tissue offered clues of the presence of a mesenchymal stem/stromal cell (MSC) population in PLA cells. Confirmation of the multilineage potential of these PLA cells was provided 1 year later by the same group (Zuk et al., 2002).

Annually, more than 200,000 elective liposuction procedures are performed in the United States and over a million worldwide (Yoshimura et al., 2006). The ease of harvest and abundance make human adipose-derived stromal/stem cells (hASCs) isolated from lipoaspirates an attractive, readily available source of adult stem cells that have become increasingly popular for use in many studies (Zanetti, Sabliov, Gimble, & Hayes, 2013). hASCs have similar self-renewal *in vitro* when compared to human bone marrow (hBMSC) or umbilical cord stem cells (hUMSC) (Zhang et al., 2009). The ability of hASCs to differentiate in other mesodermal lineages has been demonstrated on several occasions (Zuk et al., 2002, 2001), while other studies have indicated that hASCs can be reprogrammed to behave like cells of ectodermal (Kang et al., 2004; Kingham et al., 2007) and epidermal (Trottier, Marceau-Fortier, Germain, Vincent, & Fradette, 2008) lineages. Despite their apparent pluripotential nature, hASCs lack the potential of embryonic stem cell to differentiate into all embryonic and extraembryonic tissue types (Fortunel et al., 2003). Moreover, expression of surface proteins indicated that hASC differed phenotypically from hBMSC and hUMSC (Dominici et al., 2006). In general, uncommitted hASCs express the surface markers CD29, CD44, CD71, CD90, CD105/SH2, and SH3 but do not express STRO-1 and the hematopoietic markers CD34, CD45, and C117 (Bourin et al., 2013; Yoshimura et al., 2006; Zuk et al., 2002).

Since Zuk et al. (2001) first reported the PLA cells multipotentiality, a variety of terms have been used to describe the plastic-adherent cell population isolated from digests of lipoaspirates. Efforts have been made by the regenerative medicine community to refine the nomenclature of multipotent cells isolated from adipose tissue. In 2005, the Mesenchymal and Tissue Stem Cell Committee (MTSCC) of the International Society for Cellular Therapy concluded that there is not enough scientific information available to accurately support stemness in plastic-adherent populations of cells (Hutmacher, Schantz, Lam, Tan, & Lim, 2007). Therefore, the MTSCC recommended that the use of the term mesenchymal stem cells should be reserved for population of cells with proven self-renewal and multilineage differentiation potentials. For plastic-adhered population of cells that do not meet these criteria, the MTSCC strongly encouraged the replacement of the term "stem" in MSC to "stromal" (Hutmacher et al., 2007). In agreement with this recommendation, the use of the

"ASC" acronym to mean "adipose-derived stromal/stem cells" is also accepted and this terminology has been adopted recently by both the ISCT and the International Federation for Adipose Therapeutics and Science (Bourin et al., 2013).

1.2. hASCs culture, scaffolds, and osteogenesis

Since the discovery that hASCs can undergo osteogenesis (Lawrence & Madihally, 2008; Zuk et al., 2001), research has substantially progressed toward the use of hASC as a cell source for bone regeneration. Although some applications would initially involve direct administration of hASC into the target fracture site, current paradigms describing scaffolds loaded with stromal/stem cells are thought to be preferential in guiding bone regeneration by providing support for cell colonization, migration, growth, and differentiation (Lendeckel et al., 2004). Recently, several cell characterization studies have extensively described the differentiation potential and function of hASCs both *in vitro* and *in vivo*, along with the potential advantages of scaffold-directed hASC osteogenesis (Zanetti et al., 2013).

Deriving bone from hASC requires a well-defined process wherein hASCs differentiate into osteoblast precursors (osteoprogenitor and preosteoblast), followed by the maturation of osteoblasts and the formation and mineralization of a matrix (Levi & Longaker, 2011). Zuk et al. reported mRNA expression of the osteogenic transcription factor, core binding alpha factor 1 (Cbaf-1/RunX2), was first detected in hASCs after 4 days of culture in osteogenic medium consisting of DMEM, 10% fetal bovine serum (FBS), 0.01 μ*M* 1,25-dihydroxyvitamin D3 (or 0.1 μ*M* dexamethasone), 50 μ*M* ascorbate-2-phosphate, 10 m*M* β-glycerophosphate, and 1% antibiotic/antimycotic (Zuk et al., 2001). Further research indicates upregulation of RunX2 in hASC begins as early as day 1 of culture in osteogenic medium (Liu et al., 2008). Current research indicates that the molecular mechanism of MSC osteoblastogenesis resembles the mechanism observed during the development of the skeleton, where hedgehog (James et al., 2010), Wnt, and BMP (Milat & Ng, 2009; bone morphogenic protein) signaling pathways play major roles. In particular, costimulation of the Wnt and BMP pathways produces pronounced osteogenic differentiation of multipotent cells *in vitro* (Milat & Ng, 2009). BMPs stimulate the transcription of RunX2 and osterix, which in turn activates osteoblast-specific genes such as alkaline phosphatase (ALP), osteopontin (OP), osteonectin (ON), bone sialoprotein, collagen type 1 (COL1), and osteocalcin (OCN) by binding to the osteoblast-specific *cis*-acting element 2 (OSE2) in the promoter region of these genes (Milat & Ng, 2009). ALP is an early-stage marker for osteogenic differentiation, while COL1 and OCN are only expressed by mature osteoblasts (Liu et al., 2008).

Several studies describe the combination of hASCs with human autografts, acellular xenografts, first generation of synthetic polymers or ceramic scaffolds, and second and third generations of bioabsorbable and drug-eluting scaffolds (Zanetti et al., 2013). While current scaffold materials facilitate the attachment of hASC by providing an interconnected pore structure capable of supporting cell migration, proliferation, and differentiation (Lawrence & Madihally, 2008; Muller et al., 2010), the major limitations of state-of-the-art bioabsorbable scaffolds are their relatively poor mechanical properties and brittle behavior (Hutmacher et al., 2007; van Gaalen et al., 2008). The latest innovation for bioactive and fiber-reinforced biomaterials is to use both bioactive/bioabsorbable ceramics and bioabsorbable polymeric fiber reinforcement in the same composite structure (referred as second and third generations of bioabsorbable scaffolds; van Gaalen et al., 2008).

Composite ceramics/polymer systems are being explored as an alternative to address the mechanical limitations of first-generation ceramics (Garber et al., 2013; Zanetti, McCandless, Chan, Gimble, & Hayes, 2012). The polymer phase allows for tunable rheological properties while the ceramic phase contributes to osteoinduction and osteoconduction. To date, a clear trend toward the use of composite scaffolds can already be observed in some of the current models, and this methods chapter will describe common methods for hASCs culture, scaffold fabrication, and osteogenic differentiation.

2. METHODS FOR CELL CULTURE

This section describes methods for isolation, preservation, and culture of hASC. Cells are isolated from human lipoaspirates following elective liposuction surgery under an approved IRB protocol. A method for cell culture and expansion in stromal is described and cells can be either immediately used or cryopreserved for future use. Additionally, we describe methods for osteogenic differentiation of hASC. It should be noted that potentiality of adult-derived stromal/stem cells may change with passage number.

2.1. Materials

The following can be purchased from Sigma-Aldrich (St Louis, MO, USA): ascorbate, β-glycerophosphate, dexamethasone, poly-ε-caprolactone (PCL), poly-L-lactide acid (PLLA), polyglycolic acid-*co*-poly-L-lactic acid (PGLA), beta tricalcium phosphate (β-TCP) and hydroxyapatite (HA), polyvinyl alcohol (PVA), 1,4-dioxane, calcium chloride, proteinase K, TRI reagent, alizarin red S, and PVA. The following media is available from Thermo Scientific (Waltham, MA, USA): HyClone products Dulbecco's phosphate-buffered saline (DPBS), fetal bovine serum, and DMEM/F12 Hams Media. Eight percent glutaraldehyde, cacodylate buffer, and hexamethyldisilazane (HMDS) can be purchased from Electron Microscopy Sciences (Hatfield, PA, USA).

2.2. Cell isolation and culture

- Human adipose tissue lipoaspirates are processed as described in Yu et al. (2010).
- Wash the tissue in an equal volume of 37 °C phosphate-buffered saline (PBS) until all contaminating red blood cells are removed (approximately three to four washes). Perform the steps with 100 mL of lipoaspirate in a closed 250-mL centrifuge bottle.
- Digest the tissue in an equal volume of PBS containing 1% bovine serum albumin, 0.1% collagenase type I, and 2 mM $CaCl_2$ at 37 °C for 1 h with gentle agitation (rocking at 30–40 rpm).
- Centrifuge the tissue digest at 300 × g for 5 min, resuspend the pellet, and repeat the centrifugation step one additional time.
- Aspirate the floating mature adipocyte layer and supernatant fluid and resuspend the remaining SVF cell pellet in stromal medium (DMEM/Hams F12, 10% FBS, 1% antibiotic/antimycotic).
- Centrifuge the suspended SVF cells at 300 × g for 5 min, aspirate the supernatant, and resuspend the pellet obtained from 100 mL of lipoaspirate in a volume of 105 mL of stromal medium.
- Seed the T175 flasks with 35 mL of the resuspended SVF cells (3 flasks per 100 mL of lipoaspirate) and incubate overnight in a humidified, 5% CO_2 incubator at 37 °C.
- After an 18- to 24-h incubation, aspirate the supernatant on each flask, feed the cells with fresh stromal medium, and maintain the culture with fresh medium feedings every 2–3 days until 80–90% confluency is achieved.

- Harvest the adherent ASC by trypsin digestion for 5 min at 37 °C as described by Carvalho et al. (2011), wash with fresh stromal medium, and use the cells for studies directly or cyropreserved the cells for future use (Carvalho et al., 2011).

2.3. Cell cryopreservation

- Cell cryopreservation is performed using a modification of Goh, Thirumala, Kilroy, Devireddy, and Gimble (2007).
- Suspend the harvested hASC at a concentration of 10^6 cells/mL of cryopreservation medium (10% dimethyl sulfoxide, 80% calf serum, and 10% DMEM/Hams F12).
- Place the aliquots of 1 mL of resuspended ASC in labeled cryovial tubes and place within individual rack positions inside a closed plastic container surrounded by alcohol (Mr. Frosty).
- Seal the alcohol surrounded container and placed within a −80 °C freezer overnight.
- Transfer the individual cryovials to a liquid nitrogen dewar and stored until required for experiments. At that time of use, remove the vials to dry ice, thaw rapidly in a 37 °C water bath, wash with a 10 mL volume of stromal medium, and reseed at a density between 0.5 and 5×10^3 ASC per cm^2.

2.4. Osteogenic differentiation

- For osteogenic differentiation, approximately 1×10^5 hASCs were seeded in 48-well tissue culture plate (BD Falcon, Franklin Lakes, NJ, USA) or loaded on the scaffolds by methods discussed in Section 5.
- Cells were allowed to incubate for 24 h in SM at 37 °C prior to addition of osteogenic media (OM) comprised of ascorbate, β-glycerophosphate, dexamethasone, and 1,25 vitamin D3 (Gimble, Katz, & Bunnell, 2007; Zuk et al., 2001).
- Over a 2- to 4-week period *in vitro*, human adipose-derived stem cells deposit calcium phosphate mineral within their extracellular matrix and express osteogenic genes and proteins including ALP, bone morphogenic proteins, and their receptors, OCN, ON, and OP.

3. METHODS FOR SCAFFOLD PREPARATION

Scaffolds provide a convenient method to support cell proliferation, migration, and differentiation with utility both *in vitro* and *in vivo*. This section describes methods for preparing three common scaffold classes: polymer, ceramic, and composite. Poly-ε-caprolactone (PCL), PLLA, and PGLA are common biodegradable polymers used for scaffold synthesis. These materials are interchangeable in the polymer and composite scaffold synthesis protocols in the succeeding text. β-TCP and HA are commonly used ceramic phases in bone scaffolding. Here, a 40:60% ratio of β-TCP: HA powder is used for pure ceramic and composite scaffolds, but the relative composition can be adjusted as needed to modify the mechanical properties and degradation rates of the scaffold.

3.1. Unidirectional polymer precipitation method

- The scaffolds were synthesized using a unidirectional precipitation method adapted from Qureshi et al. (2012).
- Dissolve PLLA or PCL (7–10 wt% each) in dehydrated 1,4-dioxane at 50 °C until transparent.
- After cooling to room temperature (RT), pour the solution into 17 mm (height) × 10 mm (diameter) glass cylinder vials.
- Lower the glass vials into a liquid nitrogen bath at a constant rate of 2 cm/h. (The insulated liquid nitrogen bath was designed with three orifices, each only slightly larger than the glass vial, milled into the insulated lid though which the glass vials were lowered. This process minimized evaporation of the liquid nitrogen and allowed the freezing liquid nitrogen vapor to access only the portion of the vial that had transited the orifice. The vial caps were closed to prevent absorption of moisture.)
- After freezing the samples completely for 2 h, freeze-dry the solidified PLLA/PCL scaffolds in Labconco FreeZone Plus (Kansas City, MO, USA) at −80 °C for 48 h.

3.2. Molding and thermal precipitation

- Dissolve PLLA or PCL (7–10 wt% each) in dehydrated 1,4-dioxane at 50 °C until transparent.
- After cooling to RT, pour the solution into 10 mm (height) × 4 mm (diameter) cylindrical polydimethylsiloxane (PDMS) templates and transfer to −70 °C freezer overnight.
- Freeze-dry the solidified PLLA/PCL scaffolds in Labconco FreeZone Plus (Kansas City, MO, USA) at −80 °C for 48 h.

3.3. Ceramic scaffolds

- Pure ceramic scaffolds are synthesized from an adaptation of the method of Zanetti et al. (2012).
- Suspend 3 g 40:60 (wt%) β-TCP:HA powders in 1.5 g PVA aqueous solution (10 g/100 mL solution) and stir in a glass beaker to obtain a well-dispersed slurry PVA aqueous solution.
- Cut the polyurethane foam templates into the desired shape and sizes (5 mm (height) × 4 mm (diameter)) to replicate a porous scaffold.
- Immerse the prepared templates in glass beaker containing ceramic slurry and compress with glass stick to force the slurry to migrate into the pores of the foams. Incubate at 60 °C for 1 day.
- After drying, place the scaffolded materials in an open to air boat-shaped alumina crucible and heat up in a programmable box furnace from RT (at rate of 50 °C/h) to 500 °C and allow to dwell at this temperature for 5 h in order to burn off the polyurethane foam.
- After allowing the furnace to cool down to RT by furnace exhaust, heat-treat the scaffolds at 1300 °C (increasing the furnace chamber temperature from RT at rate of 60 °C/h) for 3 h and cool them down by furnace exhaust.

3.4. Composite scaffold formation

- Dissolve PLLA or PCL (7–10 wt% each) in dehydrated 1,4-dioxane at 50 °C until transparent.
- Add nanoscale 40:60 (wt%) β-TCP:HA ceramic powder to achieve a target ratio of polymer to ceramic of 25:75; 50:50; 75:25 (dry wt/dry wt) in a 5-mL glass vial and mix by vortexing.
- Add the mixtures into 5 cm (length) × 1 cm (height) × 4 cm (width) PDMS templates with twelve 6 mm (height) × 12 mm (diameter) wells, and freeze overnight at −80 °C.
- Freeze-dry the solidified PLLA/PCL scaffolds in Labconco FreeZone Plus (Kansas City, MO, USA) at −80 °C for 48 h.

4. MATERIAL CHARACTERIZATION

This section describes the characterization of the scaffold composition, mechanical properties, and morphology. The methods of composite scaffold formation described earlier rely on thermal precipitation techniques, which may result in phase exclusion of polymer from the scaffold. As such, thermal gravimetric analysis is suggested as a method to determine the organic/inorganic phase composition. Additionally, cell migration and proliferation are dependent on the morphology of the scaffold pores. An interconnected void volume with pore neck sizes greater than the cell diameter is desirable for promoting cell in growth and nutrient/waste transport. To characterize scaffold and cell morphology, two techniques are described, scanning electron microscopy and microcomputed tomography (μCT). These two complementary techniques provide information on the cell/scaffold surface morphology and an image of the scaffold internal structure.

4.1. Thermal gravimetric analysis

- Thermal gravimetric analysis (TGA) was performed on the composite scaffold foams using TA instruments TGA 2950 with a heating rate of 10 °C/min under nitrogen flow.
- Teat at least three samples for each composite material to determine the average thermal and degradation profiles and residual mass.
- Perform the scans under nitrogen flow with a temperature range of 0–600 °C at a rate of 10 °C/min.

4.2. Mechanical testing

4.2.1 Tensile testing

- Perform the tensile testing on scaffold with cylindrical geometry of 25.4 mm (height) × 28.7 mm (diameter) at RT.
- Test the tensile strength using electromechanical universal testing machine (Instron Model 5696, Canton, MA, USA) at an extension rate of 1.3 mm/min until the elongation rupture of the specimen.

4.2.2 Compressive testing

- Perform the compressive testing on scaffold with cylindrical geometry of 6 mm (height) × 12 mm (diameter) at RT.
- Test the compressive strength using electromechanical universal testing machine (Instron Model 5696) at an extension rate of 0.5 mm/min to a maximum compression strain of 90% (Garber et al., 2013).

4.3. Scanning electron microcopy

4.3.1 Scaffolds

- The morphology of the resulting specimens is observed by scanning electron microscopy (SEM).

- For SEM observations, mount the scaffolds aluminum specimen stubs and sputter coat with gold–palladium or platinum using an SPI Module sputter coater at 7 mA for 2 min and immediately examine under SEM at 10 kV (Qureshi et al., 2012).

4.3.2 Scaffolds with hASC

- Remove the scaffolds from the cell medium and fix with a solution of 8% glutaraldehyde, 2 *M* cacodylate buffer (pH=7.2), and deionized water (DI) in the ratio of 1:2:1 for 24 h.
- Rinse the fixed scaffolds in DI three times to remove any residual fixing solution.
- Dehydrate the scaffolds with 30%, 50%, 70%, 80%, 90%, and 100% ethanol for 20 min each.
- Hydrate the scaffolds with 50% and 100% HMDS solution for 30 min each and air-dry overnight.
- Mount the scaffolds on aluminum stubs and sputter coated with platinum in an Edwards S-150 sputter coater (Edwards High Vacuum Co. International, Wilmington, MA, USA) before characterization with a Jeol JSM-6610LV (Jeol USA, Peabody, MA, USA) SEM at 10 kV (Garber et al., 2013).

4.4. Micro-CT analysis

- Slice the scaffold approximately into 1–2-mm approximate cuboids of 10–15 mm height.
- Image the scaffolds with 11 keV monochromatic X-rays with 2.5 μm/px resolution at the tomography beamline. (Projections numbered 720 corresponding to $\Delta\theta = 0.25^a$; projection exposure time varied between 2 and 4 s, but reconstruction algorithms ensure normalized data.)
- The two different data sets are directly comparable, both as an aggregate dataset and as slices. Reconstruction data are 16-bit signed integer with mean air intensity scaled to zero. Pore size was measured using ImageJ 64.
- Generate volume rendering from scaffolds 3D data using Avizo 7.0.1 (Visualization Services Group). Orthogonal slices were created using ImageJ and have equivalent scale, brightness, contrast, and gray map settings (Garber et al., 2013).

5. CELL LOADING AND SCAFFOLD CULTURE

The most appropriate method for loading hASC to the scaffold is dependent primarily on scaffold dimensions and intended analysis. This section describes three methods: (1) direct pipette application, (2) spinner flask application, and (3) perfusion reactor application for the seeding of hASC. A method for quantifying the number of cells loaded to the scaffold and determining the efficiency seeding is included.

5.1. Pipette method

- Gas sterilize the scaffolds with ethylene oxide before seeding them with hASCs.
- Place the sterilized scaffolds into tissue culture plates and presoaked (prewetted) them in stromal media for 20 min.
- Pipet the aliquots of the cell suspension (10^5 cells/mL) on topside of the scaffold and incubate at 37 °C with 5% CO_2 for 30 min before pipetting the cells on the bottom surface.
- Incubate the seeded scaffolds on an orbital shaker at 5% CO_2 and 37 °C for 2 h to enhance cell infiltration into the scaffold.
- Allow the cells to attach to scaffolds for 2 h before adding 0.5 mL of stromal media into each well. Then incubate the scaffolds undisturbed overnight to allow for cell attachment.
- The following day, remove all scaffolds from the cell culture plates and grow in the appropriate culture media.
- Throughout the experiment, maintain the cells in a humidified incubator at 37 °C with 5% CO_2.
- Also prepare acellular blank scaffolds and incubate them in identical conditions.

5.2. Spinner flask

- A spinner flask induces mixing, thus increasing oxygen and nutrient delivery and facilitating the seeding of stem cells to scaffold through turbulent flow. Flasks can be disposable (plastic) or reusable (glass) and can be easily modified to accept needles for holding the scaffolds by replacing the lid with a perforated rubber stopper or by gluing needles to the existing lid.

Figure 5.1 Spinner flask assembly on magnetic hot plate/stirrer. (See color plate.)

- Sterilize the flask and scaffolds before use (preferably using ethylene oxide gas).
- In a spinner flask, mount the scaffolds at the end of metal needles in a flask of stromal media (Fig. 5.1).
- Typically, spinner flasks are around 120 mL in volume (although much larger flasks of up to 8 L have been used). It contains a magnetic stirrer that mixes the media, typically at a rate of 50–80 rpm.
- After harvesting the cells from tissue-cultured flasks, pool hASCs from multiple donors of the same passage number and add them into the stromal media in the spinner flask.
- Stir the cell solution in the flask while submerging the scaffold, on the metal needle, in the media, for 2 h. The cell density in the solution should be around 1300 cells/mL.

- After loading the cells on the scaffold, transfer the scaffolds into 48-well plates and maintain in stromal medium for the duration of the experiment.

5.3. Perfusion reactors

- Flow perfusion bioreactors, which consist of a pump and a scaffold chamber joined together by tubing, provide more homogeneous cell distribution throughout scaffolds.
- A fluid pump is used to force media flow through the cell-seeded scaffold, typically with an inline oxygenation vessel.
- Place the scaffold in a chamber that is designed to direct flow through the interior of the scaffold. Keep the scaffold in position across the flow path of the device so the media perfuse through the scaffold, thus enhancing fluid transport. Well-described examples of bioreactor design and operation can be found in Miyoshi, Ehashi, Kawai, Ohshima, and Suzuki (2010) and Partap, Plunkett, and O'Brien (2010).

5.4. Loading efficiency

- Cell loading efficiency (C_{eff}) can be calculated by cell counts of the media before (C_{init}) and after (C_{fin}) loading using the following equation: $C_{eff} = C_{fin}/C_{init} \times 100$.

6. CELL VIABILITY AND PROLIFERATION

This section describes quantitative methods for assessing cell viability and proliferation. Metabolic assays, DNA content analysis, and quantitative live/dead cell counting via flow cytometry are described. A more complete picture of cell viability and proliferation is achieved through the use of several complimentary techniques. Metabolic assays provide insight into the overall metabolic activity of the cell through enzymatic cleavage of colorimetric of fluorescent substrates. DNA assays provide a quantitative measure of cell number based on total DNA, while live/dead flow cytometry provides a quantitative measure of cells with intact versus comprised cell membranes.

6.1. Metabolic assays

6.1.1 MTT

- The cellular viability on scaffold cultures can be determined using the cytotoxicity assay (MTT proliferation kit; Life Technologies, Carlsbad, CA, USA).
- Cut the scaffolds into 2- to 4-mm-thick cylindrical sections and place it in stromal media/OM for 7, 14, and 28 days.
- On the respective day, filter the medium (extract) and pipet (100 μL/well) into a 96-well plate previously subcultured with hASC (3000 cells/well), and incubate in a CO_2 incubator at 37 °C containing 5% CO_2 for 24 h.
- After the 24-h incubation period, add 10 μL of the MTT reagent to each well, and reincubate at 37 °C in 5% CO_2 for additional 4 h.
- Then add MTT detergent (100 μL) to each well and incubate at 37 °C in 5% CO_2 overnight.
- Measure the cell viability at a wavelength of 590 nm using a plate reader (Qureshi et al., 2012).
- Perform each experiment at least in triplicate to get statistical significance.

6.1.2 AlamarBlue assay

- The growth of hASCs can be quantitatively analyzed with a metabolic activity indicator, AlamarBlue (Life Technologies).
- Plasma treat 1-mm-thick scaffold disks for 10 s to reduce the water contact angle for more facile hASCs loading.
- Load the scaffolds with hASCs (6×10^4 cells/mL) and incubate in a 12-well plate (BD Falcon) for 7, 14, and 21 days.
- On the respective day, remove the scaffolds from culture, rinse with PBS and transfer to a fresh 12-well plate (to avoid the possibility of measuring cells seeded on to the initial well bottom during the seeding process).
- Incubate the scaffolds with 0.5 mL of 10% v/v AlamarBlue in stromal medium solution for 4 h.
- Then transfer a 100-mL sample of the mixture to a 96-well plate and measure the fluorescence (excitation wavelength = 530 nm and emission wavelength = 620 nm) of AlamarBlue using a plate reader.
- Use scaffolds with no cells as a blank control, and subtract the fluorescence of these blank scaffolds from their respective scaffolds with hASC.

6.2. Cell viability

6.2.1 Flow cytometry with live/dead staining

- This method is adapted from Qureshi et al. (2011).
- Load the scaffolds with hASCs and incubate in stromal media/OM for 7, 14, and 28 days in tissue-cultured plate.
- Harvest cells:
  - Aspirate the media solution from tissue culture plates.
  - Wash the scaffolds with 1 mL PBS (with calcium and magnesium).
  - Collect the media and PBS solution containing floating cells in 15-mL tubes.
- Trypsinization:
  - Add 1 mL 0.25% trypsin for 5–7 min for cells to detach from the scaffolds.
  - Add 1 mL of stromal media and transfer trypsin and stromal media solution to the 15-mL tube containing the floating cells.
- Centrifuge:
  - Spin down 4 mL cell solution at 345 × *g* for 5 min and aspirate carefully.
  - Resuspend pellet in 1 mL PBS.
- Stain (staining steps are done under the hood with the light off):
  - Add 4 μL of 1.25 μ*M* Sytox Red (Life Technologies) and allow 15 min for cells to stain with Sytox.
- Centrifuge:
  - Spin down PBS/Sytox Red solution at 345 × *g* for 5 min.
- Fix cells:
  - Resuspend pellet in 250 μL of 1% paraformaldehyde and transfer cells in solution to a 5-mL cytometer tube.
- Run flow cytometry on BD FACSCalibur.

6.2.2 DNA content

- Total DNA content can be used to determine the number of cells on each scaffold as previously described (Liu et al., 2008).
- Briefly, crush the scaffolds and incubate with 0.5 mL proteinase K (0.5 mg/mL) at 56 °C overnight.
- Centrifuge the mixture at 108 × *g* for 5 min and mix 50 μL aliquots with 50 μL PicoGreen dye solution (0.1 g/mL; Life Technologies) in 96-well plates.

- Measure the fluorescence of the samples 480/520 nm and calculate the total DNA concentration of the sample by comparing the fluorescence to a standard curve (DNA content vs. number if hASCs).
- Subtract the fluorescence of scaffold without cells from its respective scaffold with cells.

7. ASSESSING OSTEOGENIC DIFFERENTIATION

This section covers qualitative and quantitative assays of osteogenic marker expression. ALP is often used as an early stage but nonspecific marker of osteogenesis. Both qPCR and colorimetric enzymatic activity assays are described. Alizarin red staining (ARS) for calcium deposition is a later stage marker of extracellular matrix mineralization and can be used for both qualitative and quantitative evaluation. Additionally, a qPCR method is provided for quantification of ALP, OP, ON, RunX2, and OCN mRNA expression.

7.1. ALP colorimetric enzymatic activity assay

- Perform the ALP histochemistry on hASC-loaded scaffolds at day 7. The ALP upregulation can be qualitatively measured by Millipore kits, SCR 004 (EMD Millipore Headquarters, Billerica, MA, USA) according to the manufacturer's instructions.
- Trypsinize the hASCs from scaffolds with 0.25% trypsin for 5–7 min and fix with 4% formaldehyde in DPBS for 1–2 min.
- Centrifuge the cells at $108 \times g$ for 5 min and resuspend the pellet in DPBS.
- For quantitative measurement, isolate 2×10^4 cells from each sample and react with *p*-nitrophenylphosphate (*p*-NPP) substrate provided in the kit. The absorbance of *p*-nitrophenol can then be measured at 405 nm with plate reader.

7.2. ARS assay

7.2.1 Qualitative measurement

- Wash the hASC-loaded scaffolds with PBS and fix in 10% (v/v) formaldehyde at RT for 15 min.
- After fixation, wash the scaffolds twice with excess DI prior to adding 1 mL of 40 m*M* ARS (pH 4.1) per scaffold at RT for 20 min with gentle shaking.

- After aspiration of the unincorporated dye, wash the scaffolds six times with DI while shaking for 5 min.
- Place the plates at an angle for 2 min to facilitate removal of excess water, reaspirate, and then store at −20 °C prior to dye extraction.
- Section the scaffolds using a microtome, and mineralization is visualized by phase microscopy using an inverted microscope (Nikon).

7.2.2 Quantitative measurement

- For quantification of alizarin red S staining, add 800 μL 10% (v/v) acetic acid to stained scaffold and incubate at RT for 30 min with shaking.
- The cells, now loosely attached to the scaffolds, are transferred with 10% (v/v) acetic acid to a 1.5-mL microcentrifuge tube with a wide-mouth pipette.
- After vortexing for 30 s, overlay the slurry with 500 μL mineral oil, heat to exactly 85 °C for 10 min, and transfer to ice for 5 min. Do not open the tube until fully cooled.
- Centrifuge the slurry at 345 × *g* for 15 min and remove the 500 μL of the supernatant to a new 1.5-mL microcentrifuge tube.
- Add 200 μL of 10% (v/v) ammonium hydroxide to neutralize the acid.

Read the aliquots (150 μL) of the supernatant, in triplicate, at 405 nm in 96-well plate.

7.3. qRT-PCR/PCR

- To assess the osteogenic differentiation of hASC loaded to scaffolds at different time points, extract the total RNA cell-loaded scaffolds using the TRI reagent according to the manufacturer's instructions.
- The extracted RNA can then be used to perform Q-PCR with iScript™ one-step RT-PCR kit with SYBR® Green (Bio-Rad Laboratories, Hercules, CA, USA) using MiniOpticon™ real-time PCR detection system (Bio-Rad Laboratories).
- Use water (negative), LPS-treated cells (500 μg/mL), and hASC in osteogenic medium as controls.
- Use the sequences of PCR primers (forward and backward, 5′ to 3′) listed in Table 5.1 to assess the osteogenic differentiation of hASCs.
- Normalize the samples (ΔC_t) against the housekeeping gene 18S rRNA and calculate the $-\Delta\Delta C_t$ value of ALP, OCN, and Runx2 according to the method listed in Table 5.2.

Table 5.1 Primers design of osteogenic markers

	Forward sequence	Reverse sequence
18S rRNA	AAACGGCTACCACATCCAAG	CCTCCAATGGATCCTCGTTA
ALP	AATATGCCCTGGAGCTTCAGAA	CCATCCCATCTCCCAGGAA
Osteocalcin (OCN)	GCCCAGCGGTGCAGAGT	TAGCGCCTGGGTCTCTTCAC
RunX2	GCAAGGTTCAACGATCTGAGATT	AGACGGTTATGGTCAAGGTGAAA
Osteopontin (OP)	CAGCCTTCTCAGCCAAACG	GGCAAAAGCAAATCACTGCAA
Osteonectin	GCGGGACTGGCTCAAGAAC	GATCTTCTTCACCCGCAGCTT

Table 5.2 Calculation of $\Delta\Delta C_t$ method

Sample	C_t (target gene: ALP/OCN, etc.)	C_t (reference gene: 18S rRNA)
Stromal media (control)	A1	A2
Osteogenic media	B1	B2
ΔC_t (stromal) = A1 − A2		
ΔC_t (osteogenic) = B1 − B2		
$\Delta\Delta C_t = \Delta C_t$ (stromal) − ΔC_t (osteogenic)		

REFERENCES

Bourin, P., Bunnell, B. A., Casteilla, L., Dominici, M., Katz, A. J., March, K. L., et al. (2013). Stromal cells from the adipose tissue-derived stromal vascular fraction and culture expanded adipose tissue-derived stromal/stem cells: A joint statement of the International Federation for Adipose Therapeutics and Science (IFATS) and the International Society for Cellular Therapy (ISCT). *Cytotherapy*, *15*(6), 641–648.http://dx.doi.org/10.1016/j.jcyt.2013.02.006.

Carvalho, P. P., Wu, X., Yu, G., Dietrich, M., Dias, I. R., Gomes, M. E., et al. (2011). Use of animal protein-free products for passaging adherent human adipose-derived stromal/stem cells. *Cytotherapy*, *13*(5), 594–597. http://dx.doi.org/10.3109/14653249.2010.544721.

Dominici, M., Le Blanc, K., Mueller, I., Slaper-Cortenbach, I., Marini, F., Krause, D., et al. (2006). Minimal criteria for defining multipotent mesenchymal stromal cells. The International Society for Cellular Therapy position statement. *Cytotherapy*, *8*(4), 315–317. http://dx.doi.org/10.1080/14653240600855905.

Fortunel, N. O., Otu, H. H., Ng, H. H., Chen, J., Mu, X., Chevassut, T., et al. (2003). Comment on "'Stemness': Transcriptional profiling of embryonic and adult stem cells" and "a stem cell molecular signature". *Science*, *302*(5644), 393. http://dx.doi.org/10.1126/science.1086384302/5644/393b [pii], author reply 393.

Garber, L., Chen, C., Kilchrist, K. V., Bounds, C., Pojman, J. A., & Hayes, D. (2013). Thiol-acrylate nanocomposite foams for critical size bone defect repair: A novel biomaterial. *Journal of Biomedical Materials Research. Part A*, *101*, 3531–3541. http://dx.doi.org/10.1002/jbm.a.34651.

Gimble, J. M., Katz, A. J., & Bunnell, B. A. (2007). Adipose-derived stem cells for regenerative medicine. *Circulation Research*, *100*(9), 1249–1260.

Goh, B. C., Thirumala, S., Kilroy, G., Devireddy, R. V., & Gimble, J. M. (2007). Cryopreservation characteristics of adipose-derived stem cells: Maintenance of differentiation potential and viability. *Journal of Tissue Engineering and Regenerative Medicine*, *1*(4), 322–324. http://dx.doi.org/10.1002/term.35.

Hutmacher, D. W., Schantz, J. T., Lam, C. X. F., Tan, K. C., & Lim, T. C. (2007). State of the art and future directions of scaffold-based bone engineering from a biomaterials perspective. *Journal of Tissue Engineering and Regenerative Medicine*, *1*(4), 245–260.

James, A. W., Leucht, P., Levi, B., Carre, A. L., Xu, Y., Helms, J. A., et al. (2010). Sonic Hedgehog influences the balance of osteogenesis and adipogenesis in mouse adipose-derived stromal cells. *Tissue Engineering Part A*, *16*(8), 2605–2616. http://dx.doi.org/10.1089/ten.TEA.2010.0048.

Kang, S. K., Putnam, L. A., Ylostalo, J., Popescu, I. R., Dufour, J., Belousov, A., et al. (2004). Neurogenesis of Rhesus adipose stromal cells. *Journal of Cell Science*, *117*(18), 4289–4299. http://dx.doi.org/10.1242/jcs.01264.

Katz, A. J., Llull, R., Hedrick, M. H., & Futrell, J. W. (1999). Emerging approaches to the tissue engineering of fat. *Clinics in Plastic Surgery*, *26*(4), 587–603, viii.

Kingham, P. J., Kalbermatten, D. F., Mahay, D., Armstrong, S. J., Wiberg, M., & Terenghi, G. (2007). Adipose-derived stem cells differentiate into a Schwann cell phenotype and promote neurite outgrowth in vitro. *Experimental Neurology*, *207*(2), 267–274. http://dx.doi.org/10.1016/j.expneurol.2007.06.029.

Lawrence, B. J., & Madihally, S. V. (2008). Cell colonization in degradable 3D porous matrices. *Cell Adhesion & Migration*, *2*(1), 9.

Lendeckel, S., Jodicke, A., Christophis, P., Heidinger, K., Wolff, J., Fraser, J. K., et al. (2004). Autologous stem cells (adipose) and fibrin glue used to treat widespread traumatic calvarial defects: Case report. *Journal of Cranio-Maxillofacial Surgery*, *32*(6), 370–373. http://dx.doi.org/10.1016/j.jcms.2004.06.002.

Levi, B., & Longaker, M. T. (2011). Concise review: Adipose-derived stromal cells for skeletal regenerative medicine. *Stem Cells*, *29*(4), 576–582. http://dx.doi.org/10.1002/stem.612.

Liu, Q., Cen, L., Yin, S., Chen, L., Liu, G., Chang, J., et al. (2008). A comparative study of proliferation and osteogenic differentiation of adipose-derived stem cells on akermanite and β-TCP ceramics. *Biomaterials*, *29*(36), 4792–4799. http://dx.doi.org/10.1016/j.biomaterials.2008.08.039.

Milat, F., & Ng, K. W. (2009). Is Wnt signalling the final common pathway leading to bone formation? *Molecular and Cellular Endocrinology*, *310*(1–2), 52–62. http://dx.doi.org/10.1016/j.mce.2009.06.002.

Miyoshi, H., Ehashi, T., Kawai, H., Ohshima, N., & Suzuki, S. (2010). Three-dimensional perfusion cultures of mouse and pig fetal liver cells in a packed-bed reactor: Effect of medium flow rate on cell numbers and hepatic functions. *Journal of Biotechnology*, *148*(4), 226–232. http://dx.doi.org/10.1016/j.jbiotec.2010.06.002.

Muller, A. M., Mehrkens, A., Schafer, D. J., Jaquiery, C., Guven, S., Lehmicke, M., et al. (2010). Towards an intraoperative engineering of osteogenic and vasculogenic grafts from the stromal vascular fraction of human adipose tissue. *European Cells & Materials*, *19*, 127–135.

Partap, S., Plunkett, N., & O'Brien, F. (2010). Bioreactors in tissue engineering. In A. Lazinica (Ed.), *Tissue engineering* (pp. 323–336). Vienna: IN-TECH (ISBN 978-953-7619-XX).

Qureshi, A. T., Monroe, W. T., Lopez, M. J., Janes, M. E., Dasa, V., Park, S., et al. (2011). Biocompatible/bioabsorbable silver nanocomposite coatings. *Journal of Applied Polymer Science*, *120*(5), 3042–3053. http://dx.doi.org/10.1002/app.33481.

Qureshi, A. T., Terrell, L., Monroe, W. T., Dasa, V., Janes, M. E., Gimble, J. M., et al. (2012). Antimicrobial biocompatible bioscaffolds for orthopaedic implants. *Journal of Tissue Engineering and Regenerative Medicine*. http://dx.doi.org/10.1002/term.1532.

Rodbell, M. (1966a). Metabolism of isolated fat cells. II. The similar effects of phospholipase C (Clostridium perfringens alpha toxin) and of insulin on glucose and amino acid metabolism. *The Journal of Biological Chemistry*, *241*(1), 130–139 [In Vitro].

Rodbell, M. (1966b). The metabolism of isolated fat cells. IV. Regulation of release of protein by lipolytic hormones and insulin. *The Journal of Biological Chemistry*, *241*(17), 3909–3917 [In Vitro].

Rodbell, M., & Jones, A. B. (1966). Metabolism of isolated fat cells. 3. The similar inhibitory action of phospholipase C (Clostridium perfringens alpha toxin) and of insulin on lipolysis stimulated by lipolytic hormones and theophylline. *The Journal of Biological Chemistry*, *241*(1), 140–142 [In Vitro].

Trottier, V., Marceau-Fortier, G., Germain, L., Vincent, C., & Fradette, J. (2008). IFATS collection: Using human adipose-derived stem/stromal cells for the production of new skin substitutes. *Stem Cells*, *26*(10), 2713–2723. http://dx.doi.org/10.1634/stemcells.2008-0031.

van Gaalen, S., Kruyt, M., Meijer, G., Mistry, A., Mikos, A., van den Beucken, J., et al. (2008). Chapter 19—Tissue engineering of bone. In C. van Blitterswijk, P. Thomsen, A. Lindahl, J. Hubbell, D. F. Williams, R. Cancedda, J. D. de Bruijn, & J. Sohier (Eds.), *Tissue engineering* (pp. 559–610). Burlington, CA: Academic Press.

Yoshimura, K., Shigeura, T., Matsumoto, D., Sato, T., Takaki, Y., Aiba-Kojima, E., et al. (2006). Characterization of freshly isolated and cultured cells derived from the fatty and fluid portions of liposuction aspirates. *Journal of Cellular Physiology*, *208*(1), 64–76. http://dx.doi.org/10.1002/jcp.20636.

Yu, G., Wu, X., Dietrich, M. A., Polk, P., Scott, L. K., Ptitsyn, A. A., et al. (2010). Yield and characterization of subcutaneous human adipose-derived stem cells by flow cytometric and adipogenic mRNA analyzes. *Cytotherapy*, *12*(4), 538–546. http://dx.doi.org/10.3109/14653241003649528.

Zanetti, A. S., McCandless, G. T., Chan, J. Y., Gimble, J. M., & Hayes, D. J. (2012). Characterization of novel akermanite:poly-ε-caprolactone scaffolds for human adipose-derived stem cells bone tissue engineering. *Journal of Tissue Engineering and Regenerative Medicine*. http://dx.doi.org/10.1002/term.1646.

Zanetti, A. S., Sabliov, C., Gimble, J. M., & Hayes, D. J. (2013). Human adipose-derived stem cells and three-dimensional scaffold constructs: A review of the biomaterials and models currently used for bone regeneration. *Journal of Biomedical Materials Research. Part B, Applied Biomaterials*, *101*(1), 187–199. http://dx.doi.org/10.1002/jbm.b.32817.

Zhang, Z.-Y., Teoh, S.-H., Chong, M. S. K., Schantz, J. T., Fisk, N. M., Choolani, M. A., et al. (2009). Superior osteogenic capacity for bone tissue engineering of fetal compared with perinatal and adult mesenchymal stem cells. *Stem Cells*, *27*(1), 126–137. http://dx.doi.org/10.1634/stemcells.2008-0456.

Zuk, P. A., Zhu, M., Ashjian, P., De Ugarte, D. A., Huang, J. I., Mizuno, H., et al. (2002). Human adipose tissue is a source of multipotent stem cells. *Molecular Biology of the Cell*, *13*(12), 4279–4295. http://dx.doi.org/10.1091/mbc.E02-02-0105.

Zuk, P., Zhu, M., Mizuno, H., Huang, J., Futrell, J., Katz, A., et al. (2001). Multilineage cells from human adipose tissue: Implications for cell-based therapies. *Tissue Engineering*, 7(2), 211–228.

CHAPTER SIX

Analysis of Adipose Tissue Lipid Using Mass Spectrometry

Rodney C. Baker*, Yana Nikitina†, Angela R. Subauste†,[1]

*Department of Pharmacology and Toxicology, University of Mississippi Medical Center, Jackson, Mississippi, USA

†Division of Endocrinology, Department of Medicine, University of Mississippi Medical Center, Jackson, Mississippi, USA

[1]Corresponding author: e-mail address: asubauste@umc.edu

Contents

Abstract

Mass spectrometry technology has enabled significant advances in detailing the alterations of the lipidome in response to pathological conditions or experimental manipulations. Lipids comprise a wide range of compounds with functions that include structural, intracellular signaling, trafficking, and storage. Characterization of lipid species has evolved significantly over recent years due to the progress made in the area of mass spectrometry. This chapter details the methods used for the analysis of lipids tailored to the intrinsic characteristics of adipose tissue. Particular attention is given to the analysis of triglycerides, diacylglycerols, and phospholipid.

Methods in Enzymology, Volume 538
ISSN 0076-6879
http://dx.doi.org/10.1016/B978-0-12-800280-3.00006-2

1. INTRODUCTION

Recent developments in mass spectrometry technology have enabled the comprehensive analysis of the lipidome, fostering significant advances in the molecular biology of metabolism-related diseases. Lipidome profiling by mass spectrometry can be preceded by the chromatographic separation of lipid species. The former identifies and quantifies lipid species preseparated by normal or reversed-phase chromatography coupled online to a mass spectrometer, which then leads to the acquisition of MS or MS/MS spectra. An alternative approach is shotgun lipidomics, a total lipid extract is infused directly into a mass spectrometer, and the molecular characterization of lipid species relies either on the accurately determined mass–charge ratio (m/z) of precursor ions or on the detection of specific fragment ions or neutral losses in tandem mass spectrometric experiments.

The analysis of the different signaling lipid species by mass spectrometry is particularly challenging in adipose tissue given the high abundance of neutral lipids, which leads to massive ion suppression. Also, high concentrations of neutral lipids are known to accumulate in the column or precolumn, interfering with the separation of more polar compounds. In this chapter, we will focus on the analysis of glycerolipids and phospholipids using a combination of direct infusion and HPLC separation in line with an MS or MS/MS system.

Key in the development of this protocol is the extraction protocol, which is aimed toward the separation of neutral lipids and phospholipids to increase the sensitivity in the detection of small-abundance lipids while simplifying the analysis. Figure 6.1 shows a scheme of the sample analysis used in this chapter.

2. NEUTRAL LIPIDS

Neutral lipids are the most abundant lipids in adipose tissue and include triacylglycerols (TAG), diacylglycerols (DAG), cholesteryl esters, and cholesterol. Another layer of complexity is added by the fact that they exist as a mixture of different species that differ in the fatty acyl group; this is of particular importance in the glycerolipid group. Discriminating glycerolipids at this level of detail has already demonstrated the novel regulatory roles of specific lipids in metabolism. For example, there is already evidence of a graded relationship between specific TAGs and insulin resistance. In plasma, TAGs of lower carbon number and double-bond content

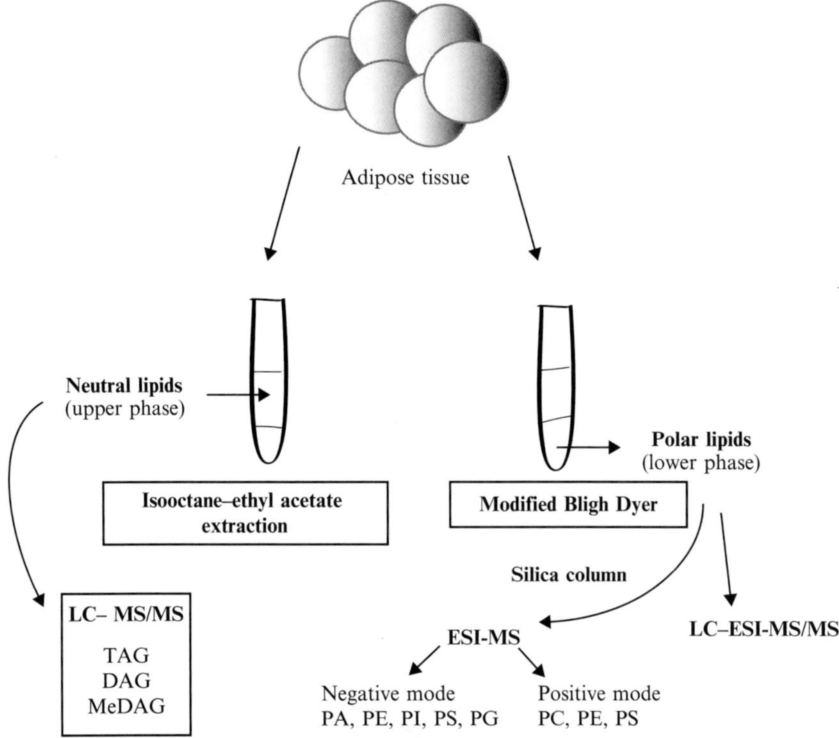

Figure 6.1 Scheme showing the extraction and analysis followed for adipose tissue.

were associated with an increased risk of type 2 diabetes, whereas TAGs of higher carbon number and double-bond content were associated with a decreased risk of type 2 diabetes (Rhee et al., 2011). In this section, we will detail the methods for analyzing neutral lipids in adipose tissue samples placing particular attention on the identification of DAGs and TAGs.

Reagents

- Phosphate-buffered saline, pH 7.4 (PBS) (Catalog No. 10010-023; Life Technologies, Carlsbad, CA)
- Isooctane (Catalog No. O296-1; Fisher Scientific, Waltham, MA)
- Ethyl acetate (Catalog No. E195-1; Fisher Scientific, Waltham, MA)
- 14:0-16:1-14:0 D5 TAG (Catalog No. 110541; Avanti Polar Lipids, Alabaster, AL)
- 1,3-14:0 D5 DAG (Catalog No. 110535; Avanti Polar Lipids, Alabaster, AL)
- Methyl tert-butyl ether (MTBE) (Catalog No. 650560, Sigma-Aldrich, St Louis, MO 63103)

- Hexane (Catalog No. 650552, Sigma-Aldrich, St Louis, MO 63103)
- Ammonium acetate (Catalog No. A7330, Sigma-Aldrich, St Louis, MO 63013)
- 2-Propanol (Catalog No. 34965, Sigma-Aldrich, St Louis, MO 63013)
- Acetonitrile (Catalog No. 34967, Sigma-Aldrich, St Louis, MO 63013)
- Dichloromethane (Catalog No. 650463, Sigma-Aldrich, St Louis, MO 63103)
- Water LC–MS (Catalog No. W6-4, Fisher Scientific, Hanover Park, IL 60135)

2.1. Extraction protocol

The analytical procedure depends on the completeness of the extraction of lipids and the simplification of the sample. All major classes of lipids could be recovered via chloroform/methanol extraction, typically according to the Bligh and Dyer recipes (Bligh & Dyer, 1959), in which they are mostly enriched in the chloroform phase (lower phase). One of the limitations of this approach is that the analysis of a more complex sample requires a very involved separation process, which could limit its widespread application. The extraction protocol described here aims at the separation of neutral lipids from polar lipids. This approach allows for a more complete and simplified analysis of the different neutral lipids present in adipose tissue. It is important to note that while TLC methods have provided a good separation of neutral lipids based on polarity, its use in lipidomics is limited by the fact that lipids are prone to oxidation during TLC (DeLong, Baker, Samuel, Cui, & Thomas, 2001).

The extraction protocol is based on the work of Hutchins, Barkley, and Murphy (2008) utilizing isooctane/ethyl acetate extraction. This less polar immiscible solvent inefficiently extracts the more polar phospholipids and sphingolipids while extracting over 90% of the neutral lipids.

Human adipose samples were obtained through needle aspiration of subcutaneous fat from the lower abdomen. Through this procedure, an average yield of 20–50 mg of adipose tissue is obtained. Immediately after the sample is obtained, the tissue is cleaned from any blood with cold PBS. Samples should be quickly frozen by immersion in liquid nitrogen and stored at −80 °C if not immediately processed. As a general rule, samples should be kept cold during the entire procedure. This is to minimize the potential oxidation of unsaturated fatty acids. It is important to take into consideration that samples intended for lipid analysis should not be stored long term at −80 °C without further extraction. Previous work has demonstrated that after 1 year of storage of plasma samples at −80 °C, there are already alterations in some lipid parameters. In contrast, there is evidence that lipid extracts are stable for nearly 4 years when stored at −80 °C (Hodson, Skeaff, Wallace, & Arribas, 2002). The use of butylated hydroxytoluene should be considered when storing samples to minimize the potential oxidation of the sample. The usual concentration is 1 mg per 10 ml added to the solvent used for storage purposes.

Samples are then placed in 2 ml of ice-cold PBS. Tissue is then homogenized in a tight-fit glass Dounce homogenizer (Kimble/Kontes Glass Co., Vineland, NJ) for approximately 1 min. Samples are kept on ice during homogenization. Then, isooctane/ethyl acetate (75:25) 4 ml containing internal standards is added. In general, solvent to sample ratio should be at least 20 ml:1 g. The larger the ratio, the better the extraction efficiency. Internal standards were added in the following amounts: a mixture of TAG (14:0–16:1–14:0 D_5 TG) and DAG (1:3–14:0 D_5 DG). Internal standard is added to a final concentration of 1 μg/sample. Vortex on a tabletop touch mixer until sample and solution are fully mixed. Centrifuge for 10 min at 4 °C and 900 × *g*. Collect the organic fraction (top). Repeat the extraction on the bottom fraction with 4 ml of 75:25 isooctane/ethyl acetate and combine both organic fractions.

2.2. LC separation of neutral lipids

While neutral lipid samples can be analyzed by direct infusion, the most accurate data are obtained by chromatographic separation. This is particularly the case for TAG analysis as the presence of monoalkyldiacylglycerols can potentially have the same mass as TAGs containing one odd-chain fatty acyl (Hutchins et al., 2008). These two lipids can be easily separated by HPLC.

A typical method for the separation of neutral lipid classes by HPLC employs normal-phase chromatography using either silica, amino, or diol columns eluted with normal-phase solvents; the HPLC eluent is doped with ammonium acetate just before introduction into the mass spectrometer. Hutchins et al. (2008) reported the class separation of cholesterol esters and TAG and the separation of 1,3- and 1,2-diacylglycerols. For separation by NP-LC, the prepared sample solution (40 μl) is loaded onto a 250 × 4.6 mm, 5 μm, silica column fitted with a 2 × 4 mm silica guard cartridge. The neutral lipid classes are eluted with a 1 ml/min gradient of

MTBE in hexane as follows: 4.5% MTBE is isocratic from 0 to 10 min; from 10 to 20 min, MTBE is ramped to 45%, where it remains for 1 min; and from 21 to 22 min, MTBE is returned to 4.5%, where it remains until the end of the run at 27 min. Ten percent of the HPLC eluent (100 μl) is introduced into the mass spectrometer after mixing with 30 μl of 10 m*M* ammonium acetate in 45:45:5:5 isopropanol–acetonitrile–water–dichloromethane. The ammonium acetate solution is introduced into the HPLC eluent by a second pump.

2.3. MS/MS of triglycerides and diacylglycerides

In general, the characterization of a complex sample requires the acquisition of a positive or negative lipid charge. For TAGs and DAGs, this is achieved through the generation of NH_4^+ adduct ions (Han & Gross, 2001; Hsu & Turk, 1999). The analysis is carried out with neutral loss experiments. A typical method is outlined here: triglycerides and diacylglycerides are detected using a hybrid quadrupole/linear ion trap mass spectrometer (AB SCIEX 4000 QTRAP). Individual compounds are identified by trapping and scanning selected ions from the ion trap (enhanced product ion experiments). The mass spectrometer is operated in the positive mode with a spray voltage to 5000, a scan rate of 4000 Da/s, and mass range from *m/z* 500 to 1000. The instrument is operated at high collision energy, temperature at 200 °C, and declustering potential at 100. The settings for the enhanced product ion experiment are as follows: positive mode, scan rate 4000 Da/s unit Q1 resolution, and dynamic ion trap fill time. Sample is introduced in 45:45:10 methanol–acetonitrile–ammonium acetate over a 10 min period at a flow rate of 10 μl/min. Data are collected from *m/z* 500 to 1200.

The mass spectral analysis of $[M+NH_4]+$ from the neutral lipid extract revealed a complex mixture of both TAGs and DAGs. Figure 6.2B depicts the TAG spectra. For each reported *m/z*, there are numerous isobaric species, which limit the successful identification of each triglyceride (McAnoy, Wu, & Murphy, 2005). For example, for the TAG ion corresponding to 48:1, a total of 23 isobaric species can be identified. MS/MS of $[M+NH_4]^+$ generates three DAG-type ions. The difference in mass (neutral loss) between the TAG and the DAG ion reveals the fatty acyl chain (Fig. 6.2A). For example, as shown in Fig. 6.2B, MS/MS of the *m/z* (872.8) revealed the abundant loss of 273, 299, 297, and 295 at *m/z* 599, 575, 573, and 572 (corresponding to fatty acyl groups 16:0, 18:1, 18:2,

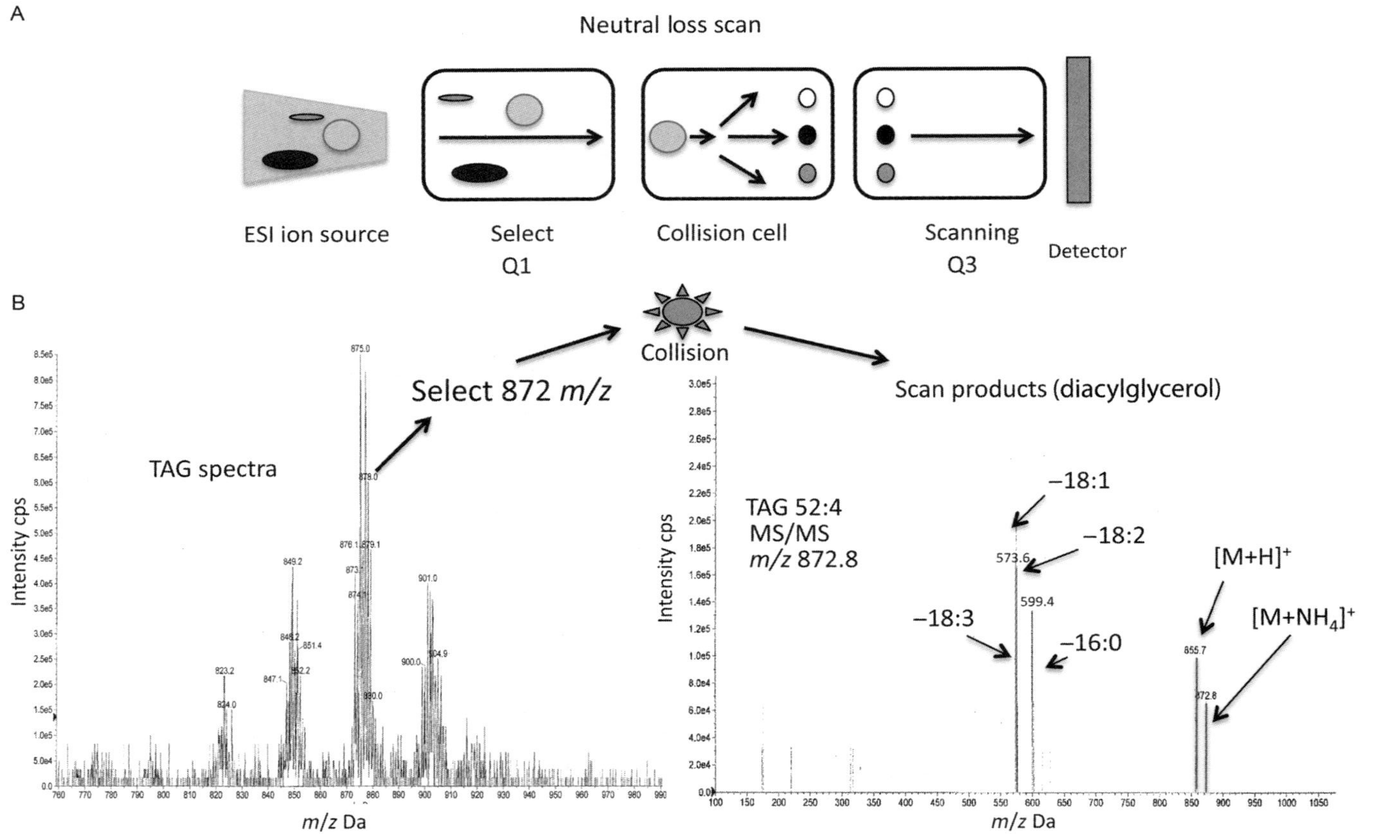
A
Neutral loss scan
ESI ion source
Select
Q1
Collision cell
Scanning
Q3
Detector
B
Collision
Select 872 m/z
Scan products (diacylglycerol)
TAG spectra
Intensity cps
m/z Da
TAG 52:4
MS/MS
m/z 872.8
−18:1
−18:2
−18:3
−16:0
[M+H]+
[M+NH4]+
573.6
599.4
855.7
875.0
849.2
901.0
823.2

Figure 6.2 (A) Scheme of a neutral loss scan. (B) Positive-ion electrospray ionization of human adipose tissue neutral lipid fraction extracted from approximately 20 mg of tissue. No D5 internal standards were present in this sample. The electrospray mobile phase contained 10 m*M* NH_4OAc so that the ammonium adduct ion for each molecular species of TAGs predominated. Collision-induced dissociation of *m/z* 872.8 corresponding to the $[M+NH_4]+$ for 52:4 TAG molecular species present in the adipose tissue lipid extract. Each product ion corresponds to neutral loss of each fatty acyl group plus ammonia (NH_3) present in these isobaric molecular species of TAGs.

and 18:3, respectively). Based on the fatty acyl profile, a list of possible TAGs can be determined. For our particular example, *m/z* 872.8 corresponds to TAG 52:4. One can then infer that the possible combinations include (16:0/18:1/18:3) and (16:0/18:2/18:2). It is important to keep in consideration that the only way to ascertain the unequivocal identification of TAGs requires MS/MS/MS (McAnoy et al., 2005). In the case of DAGs, a single neutral loss provides enough information to identify the both fatty acyl substituents in a molecular species. The samples can be systematically analyzed through automatic identification with software as Lipid Profiler or LipidView™ AB SCIEX. Table 6.1 shows the expected mass loss for each fatty acyl component during MS/MS.

3. PHOSPHOLIPIDS

Phospholipids are the major component of cellular membranes and have an important role in its physical and chemical properties. In regard to adipose tissue, recent literature has determined that obesity has a significant effect on the phospholipid composition of adipose tissue. These changes affect membrane fluidity and allow for the appropriate adipose expansion while maintaining adequate cellular homeostasis (Pietilainen et al., 2011). Furthermore, phospholipids have also been linked to signal transduction, as is the case for phosphatidic acid (PA), which is a known activator of mTOR (Foster, 2013). These signaling lipids are characterized by low concentrations, rapid synthesis and degradation, and specific spatial localization, consistent with the highly specific responses required for signaling.

Electrospray ionization (ESI) has become a very valuable tool in the analysis of phospholipids. ESI ionization is based on the differential propensity of each lipid class to acquire either a positive or a negative charge (Brugger, Erben, Sandhoff, Wieland, & Lehmann, 1997). This allows for the resolution of lipid classes directly from unprocessed lipid extracts without prior chromatographic separation. In other words, lipid classes can be separated through their endogenous electric potential, obviating chromatographic procedures. This is a very useful method in particular when determining qualitative changes in the lipidome under different conditions. In general, the quantitative analysis of phospholipids is best done after chromatographic separation by class to avoid the mass overlap of phospholipids from a complex biological sample.

Table 6.1 Mass losses observed for MS/MS experiments of TAG and DAG

	Neutral loss (MS/MS)
	$RCOOH + NH_3$
14:0	245
14:1	243
16:0	273
16:1	271
16:2	269
17:0	287
18:0	301
18:1	299
18:2	297
18:3	295
18:4	294
20:0	329
20:1	327
20:2	325
20:3	323
20:4	321
20:5	319
20:6	317
22:0	357
22:1	355
22:2	353
22:3	351
22:4	349
22:5	347
22:6	345

As mentioned previously, one of the challenges when working with adipose tissue is the high concentration of neutral lipids. Direct infusion of adipose lipid extracts obtained from a methylene chloride–methanol–water extraction can give significant data when analyzing highly abundant phospholipids (i.e., phosphatidylcholine (PC), phosphatidylethanolamine (PE)). If the array aims at the identification of lower-abundant lipids, the extraction protocol should seek the isolation of the phospholipid fraction. In this section, we include the separation of phospholipids with a silica column targeted to the phospholipid fraction. This approach has been successful in the study of the phospholipid composition of lipid droplets (Tauchi-Sato, Ozeki, Houjou, Taguchi, & Fujimoto, 2002). We describe the value of ESI–MS for the separation of phospholipids under different classes. We also describe the targeted analysis of low-abundance phospholipids involved in PPARγ activation using LC–ESI–MS/MS.

Reagents

- PBS (Catalog No. 10010-023; Life Technologies, Carlsbad, CA)
- Methylene chloride (Catalog No. 650463, Sigma-Aldrich, St Louis, MO 63013)
- Methanol (Catalog No. 14262, Sigma-Aldrich, St Louis, MO 63013)
- Acetic acid (Catalog No. BP1185-500 Fisher Scientific, Hanover Park, IL 60135)
- Hexane (Catalog No. 650552, Sigma-Aldrich, St Louis, MO 63103)
- Silica gel (Catalog No. 7024-5 J.T. Baker Co., Phillipsburg, NJ)
- Solid-phase extraction column (Catalog No. 7121-03 J.T. Baker Co., Phillipsburg, NJ)

3.1. Extraction protocol

Sample from adipose biopsy is quickly frozen by immersion in liquid nitrogen. The frozen sample is then placed in a tight-fit glass homogenizer with 1.0 ml of PBS and homogenized for 1 min while keeping the sample cold. Methylene chloride (1 ml) and methanol containing 2% acetic acid (2 ml) are added, which yield a monophase. The sample is mixed vigorously and held at room temperature for 10 min. The organic and methanol/water phases are separated by adding 1 ml of PBS and 1 ml of methylene chloride, followed by centrifugation for 10 min. The organic lower layer is dried under nitrogen and stored in methylene chloride–methanol (2/1) or transferred to solid-phase extraction column (silica gel, SPE) using approximately 1 ml of hexane five times.

Neutral and phospholipid fractions are separated using a 0.3 g 40 μ*M* silica gel in a 3 ml SPE column. The column is conditioned with hexane containing 2% methylene chloride. The neutral lipid fraction is eluted with 5 ml of methylene chloride and the phospholipid fraction with methanol. To assure all phospholipids are transferred to the solid-phase extraction column, the original sample tube is washed with 1.0 ml of 2:1 methylene chloride–methanol followed by 1.0 ml of methanol, and each wash is transferred to the SPE column prior to eluting the phospholipid fraction. Phospholipid and neutral fractions are dried under nitrogen. The cholesterol ester fraction is then eluted with chloroform (5 ml) followed by a neutral lipid fraction eluted with acetone (20 ml) and a phospholipid fraction eluted with methanol (5 ml). The phospholipid fraction is dried under nitrogen and suspended in methylene chloride–methanol. Methylene chloride is used for the suspension of the neutral fraction. The use of acetic acid in this particular lipid extraction protocol is aimed at improving the extraction of more polar lipids such as PA and lysophosphatidic acid (LPA). Plasmalogens can be degraded by strong acids but are stable to acetic acid at moderate concentrations.

3.2. Direct infusion

All phospholipid classes can be detected and identified in a single run; however, in order to detect and identify low-abundant molecular species, samples are generally analyzed in both negative and positive modes. Each phospholipid class has a specific fragmentation pattern that can be used during the identification process.

Phospholipid samples can acquire either a positive or a negative charge during ESI. This allows for better identification of species even in the presence of a relatively complex sample. Positive mode can detect PC, phosphatidylserine (PS), and PE. Six lipid classes and their lyso variants can be detected in negative mode: phosphatidylinositol (PI), PS, PA, PE, and phosphatidylglycerol. Fully annotated MS/MS spectra for all six major glycerophospholipids classes are available at LIPID MAPS public website: http://www.lipidmaps.org/data/standards/standards.php?lipidclass=LMGP.

Phospholipids were detected using a hybrid quadrupole/linear ion trap mass spectrometer (AB SCIEX 4000 QTRAP) equipped with a Harvard Apparatus syringe pump (Harvard Apparatus, Holliston, MA) and an electrospray source. Samples are analyzed at an infusion rate of 10 ml/min in both positive and negative ionization modes over the range of m/z 400 to 1200. The instrument was operated with the following settings: spray voltage, 5500; temperature, 200 °C; declustering potential, 100; and collision energy, 40. Instrument parameters are optimized with 1,2-dioctanoyl-sn-glycero-3-phosphotidylethanolamine (16:0 PE) prior to analysis. This particular phospholipid class results in adequate spectra for both positive and negative scans.

In precursor ion scans, also known as parent ion scans, the precursor ion collides to create fragments or product ions. Only those compounds that give a specific fragment ion are detected (Fig. 6.3A). The software produces the spectrum of compounds, which produced the selected precursor ion. For example, a positive-ion mode precursor ion scan of m/z 184 is specific for phosphocholine-containing phospholipids, and the spectra originated

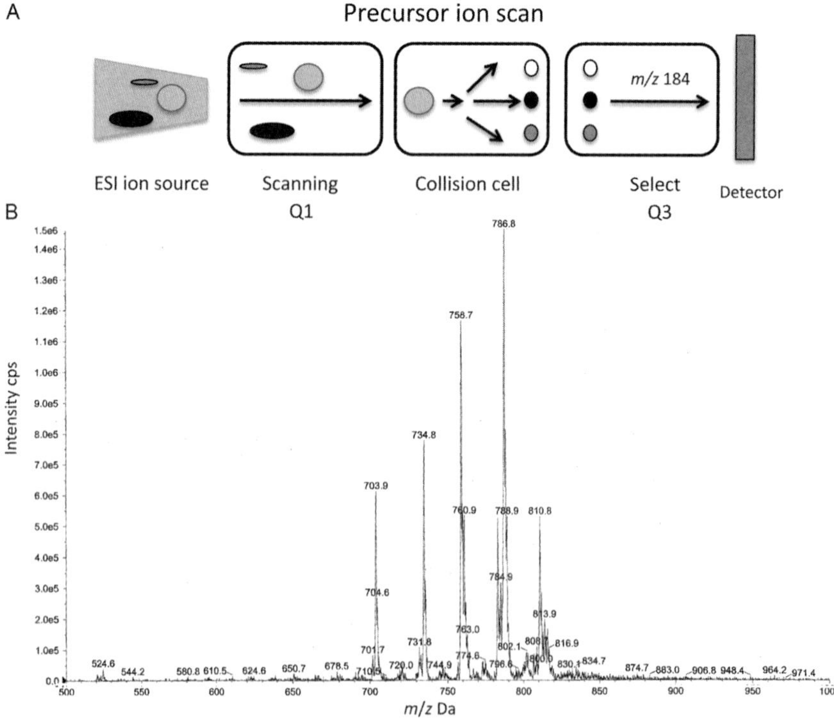

Figure 6.3 (A) Scheme showing reconstruction of the precursor ion. (B) Example of phosphatidylcholine spectra from adipose tissue using the ion precursor scan. In this example, the mass spectrometer is set for scanning precursor ions that produce 184 Da fragment ions (phosphocholine) in the third quadrupole in positive mode.

	Specificity	Fragment structure	Scan mode	Fragment type
Negative mode	All [M-H]⁻ ions of glycerophospholipids	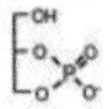	Precursor of 153	glycero phosphate-H_2O
	PI	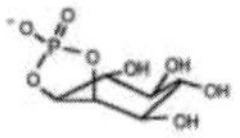	Precursor of 241	Head group -H_2O
	PS	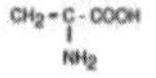 CH_2-C-COOH NH_2	Precursor of 87	Head group -H_3PO_4
	PE	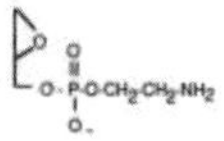 O-P-O-CH_2-CH_2-NH_2	Precursor of 196	dilyso -H_2O
	Sphingomyelin	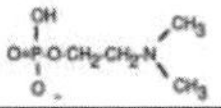 O=P-O-CH_2-CH_2-N(CH_3)CH_3	Precursor of 168	Head group (demethylated)
Positive mode	PC and sphingomyelin	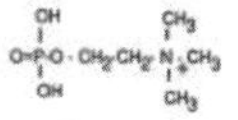 O=P-O-CH_2-CH_2-N^+(CH_3)$_3$	Precursor of 184	Head group
	PE	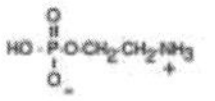 HO-P-O-CH_2-CH_2-NH_3^+	Neutral loss of 141	Head group
	PS	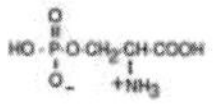 HO-P-O-CH_2-CH-COOH $^+NH_3$	Neutral loss of 185	Head group

Figure 6.4 Phospholipid class-specific scan modes available in positive- and negative-ion scan. *Modified from Brugger et al. (1997).*

from adipose tissue are shown in Fig. 6.3B. Brugger et al. (1997) described the general principles of phospholipid fragmentation. A summary of the specific fragments for phospholipid analysis is shown in Fig. 6.4.

3.3. LC–MS analysis of phospholipids

In general, low-abundance phospholipids, usually involved in signaling, are often not detected when analyzed in the presence of high-abundance phospholipids, due to the effect of ion suppression. The chromatographic separation of phospholipids by class prior to mass analysis increases the detection of low-abundance phospholipids. Moreover, the presence of many isomers with a narrow range of mass and isotopic ions makes quantitative measurement of phospholipids difficult. By using liquid chromatography, the information regarding retention time will enable the separate identification of isobaric molecular species. Similarly, the separation of low-abundance lipid species from high-abundance lipid species by LC also prevents the occurrence of ion suppression in MS.

Of significant interest in adipose tissue biology is the study of lipid intermediates described as endogenous PPAR γ ligands. LPA and cyclic phosphatidic acid, two closely related lipids, have been described as PPARγ agonist (LPA) and antagonist (CPA) (McIntyre et al., 2003; Tsukahara et al., 2010). In this section, we will describe the protocol for targeted analysis of these lipids. The protocol will detail the lipid extraction from cells and subsequent LC–MS/MS analysis (Subauste et al., 2012).

Reagents

- Methanol (Catalog No. 14262, Sigma-Aldrich, St Louis, MO 63013)
- Acetic acid (Catalog No. W6-4, Fisher Scientific, Hanover Park, IL 60135)
- LPA 16:0 (CAS 22002-85-3, Cayman Chemical, Ann Arbor, MI 48108)
- Chloroform (Catalog No. C607-1, Fisher Scientific, Hanover Park, IL 60135)
- Ammonium acetate (Catalog No. A-114-50, Fisher Scientific, Hanover Park, IL 60135)
- Ammonium hydroxide (Catalog No. A- 470-500, Fisher Scientific, Hanover Park, IL 60135)
- Acetonitrile (Catalog No. A956-4, Fisher Scientific, Hanover Park, IL 60135)
- 2-Propanol (Catalog No. A464-1, Fisher Scientific, Hanover Park, IL 60135)

3.3.1 Lipid extraction from cultured cells

Cell culture plates (10 cm dishes) are aspirated to remove media, rapidly rinsed with water, and immediately quenched by rapid addition of liquid nitrogen directly to the culture plate. The entire rinse–quench procedure has been shown not to induce significant changes in the metabolite profile of cultured cells (Lorenz, Burant, & Kennedy, 2011). Two milliliters of methanol containing 2% acetic acid and 1 nM $^{13}C_{16}$ 16:0 LPA are then added to the quenched culture plate, and cells are released from the plate surface by scraping. The cell suspension is transferred to a glass tube and 1 ml of chloroform is added, to yield the 2:1 composition of methanol–chloroform. The monophase is mixed and kept at room temperature for 10 min. The bottom organic layer is then transferred to another tube and dried under nitrogen. The dried lipid is dissolved in 200 μl of 0.1 M ammonium acetate in methanol. This sample is transferred to a microcentrifuge tube, centrifuged at 10,000 × g for 5 min at 4 °C, and transferred to an autosampler vial for further analysis.

3.3.2 LC–ESI–MS/MS of cyclic phosphatidic acid and LPA

Initial screening of PA and LPA can be carried out using normal-phase columns to isolate the two lipid classes. Identification of both PA and LPA molecular species can be accomplished by precursor ion scanning (negative precursors of *m/z* 153). If the concentration of a specific molecular species of either PA or LPA is to be measured, reverse-phase HPLC and multiple reaction monitoring (MRM) are a better choice of methods compared to product ion scans and normal-phase separation. A typical method as reported by Subauste et al. (2012) using an Agilent 6410 triple quadrupole mass spectrometer (Agilent Technologies, Santa Clara, CA) is outlined later.

HPLC separation is performed using an Xbridge C18 column (50 × 2.1 mm, 3 μm, Waters Corp., Milford, MA). Mobile phase A consists of 5 m*M* ammonium acetate in water, adjusted to pH 9.9 with ammonium hydroxide. Mobile phase B is 60% acetonitrile and 40% 2-propanol. The gradient begins with a linear increase from 20% B to 99.5% B over 15 min, followed by a 5 min hold at 99.5% B. The mobile phase is then returned to 20% B and the column is flushed for 8 min prior to beginning the next run. The sample injection volume is 40 μl. Online tandem MS analysis is performed using negative-ion electrospray in MRM mode. MRM is a quantitative target analyte scan. MS parameters are gas temperature 325 °C, drying gas flow 10 l/min, nebulizer 40 psig, and capillary voltage 4000 V. Optimum MRM parameters are determined by flow-injection analysis of authentic standards and are listed in Table 6.2 (Subauste et al., 2012). The parent ion for LPA and CPA species is the deprotonated molecular anion (M-H). Daughter ions for all LPA species are a common fragment at *m/z* 153.1. cPA species give daughter ions that are characterized by a common loss of 136. For all transitions, unit mass

Table 6.2 MRM transitions for LC–ESI–MS/MS analysis of CPA and LPA species

Compound name	Parent ion *m/z*	Daughter ion *m/z*
$^{13}C_{16}$ LPA 16:0 (internal standard)	425.2	153.1
cPA16:0	391.2	255.2
cPA18:0	419.3	283.3
cPA18:1	417.3	281.3
LPA18:0	437.3	153.1
LPA18:1	435.3	153.1

resolution was used for both MS1 and MS2, the dwell time was 35 ms per transition, the fragmentor voltage is 140, and the collision energy is 20.

4. SUMMARY

The study of the lipidome in adipose tissue offers a great potential for advances in our understanding of the regulation of metabolism in this tissue. In this chapter, we summarized approaches focused toward the qualitative and quantitative study of lipids. The simplification of the sample for the study of neutral lipids and phospholipids offers significant advantages when approaching studies involving adipose tissue given the high concentration of neutral lipids. We consider of particular importance the study of low-abundance phospholipids, as they are usually involved in regulatory processes. The use of a simple silica column offers an improvement in detection by isolating phospholipids to avoid the interference from the highly abundant neutral lipids. The use of LC–ESI–MS/MS has proven capable of detecting these intermediates in the picomolar range. Another variable that will need to be considered when investigating the role of lipid intermediates in metabolism is subcellular localization. As mentioned previously, the regulatory role of these intermediates also has a spatial variable governing its effect. Cellular fractionation and the use of labeled lipid intermediates are other valuable tools specially when combined to the high sensitivity and wide range of data offered by mass spectrometry.

ACKNOWLEDGMENTS

The authors thank Chris Purser for the excellent technical assistance with various aspects for this work.

REFERENCES

Bligh, E. G., & Dyer, W. J. (1959). A rapid method of total lipid extraction and purification. *Canadian Journal of Biochemistry and Physiology*, *37*(8), 911–917.

Brugger, B., Erben, G., Sandhoff, R., Wieland, F. T., & Lehmann, W. D. (1997). Quantitative analysis of biological membrane lipids at the low picomole level by nano-electrospray ionization tandem mass spectrometry. *Proceedings of the National Academy of Sciences of the United States of America*, *94*(6), 2339–2344.

DeLong, C. J., Baker, P. R., Samuel, M., Cui, Z., & Thomas, M. J. (2001). Molecular species composition of rat liver phospholipids by ESI-MS/MS: The effect of chromatography. *Journal of Lipid Research*, *42*(12), 1959–1968.

Foster, D. A. (2013). Phosphatidic acid and lipid-sensing by mTOR. *Trends in Endocrinology and Metabolism*, *24*(6), 272–278.

Han, X., & Gross, R. W. (2001). Quantitative analysis and molecular species fingerprinting of triacylglyceride molecular species directly from lipid extracts of biological samples by electrospray ionization tandem mass spectrometry. *Analytical Biochemistry*, *295*(1), 88–100. http://dx.doi.org/10.1006/abio.2001.5178, S0003-2697(01)95178-4 [pii].

Hodson, L., Skeaff, C. M., Wallace, A. J., & Arribas, G. L. (2002). Stability of plasma and erythrocyte fatty acid composition during cold storage. *Clinica Chimica Acta*, *321*(1–2), 63–67, S0009898102001006 [pii].

Hsu, F. F., & Turk, J. (1999). Structural characterization of triacylglycerols as lithiated adducts by electrospray ionization mass spectrometry using low-energy collisionally activated dissociation on a triple stage quadrupole instrument. *Journal of the American Society for Mass Spectrometry*, *10*(7), 587–599. http://dx.doi.org/10.1016/S1044-0305(99)00035-5, S1044-0305(99)00035-5 [pii].

Hutchins, P. M., Barkley, R. M., & Murphy, R. C. (2008). Separation of cellular nonpolar neutral lipids by normal-phase chromatography and analysis by electrospray ionization mass spectrometry. *Journal of Lipid Research*, *49*(4), 804–813. http://dx.doi.org/10.1194/jlr.M700521-JLR200, M700521-JLR200 [pii].

Lorenz, M. A., Burant, C. F., & Kennedy, R. T. (2011). Reducing time and increasing sensitivity in sample preparation for adherent mammalian cell metabolomics. *Analytical Chemistry*, *83*(9), 3406–3414. http://dx.doi.org/10.1021/ac103313x.

McAnoy, A. M., Wu, C. C., & Murphy, R. C. (2005). Direct qualitative analysis of triacylglycerols by electrospray mass spectrometry using a linear ion trap. *Journal of the American Society for Mass Spectrometry*, *16*(9), 1498–1509. http://dx.doi.org/10.1016/j.jasms.2005.04.017, S1044-0305(05)00363-6 [pii].

McIntyre, T. M., Pontsler, A. V., Silva, A. R., St Hilaire, A., Xu, Y., Hinshaw, J. C., et al. (2003). Identification of an intracellular receptor for lysophosphatidic acid (LPA): LPA is a transcellular PPARgamma agonist. *Proceedings of the National Academy of Sciences of the United States of America*, *100*(1), 131–136. http://dx.doi.org/10.1073/pnas.0135855100, 0135855100 [pii].

Pietilainen, K. H., Rog, T., Seppanen-Laakso, T., Virtue, S., Gopalacharyulu, P., Tang, J., et al. (2011). Association of lipidome remodeling in the adipocyte membrane with acquired obesity in humans. *PLoS Biology*, *9*(6), e1000623. http://dx.doi.org/10.1371/journal.pbio.1000623, 10-PLBI-RA-8986R3 [pii].

Rhee, E. P., Cheng, S., Larson, M. G., Walford, G. A., Lewis, G. D., McCabe, E., et al. (2011). Lipid profiling identifies a triacylglycerol signature of insulin resistance and improves diabetes prediction in humans. *Journal of Clinical Investigation*, *121*(4), 1402–1411. http://dx.doi.org/10.1172/JCI44442, 44442 [pii].

Subauste, A. R., Das, A. K., Li, X., Elliott, B. G., Evans, C., El Azzouny, M., et al. (2012). Alterations in lipid signaling underlie lipodystrophy secondary to AGPAT2 mutations. *Diabetes*, *61*(11), 2922–2931. http://dx.doi.org/10.2337/db12-0004, db12-0004 [pii].

Tauchi-Sato, K., Ozeki, S., Houjou, T., Taguchi, R., & Fujimoto, T. (2002). The surface of lipid droplets is a phospholipid monolayer with a unique fatty acid composition. *Journal of Biological Chemistry*, *277*(46), 44507–44512. http://dx.doi.org/10.1074/jbc.M207712200, M207712200 [pii].

Tsukahara, T., Tsukahara, R., Fujiwara, Y., Yue, J., Cheng, Y., Guo, H., et al. (2010). Phospholipase D2-dependent inhibition of the nuclear hormone receptor PPARgamma by cyclic phosphatidic acid. *Molecular Cell*, *39*(3), 421–432. http://dx.doi.org/10.1016/j.molcel.2010.07.022, S1097-2765(10)00569-1 [pii].

CHAPTER SEVEN

Measurement of Long-Chain Fatty Acid Uptake into Adipocytes

Elena Dubikovskaya[*,1], **Rostislav Chudnovskiy**[†,1], **Grigory Karateev**[*], **Hyo Min Park**[†], **Andreas Stahl**[†,2]

[*]Department of Chemistry, École Polytechnique Fédérale de Lausanne, Lausanne, Switzerland
[†]Department of Nutritional Science and Toxicology, UC Berkeley, Berkeley, California, USA
[1]Authors contributed equally to manuscript.
[2]Corresponding author: e-mail address: astahl@berkeley.edu

Contents

Methods in Enzymology, Volume 538
ISSN 0076-6879
http://dx.doi.org/10.1016/B978-0-12-800280-3.00007-4

Abstract

The ability of white and brown adipose tissue to efficiently take up long-chain fatty acids is key to their physiological functions in energy storage and thermogenesis, respectively. Several approaches have been taken to determine uptake rates by cultured cells and primary adipocytes including radio- and fluorescently labeled fatty acids. In addition, the recent description of activatable bioluminescent fatty acids has opened the possibility for expanding these *in vitro* approaches to real-time monitoring of fatty acid uptake kinetics by adipose depots *in vivo*. Here, we will describe some of the most useful experimental paradigms to quantitatively determine long-chain fatty acid uptake by adipocytes *in vitro* and provide the reader with detailed instruction on how bioluminescent probes for *in vivo* imaging can be synthesized and used in living mice.

1. INTRODUCTION

1.1. Fatty acid uptake by adipocytes

Long-chain fatty acid (LCFA) uptake by adipocytes plays an important role in maintaining lipid homeostasis. Adipose tissue produces lipoprotein lipase (Wang & Eckel, 2009), which can generate LCFAs in the local vasculature through its action on triacylglycerol (TAG)-rich lipoprotein particles. Following transition across the endothelium (Hagberg et al., 2010), interstitial albumin-bound LCFAs can interact with adipocytes. This interaction is thought to involve dissociation from albumin, fatty acid translocation across the plasma membrane, and interaction with cytosolic fatty acid-binding proteins and/or activation to acyl-CoA (Bernlohr, Coe, & LiCata, 1999), which subsequently can participate in a variety of metabolic processes such as mitochondrial β-oxidation, particularly in brown adipose tissue (BAT), and TAG synthesis, particularly in white adipose tissue (WAT). Depending on the assay system and physiological condition, several, if not all, of these steps can become rate-limiting for the net flux of exogenous LCFAs into adipocytes. However, particular attention has been devoted to identifying proteins involved in the binding and transfer of LCFAs on the plasma membrane, leading to the identification of the solute carrier family 27 (fatty acid transport proteins, FATPs) (Anderson & Stahl, 2013), the scavenger receptor CD36 (Su & Abumrad, 2009), and the mitochondrial aspartate amino transferase (FABPpm) (Isola et al., 1995). Of these potential membrane LCFA transporters, WAT has been shown to express robust levels of FATP1, FATP4, CD36, and FABPpm (Hui & Bernlohr, 1997). Importantly, physiological stimuli such as insulin stimulation of white adipocyte (Wu et al.,

2006b) and adrenergic stimulation of brown adipocyte (Wu et al., 2006a) cell lines can drastically change cellular LCFA uptake rates. Also, genetic loss- and gain-of-function models for membrane transporters such as FATPs (Doege & Stahl, 2006) and CD36 (Coburn et al., 2000) have shown the expected alterations in uptake rates in a variety of tissues including the liver, heart, WAT, and BAT. Specifically, loss of FATP1 function *in vitro* and *in vivo* results in the loss of insulin-stimulatable, but not basal, LCFA uptake by WAT (Wu, Ortegon, et al., 2006b) and 3T3-L1 adipocytes (Lobo, Wiczer, Smith, Hall, & Bernlohr, 2007) and in diminished LCFA uptake by BAT, resulting in severe cold sensitivity (Wu, Kazantzis, et al., 2006a). Conversely, overexpression of the same transporter in the heart leads to TAG accumulation and symptoms of diabetic cardiomyopathy (Chiu et al., 2001). Homozygote CD36-null mutations have been created in *Mus musculus* (Febbraio et al., 1999) and also occur spontaneously in humans (Hirano et al., 2003) with resulting profound alterations in fatty acid uptake rates by various tissues including adipose. Given the clear evidence that cellular fatty acid uptake rates can be dynamically regulated and the importance of free fatty acid uptake for cellular energetics and insulin sensitivity (Samuel, Petersen, & Shulman, 2010), several experimental routes have been taken to determine LCFA uptake kinetics by WAT and other tissues both *in vitro* and *in vivo*.

1.2. *In vitro* fatty acid uptake assays

A variety of approaches have been developed to determine LCFA uptake that fall into the general categories of either tracing labeled fatty acids or indirect detection of transition of fatty acids across the plasma membrane. The most common indirect approach to determine fatty acid uptake is the measurement of cellular TAG stores, which is technically straightforward and can be easily done *in vivo* and *in vitro* but has the obvious shortcoming of being driven not only by cellular LCFA uptake but also by rates of lipolysis, fatty acid catabolism, LCFA efflux, and *de novo* synthesis from glucose and other substrates. Additional indirect approaches taken include the use of fluorescent intracellular fatty acid-binding proteins (Kampf & Kleinfeld, 2004), pH indicators (Berk & Stump, 1999), and growth assays of yeast cells plated on oleate media following the expression of murine FATPs on oleate media (Dirusso et al., 2000). Use of intracellular pH indicators for LCFA uptake assays assumes that fatty acids translocate across the plasma membrane as a protonated species and that fatty acid-induced proton fluxes across the plasma membrane relate to actual LCFA fluxes. However, recent studies of the function of the mitochondrial uncoupling protein 1 (UCP1), found in BAT mitochondria, have shown that unprotonated fatty acids can also be moved across membranes (Fedorenko, Lishko, & Kirichok, 2012). Forcing yeast to grow on oleate as their primary carbon source following the ablation of the endogenous yeast FATP has demonstrated that certain mammalian FATPs, including FATP1 and FATP4, which are highly expressed by adipocytes, can functionally rescue growth and thus open the door to rapid screening of point mutations (DiRusso et al., 2005). However, yeast is not able to functionally express all human FATPs and other transporters such as CD36 have not been tested. As is the case for all reductionist approaches, focusing on one adipocyte protein expressed ectopically in a model cell system also carries the risk of missing important complex interactions that could be present at the adipocyte surface to mediate efficient LCFA uptake. An alternative strategy is the detection of intracellular fatty acid concentrations using a fluorescent fatty acid-binding protein (ADIFAB) (Kampf, Parmley, & Kleinfeld, 2007). As with TAG levels, fluorescent-binding proteins are unable to distinguish exogenous from endogenous LCFAs, an important point particularly in adipocytes, which are able to mobilize large numbers of fatty acids during induction of lipolysis. In addition, this approach requires the microinjection of cells and cannot be translated to *in vivo* measurements.

Uptake assays with labeled fatty acids have either relied on radiolabeled fatty acids or used fatty acid analogs conjugated to fluorescent or bioluminescent probes. A variety of ^{14}C and ^{3}H radiolabeled fatty acids are commercially available and have the additional advantage of faithfully mimicking the biochemical properties of natural fatty acids. Thus, radiolabeled LCFAs have been extensively used for LCFA uptake assays in many *in vitro* cell systems including brown (Wu, Kazantzis, et al., 2006a) and white adipocytes (Stahl, Evans, Pattel, Hirsch, & Lodish, 2002). They also have found use *in vivo* after delivery by gavage or injection (Doege et al., 2008) and have even been used for high-throughput screens (HTS) aimed at identifying FATP4 inhibitors (Blackburn et al., 2006). Disadvantages of this approach include the use of radioactive materials, the lack of dynamic monitoring of cellular uptake, and high costs that frequently prevent their use in HTS applications. Fluorescently labeled fatty acid analogs, particularly C1-BODIPY-C12, have proved to be suitable alternatives to radiolabeled LCFAs. Uptake kinetics of C1-BODIPY-C12 have been determined for several cell types including adipocytes (Fig. 7.1) and the fluorescent fatty acids are capable of

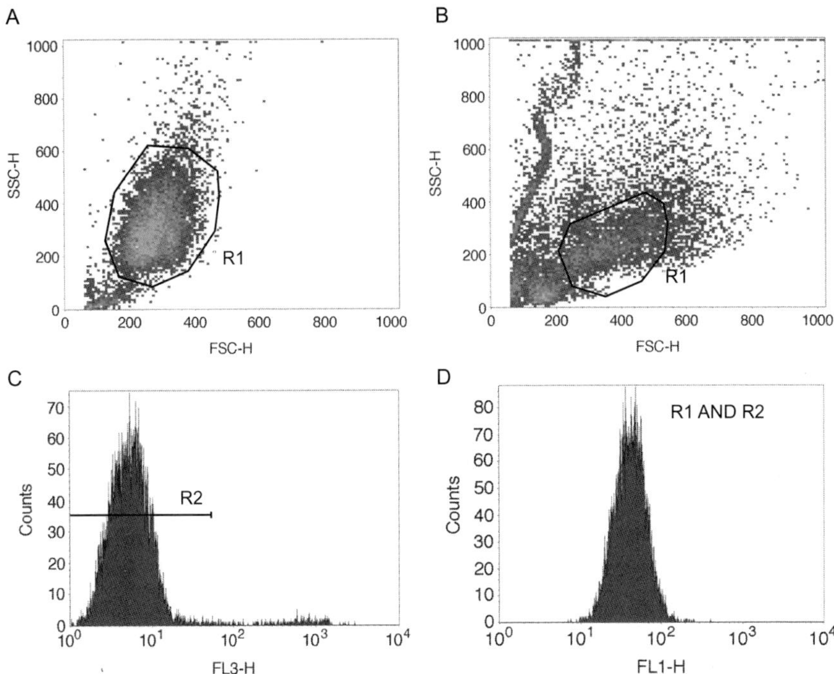

Figure 7.1 FACS-based fatty acid uptake assays. (A) Forward and side scatter dot blot of mature 3T3-L1 adipocytes. (B) Forward and side scatter dot blot of primary WAT-derived adipocytes. Gate R1 identifies intact adipocytes. (C) PI staining of 3T3-L1 adipocytes as recorded in FL3. Gate R2 identifies viable cells. (D) C1-BODIPY-C12 uptake of 3T3-L1 adipocytes with the logical gate R3 = R1 AND R2.

participating in downstream metabolic reactions following uptake (Kasurinen, 1992). While C1-BODIPY-C12 has been the most widely used fluorescent fatty acid, BODIPY-conjugated fatty acids are available with acyl chain lengths ranging from C5 to C15. In side-by-side comparisons, we found that all BODIPY-LCFAs but not the medium-chain BODIPY-C5 are readily taken up by 3T3-L1 adipocytes (Fig. 7.2), which is congruent with studies demonstrated that FATPs have a substrate preference for $C \geq 8$ fatty acids (Stahl et al., 1999). Fluorescent LCFA uptake can be quantitated using either plate readers or fluorescence-activated cell scanners (FACS). FACS has the advantage of being able to assess additional parameters such as cell size, surface markers, or cell viability and thus lends itself to applications for heterogeneous cell populations such as isolations of primary BAT or WAT cells. In contrast, plate reader-based assays offer a higher throughput. Traditionally, both approaches have relied on endpoint

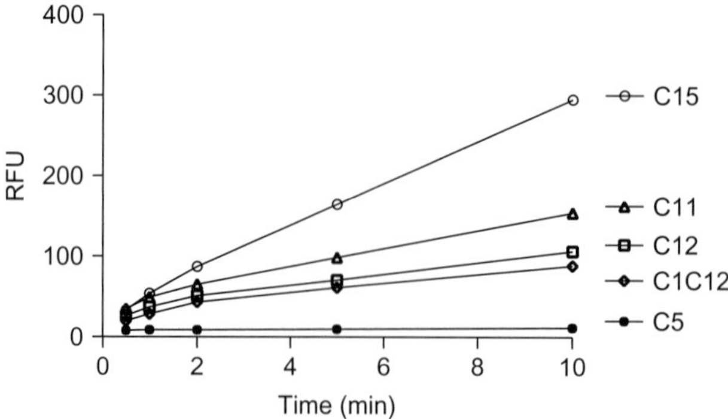

Figure 7.2 FACS-based fatty acid uptake assays with different BODIPY-LCFAs. Mature 3T3-L1 adipocytes were incubated with 2 μ*M* of either BODIPY C15 (Cat. # 3821), BODIPY C11 (Cat. # 3862), BODIPY C12 (Cat. # 3822), BODIPY C1C12 (Cat. # 3823), or BODIPY C5 (Cat. # 3834) for the indicated amount of time. Each time point was performed in duplicate.

assays with each time point requiring the removal of the uptake solution, washing of cells, and subsequent analysis. A convenient alternative to this approach is the combination of water-soluble, cell-impermeable, quenchers with C1-BODIPY-C12 fatty acids, which provides a homogeneous assay system that allows for the continuous real-time assessment of fatty acid uptake by adherent cells (Liao, Sportsman, Harris, & Stahl, 2005).

1.3. *In vivo* fatty acid uptake assays

Given the recent surge in interest in imaging BAT and its activation, there is also a need to develop suitable clinical and preclinical assays to determine LCFA uptake rate by BAT *in vivo*. Current approaches have mainly relied on PET using ^{18}F-2deoxyglucose (Cypess & Kahn, 2010); however, LCFAs are the predominant substrates for BAT uncoupled respiration (Ma & Foster, 1986). PET imaging approaches are costly and, due to the short half-life of ^{18}F (~110 min), do not lend themselves for imaging of slow metabolic changes nor longitudinal studies. As a suitable alternative, we recently developed a bioluminescent approach based on an activatable LCFA probe that can generate luciferin following cellular uptake. Thus, in the presence of luciferase, light generation is proportional to cellular fatty acid uptake (Henkin et al., 2012). Cellular uptake of this probe was shown to occur via physiological pathways and could be monitored in live animals in real time using bioluminescent imagers (Henkin et al., 2012). Importantly, using whole-body luciferase-expressing mice, uptake of fatty acids by classical

BAT pads could be quantitatively determined. The generation of tissue-specific luciferase-expressing animals, for example, utilizing adipose-specific promoters, should further expand the utility of this approach.

2. ADIPOCYTE SOURCES

There are several white, for example, OP9, and brown, for example, HIB1B, adipocyte lines available, but 3T3-L1 preadipocytes remain the most widely used *in vitro* system. Primary adipocytes can also be isolated from a variety of adipose depots and are particularly useful for the characterization of fatty acid uptake alterations in transgenic animals.

2.1. 3T3-L1 adipocyte differentiation

2.1.1 Materials required

Cell culture dishes (Corning Inc., Corning, NY; Cat. # 430167)
Medium 1 (DMS)

Materials	
DMEM (Life Technologies, Carlsbad, CA; Cat. # 11995-065)	500 ml
Fetal bovine serum (Life Technologies, Carlsbad, CA; Cat. # 10437-028)	50 ml (or 10%)
Pen/strep/glutamine (Life Technologies, Carlsbad, CA; Cat. # 10378-016)	5 ml (or 1%)
Combine and sterile filter in a tissue culture (TC) hood	

Medium 2 (DM1)

Materials	
DMS (medium 1)	500 ml
Dexamethasone (Sigma-Aldrich, St. Louis, MO; Cat. # D4902)	500 μl of 0.25 m*M*
IBMX (Sigma-Aldrich, St. Louis, MO; Cat. # I5879)	5 ml of 50 m*M*
Insulin (Sigma-Aldrich, St. Louis, MO; Cat. # I2643)	100 μl of 5 mg/ml
Combine and sterile filter in TC hood	

Medium 3 (DM2)

Materials	
DMS (medium 1)	500 ml
Insulin	100 μl of 5 mg/ml
Combine and sterile filter in TC hood	

2.1.2 Protocol

All cell culture work should be done in a TC hood under sterile conditions.

Cell maintenance

1. For propagation, cells must be kept at a low confluency without cells touching each other. For splitting cells, remove DMS from dish and add 0.5 ml of trypsin (Life Technologies, Carlsbad, CA; Cat. # 15400-054).
2. Swirl trypsin onto the entire dish and put the dish into a 37 °C 5% CO_2 incubator for 1–3 min.
3. Remove cells from incubator and add 10 ml of DMS to dish. Pipette up and down to detach cells from the dish.
4. Pipette the contents of the dish into a 15-ml conical vial and centrifuge the vial at 600 × *g* for 2 min at room temperature. A cell pellet should be visible at the bottom of the vial after the spin is complete.
5. Carefully aspirate the media as close to the pellet as possible without sucking up the pellet. Resuspend the pellet in DMS. Cells used for maintenance should be split every other day at a ratio of 1:10.

Cell differentiation

1. Seed 200,000 cells evenly onto a 10-cm cell culture dish and add 10 ml of DMS to the dish.
2. Exchange the DMS every other day until cells reach 100% confluency.
3. Upon confluency, replace DMS with 10 ml DM1 to initiate differentiation. Once DM1 is added, cells are in day 0 of differentiation.
4. On day 2, replace DM1 with 10 ml of DM2.
5. On day 4, replace DM2 with DMS and continue replacing DMS every other day until day 8.
6. By day 8, adipocytes should have visible lipid droplets and are ready for uptake assay.

2.2. Primary adipocytes

2.2.1 Materials required

Sterile cheesecloth
Shaking water bath

Solutions:	
Transfer solution: 5% BSA in Ringer	50 ml
Wash solution: 1% BSA in Ringer	100 ml
0.1% Fatty acid-free BSA in 1 × HBS	50 ml

Collagenase solution: Dissolve 1 g of type I collagenase (Life Technologies, Carlsbad, CA; Cat. # 17100-017) in 1 l of DMEM with 50 g of fatty acid-free BSA. Sterile filter, aliquot, and store at −20 °C.

2.2.2 Procedure

1. Prepare small petri dish (60 × 15 mm) containing 2.25 ml of transfer solution on ice.
2. Euthanize mice using CO_2 or isoflurane + cervical dislocation.
3. Remove desired adipose depots (using sterile tools and alcohol) and do not allow hair to mix with sample.
4. Place fat pads into petri dish on ice and transfer into sterile tissue culture cabinet.
5. Chop gently with scissors and add 0.25 ml collagenase solution.
6. Transfer cells using 25 ml pipette into a 15 ml conical.
7. Incubate at 37 °C while gently shaking for 55 min.
8. Place cheesecloth atop a 50-ml conical, decant cells in collagenase into 50 ml conical, and rinse cheesecloth twice with 10 ml of wash solution.
9. Remove cheesecloth and pull up the entire volume of collagenase/wash solution into a 25-ml pipette.
10. Let the pipette stand upright for 10 min, and cells will slowly rise to the top.
11. Slowly release cell-free solution until cell layer approaches end of pipette.
12. Wash cells by pulling up 10 ml of fresh wash solution and resuspend cells by bubbling air through the pipette.
13. Repeat washing procedure twice by going back to step 9.
14. Wash once in 5–10 ml of 0.1% fatty acid-free BSA in 1 × HBS and collect cells into a 1.5-ml centrifuge tube after discarding the cell-free solution.

3. UPTAKE ASSAY WITH RADIOACTIVE FATTY ACIDS

This assay can be performed with a variety of ^{3}H- and ^{14}C-labeled fatty acids and is useful for both cultured and primary adipocytes.

3.1. Materials required

Scintillation counter (Beckman LS6500).
Scintillation cocktail (EcoLume™; MP Biomedicals, Solon, OH; Cat. #0188247001).
10 × Dulbecco's phosphate-buffered saline (Life Technologies, Carlsbad, CA; Cat. # 14200-075).
Fatty acid-free BSA (Sigma-Aldrich, St. Louis, MO; Cat. # A7511).
EDTA (Sigma-Aldrich, St. Louis, MO; Cat. # E5134).
Sodium-deoxycholate (Sigma-Aldrich, St. Louis, MO; Cat. # D6750).
NP-40 (Calbiochem; Cat. #492016).
Sodium dodecyl sulfate (Sigma-Aldrich, St. Louis, MO; Cat. # L3771).
Oleic acid [1-^{14}C] (American Radiolabeled Chemicals; Cat. #ARC0297).
Octanoic acid [1-^{14}C] (American Radiolabeled Chemicals; Cat. #ARC0149).
Arachidonic acid [1-^{14}C] (American Radiolabeled Chemicals; Cat. #ARC0290).
Palmitic acid [1-^{14}C] (American Radiolabeled Chemicals; Cat. #ARC0172A).
24-Well cell culture dish (BD Falcon, Franklin Lakes, NJ; Cat. # 353043).

Solutions

Radiolabeled fatty acid solution	
Fatty acid stock solution (55 mCi/mmol, 1 mCi/ml)	5 μl
0.1% Fatty acid-free BSA PBS	445 μl
Serum-free DMEM	450 μl
Final fatty acid concentration will be 0.1 m*M*	

Control solution	
100% Ethanol	5 μl
0.1% Fatty acid-free BSA PBS	450 μl
Serum-free DMEM	450 μl

RIPA buffer	
50 m*M* Tris–HCL, pH 7.4	25 ml of 1 *M*
1% NP-40	5 ml
0.5% Na-deoxycholate	2.5 g
0.1% SDS	0.5 g
150 m*M* NaCl	15 ml of 5 *M*
2 m*M* EDTA	2 ml of 0.5 *M*
in 500 ml distilled water	

3.2. Protocol for attached cells

1. Fully differentiate 3T3-L1 fibroblasts into adipocytes (see differentiation protocol in the preceding text) in a 24-well cell culture dish. Perform experiment on day 8 of differentiation. Alternatively, use primary cells.
2. If treating the cells, for example, with insulin, serum starve the cells for 3–8 h to establish basal conditions. Rinse cells with serum-free medium three times, treat with desired hormone/growth factor, etc.
3. Add 20 μl of radiolabeled fatty acid solution or control solutions to cells. (The final concentration of each fatty acid should be 1 μ*M*.)
4. After 1–60 min, remove media and carefully wash cells twice in 2 ml ice-cold 0.1% fatty acid-free BSA in PBS.
5. Add 200 μl of ice-cold RIPA buffer and incubate plates on ice for 5 min.
6. Scrape the cells into RIPA buffer, transfer cell lysates into microcentrifuge tubes, and centrifuge at 16,000 × *g* in a tabletop centrifuge for 10 min.
7. Use 50 μl of supernatant for protein assay of your choice and add 100 μl of lysate to 4 ml of scintillation fluid.

8. Using a beta scintillation counter, determine activity and calculate the uptake rate, taking into account protein amount and incubation time (pmol/min/mg).

4. UPTAKE ASSAY WITH FLOW CYTOMETER

This assay relies on determining the uptake of a fluorescently labeled fatty acid analog and is particularly useful in combination with primary cells as nonadipocytes and dead cells can easily be excluded. However, the assay can also be performed as an endpoint assay for adherent cells.

4.1. Materials required

Flow cytometer (BD FACSCalibur™, BD Franklin Lakes, NJ, or equivalent).
FACS tubes (USA Scientific, Ocala, FL; Cat. # 1450 2000).
BODIPY fatty acids (Invitrogen—Life Technologies, Carlsbad, CA; Cat. #'s D-3821 D-3822 D-3823 D-3834 D-3862).
Propidium iodide (Invitrogen—Life Technologies, Carlsbad, CA; Cat. # P3566).
10 × Hank's balanced salt solution (HBSS, Gibco—Life Technologies, Carlsbad, CA; Cat. # 14065-056).
Fatty acid-free BSA (Sigma-Aldrich, St. Louis, MO; Cat. # A7511).
EDTA (Sigma-Aldrich, St. Louis, MO; Cat. # E5134).

Solutions	
Cell buffer	Fatty acid-free 0.1% BSA in 1 × HBSS
BODIPY solution	2.0 μ*M* final concentration of fatty acid BODIPY in cell buffer
	Vortex, sonicate, and keep at 37 °C in a water bath
FACS buffer	10% Fetal bovine serum
	10 m*M* EDTA
	50 μg/ml propidium iodide
	In 1 × HBSS
	Keep on ice
Stop solution	0.2% BSA in 1 × PBS
	Keep on ice

4.2. Protocol for detached cells

1. Fully differentiate 3T3-L1 fibroblasts into adipocytes in a 10 cm cell culture dish (see differentiation protocol in the preceding text). Perform experiment with day 8 differentiated cells. Alternatively, use primary adipocytes and go to step 7.
2. If treating the cells, for example, with insulin, serum starve the cells for 3–8 h to establish basal conditions. Rinse cells with serum-free medium three times, treat with desired hormone/growth factor, etc.
3. Aspirate and remove media from the 10-cm dish. Add 0.5 ml of trypsin to the dish. Swirl around for a few seconds so that trypsin covers the dish.
4. Carefully aspirate trypsin without detaching cells.
5. Add 1 ml trypsin and place the cells in a 37 °C 5% CO_2 incubator for 3 min.
6. Carefully aspirate the trypsin. Tap on edge of dish to dislodge adipocytes. Cells should all detach and slide off.
7. Add 9.5 ml of cell buffer to dish and gently pipette up and down so that cells are evenly mixed.
8. Transfer the 9.5 ml of cell buffer with the cells to a 15-ml conical tube and put the tube in a 37 °C water bath.
9. Prepare a 2-μ*M* fatty acid BODIPY solution in 0.5 ml of cell buffer and warm to 37 °C.
10. Prepare and label FACS tubes and place them on ice.
11. Pipette 4 ml of stop solution into each FACS tube on ice.
12. Add 0.5 ml of fatty acid BODIPY solution to 9.5 ml of cells in a 37 °C water bath and invert gently. This is the beginning of your time course.
13. At appropriate time point (there is enough solution for 20 FACS assays, for example, quadruplicates at 30 s, and 1, 2, 5, 10 min), gently invert solution with cells and remove 500 μl of cells and add them to the appropriate FACS tube.
14. After completing the time course, spin down FACS tubes at 600 × *g* for 10 min in a tissue culture centrifuge with appropriate adaptors.
15. Cell pellet should be visible at the end of the spin. Carefully decant or aspirate supernatant and discard.
16. Add 500 μl FACS buffer with propidium iodide to each FACS tube with cells.
17. Resuspend the cells by gently pipetting up and down several times and briefly vortex each FACS tube immediately before collecting data.
18. Run FACS by setting one gate in a forward/side scatter plot to include single-cell mature adipocytes (see Fig. 7.1A and B for 3T3 L1 and

primary adipocytes respectively) and set a second gate in the FL3 channel to exclude propidium iodide-positive cells (see Fig. 7.1C).

19. Determine fatty acid uptake as fluorescence (in arbitrary units) in the FL1 channel (see Fig. 7.1D).

4.3. Protocol for attached cells

1. Fully differentiate 3T3-L1 fibroblasts into adipocytes (see differentiation protocol in the preceding text) in a 12-well cell culture dish (BD Falcon, Franklin Lakes, NJ; Cat. # 353043). Perform experiment on day 8 differentiated cells.
2. Prepare 12 ml of cell buffer with a final concentration of 2.0 μ*M* BODIPY.
3. Warm BODIPY solution to 37 °C.
4. Prepare and label FACS tubes and place them on ice (duplicates).
5. Prepare 2 ml of stop solution per well (24 ml total) and place on ice.
6. Prepare FACS buffer with propidium iodide and keep on ice.
7. Aspirate and remove media from the 12-well dish.
8. Add 1 ml BODIPY solution to each of the wells in the 12-well dish.
9. At each of the time points (e.g., 30 s, and 1, 2, 5, or 10 min), add 2 ml of cold stop solution to the appropriate wells.
10. After completing the time course, aspirate and discard the stop solution.
11. Pipette 100 μl of trypsin per well into each of the 12 wells. Swirl gently and aspirate and discard the trypsin.
12. Pipette 200 μl of trypsin into each of the 12 wells. Place the dish of cells in a 37 °C 5% CO_2 incubator for 3 min.
13. Carefully aspirate trypsin. Tap on dish gently to shake cells loose. Cells should all detach and slide off.
14. Add 500 μl FACS buffer per well.
15. Pipette up and down to detach the rest of the cells. Pipette the 500 μl of FACS buffer and cells into the respective FACS tube.
16. Resuspend the cells by gently pipetting up and down several times with a 100-μl pipette.
17. Vortex each FACS tube with cells right before collecting data.
18. Perform FACS run as described earlier.

4.4. Variation of BODIPY-FFA chain length

Alternatively, experiments B and C in the preceding text can be performed with BODIPY-fatty acids of different chain length, such as C5, C11, C12, and C15, to determine chain length specificity of uptake. A representative experiment using 3T3-L1 adipocytes is shown in Fig. 7.2.

5. QUENCHER-BASED LCFA UPTAKE ASSAY

Plate reader-based assays have the advantage of allowing for high throughput and convenience but rely on homogeneous cell populations and high well-to-well consistency in viability and cell number. Importantly, through the addition of a water-soluble, cell-impermeable quencher to the uptake solution, this assay can be used to ascertain live uptake kinetics for fatty acids into adipocytes.

5.1. Materials required

Bottom-read-capable spectrophotometer (Molecular Devices, SpectraMax Gemini EM)
QBT™ fatty acid uptake assay kit (Molecular Devices; Cat. # R8132)
10 × Hank's balanced salt solution (Gibco; Cat. # 14065-056)
Fatty acid-free BSA (Sigma-Aldrich, St. Louis, MO; Cat. # A7511)
1 *M* HEPES buffer solution
Insulin (Sigma-Aldrich, St. Louis, MO; Cat. # I2643)

Solutions

Modified HBSS buffer.
1 g BSA.
10 ml of 1 *M* HEPES buffer solution.
50 ml 10 × HBSS.
Add 440 ml of sterile TC quality water to make up 500 ml.

5.2. Protocol

1. Fully differentiate 3T3-L1 fibroblasts into adipocytes (see differentiation protocol in the preceding text) in a 12-well cell culture dish. Perform experiment on day 8 of differentiation. If treating the cells, for example, with insulin, serum starve the cells for 3–8 h to establish basal conditions. Rinse cells with serum-free medium three times, treat with desired hormone/growth factor, etc.

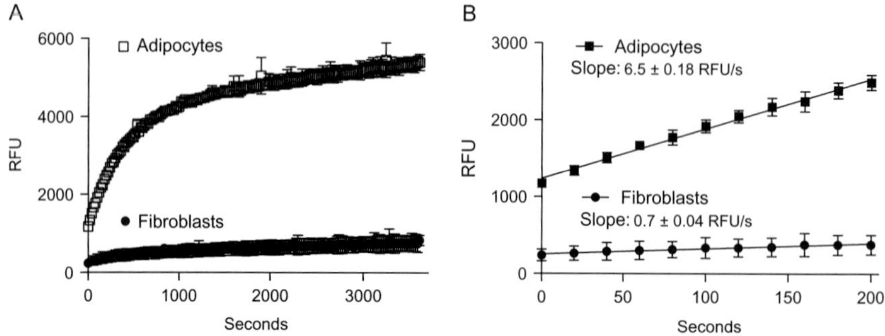

Figure 7.3 Quencher-based real-time fatty acid uptake assay. Mature 3T3-L1 adipocytes or undifferentiated fibroblasts were treated with BODIPY-fatty acid quencher mix as described under protocol 5. (A) Real-time uptake over 1 h. (B) Linear regression through initial time points to determine uptake velocities.

2. Prepare a 1 × loading buffer by adding 10 ml of HBSS buffer to 1 vial from the QBT fatty acid uptake kit.
3. Keep the 1 × loading buffer at 37 °C until ready to add to cells. Vortex before adding.
4. Aspirate the media from the wells and put 1 ml of serum-free media (warmed to 37 °C) into each of the 12 wells. Incubate for 1 h in the 37 °C 5% CO_2 incubator.
5. Aspirate and discard the serum-free medium.
6. Add 800 µl of the 1 × loading buffer per well and immediately transfer the dish to the fluorescent spectrophotometer (set to 37 °C).
7. Start kinetic assay.
8. Kinetic assay settings:
   a. Measure fluorescence with bottom-read mode.
   b. Excitation = 485 nm. Emission = 515 nm. Cutoff = 495 nm.
   c. Read every 20 s for 60 min.

Representative results for short- and long-term uptake assays with undifferentiated and differentiated 3T3-L1 adipocytes are shown in Fig. 7.3.

6. BIOLUMINESCENT FATTY ACID UPTAKE ASSAYS

Assays with bioluminescent fatty acids can be performed *in vitro* using cells or tissues and *in vivo*. In both cases, a sensitive optical detection system, for example, an IVIS Spectrum, is required. Since the bioluminescent LCFA analogs are, at the time of this writing, not commercially available in the United States, we will also provide the reader with instruction for probe synthesis.

6.1. Probe synthesis

6.1.1 Required materials

Aldrithiol-2 (Sigma-Aldrich, St. Louis, MO; Cat. # 143049)
3-Mercapto-1-propanol (Sigma-Aldrich, St. Louis, MO; Cat. # 405736)
Triphosgene (Fluorochem West Columbia, SC; Cat. # M03560)
Pyridine (Sigma-Aldrich, St. Louis, MO; Cat. # 270970)
2-Cyano-6-hydroxybenzothiazole (Shanghai Chemical Pharm-Intermediate Tech. Co., Ltd.)
16-Mercaptohexadecanoic acid (Sigma-Aldrich, St. Louis, MO; Cat. # 674435)
Methyl-16-mercaptohexadecanoate (Asemblon Inc.)
D-cysteine hydrochloride (AnaSpec Inc., Fremont, CA; Cat. # 61814-5)
Potassium carbonate (Sigma-Aldrich, St. Louis, MO; Cat. # 209619)
Triethylamine (Acros Organics—Thermo Fisher, Waltham, MA; Cat. # 219510500)
Dichloromethane (Sigma-Aldrich, St. Louis, MO; Cat. # 270997)
Methanol (Sigma-Aldrich, St. Louis, MO; Cat. # 322415)
Hexane (Sigma-Aldrich, St. Louis, MO; Cat. # 32293)
Ethyl acetate (Sigma-Aldrich, St. Louis, MO; Cat. # 45760)
Tetrahydrofuran (Sigma-Aldrich, St. Louis, MO; Cat. # 401757)
Dimethylformamide (Sigma-Aldrich, St. Louis, MO; Cat. # 227056)
Acetic acid (Sigma-Aldrich, St. Louis, MO; Cat. # 320099)
Chloroform-d (Sigma-Aldrich, St. Louis, MO; Cat. # 431915)
Methan(ol-d) (Sigma-Aldrich, St. Louis, MO; Cat. # 151939)
Acetone-d^6 (Sigma-Aldrich, St. Louis, MO; Cat. # 175862)
Deionized water (obtained from Milli-Q purification system)
Hydrochloric acid 1 *M* solution in water (Sigma-Aldrich, St. Louis, MO; Cat. # 71763)
Silica gel (Silicycle SiliaFlash P60 230-400 mesh; Cat. # R12030B)
Flash chromatograph (Biotage FLASH + column Biotage Si 12 + M or equivalent)
Glass-backed silica gel 60 Å F254 plates for analytical thin layer chromatography (Silicycle; Cat. # TLG-R10011B-333)

Reversed-phase analytical LC–MS system (Agilent Technologies 6120 Quadrupole LC–MS with 1260 Series HPLC system, Agilent Technologies, Santa Clara, CA, or equivalent)

Reversed-phase preparative high-performance liquid chromatography system (RP-HPLC Varian Pro Star + UV–Vis detector model 330 + preparative column Microsorb C-18 (21.4 × 250 mm or equivalent)

NMR spectrometer (AVQ-400 MHz or equivalent)

6.1.2 Synthesis of FFA-SS-luc probe

The synthesis of FFA-SS-luc probe is outlined in Scheme 7.1.

3-Mercaptopropanol **1** is converted into the activated disulfide derivative **3** by reaction with aldrithiol-2 **2**. In a separate route, 2-cyano-6-hydroxybenzothiazole **4** is transformed *in situ* into chloroformate **5**, which gives carbonate **6** after coupling with the compound **3**. Subsequent reaction of **6** with 16-mercaptohexadecanoic acid leads to the displacement of the thiopyridyl moiety to produce the conjugate **7**. In the last step, condensation of **7** with D-cysteine formed *in situ* from D-cysteine hydrochloride and K_2CO_3 results in the FFA-SS-luc probe **8**.

6.1.2.1 Synthesis of 3-(pyridin-2-yldisulfanyl)propan-1-ol (**3**)

1. Place 3.84 g (17.44 mmol, 3 equiv.) of aldrithiol-2 under nitrogen in an oven-dried 25-ml flask equipped with a stir bar. Add 12 ml of methanol previously purged with nitrogen for 20 min.
2. Add to the reaction mixture 0.50 ml (5.81 mmol, 1 equiv.) of 3-mercapto-1-propanol dropwise.
3. Allow the solution to stir for 3 h at room temperature.
4. Evaporate the solvent *in vacuo*.
5. Purify the residue by flash silica gel chromatography using hexane/ethyl acetate 1:1 (v/v) as an eluent. Determine which fractions contain the product by TLC and evaporate the solvent *in vacuo*.
6. Prove the structure and check the purity of the product after purification by ^{1}H and ^{13}C NMR using CD_3OD as a solvent for NMR sample preparation and compare the spectra with those previously reported (Henkin et al., 2012).

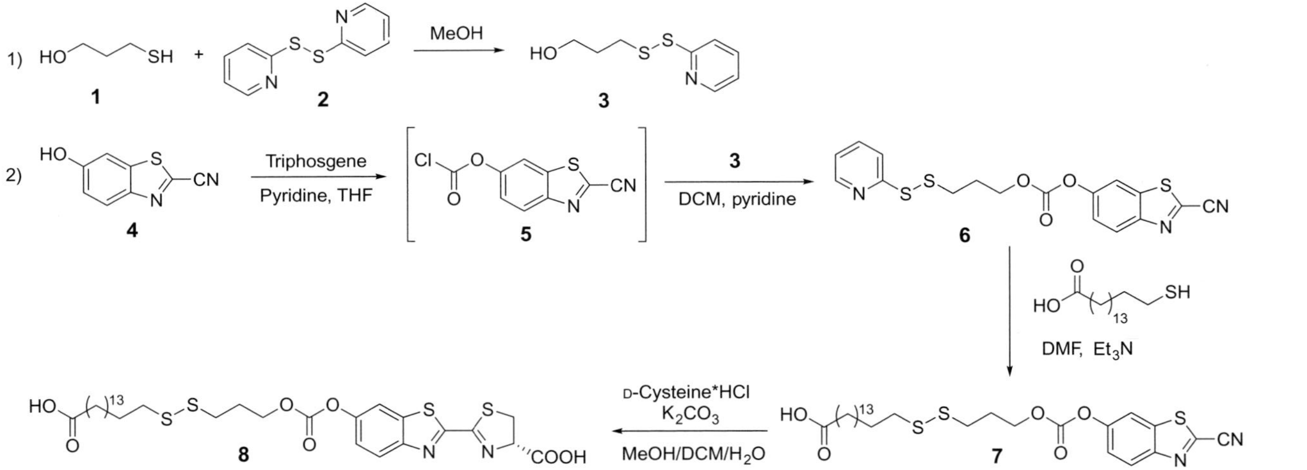

1) HO–(CH2)3–SH **1** + 2,2′-dipyridyl disulfide **2** → (MeOH) → **3**

2) **4** → (Triphosgene; Pyridine, THF) → [**5**] → (**3**; DCM, pyridine) → **6**

6 → (HO(O)C–(CH2)13–CH2CH2–SH; DMF, Et_3N) → **7**

7 → (D-Cysteine*HCl, K_2CO_3; $MeOH/DCM/H_2O$) → **8**

Scheme 7.1 Synthesis of FFA-SS-luc probe.

6.1.2.2 Synthesis of 16-((3-((2-cyanobenzo[*d*]thiazol-6-yloxy)carbonyloxy) propyl)disulfanyl)hexadecanoic acid (**7**)

1. Place 376 mg (0.93 mmol, 1 equiv.) of the compound **6** and 269 mg (0.93 mmol, 1 equiv.) of 16-mercaptohexadecanoic acid under nitrogen in an oven-dried 100-ml flask equipped with a stir bar. Add 40 ml of dry DMF and 207 μl (1.49 mmol, 1.6 equiv.) of dry triethylamine to the reaction flask.
2. Allow the reaction mixture to stir at room temperature for 2 h.
3. Evaporate the solvent *in vacuo*.
4. Purify the residue by flash silica gel chromatography using hexane/ethyl acetate 4:1 (v/v) as an eluent. Determine which fractions contain the product by TLC and evaporate the solvent *in vacuo*.
5. Prove the structure and check the purity of the product after purification by ^{1}H and ^{13}C NMR using $CDCl_3$ as a solvent for NMR sample preparation and compare the spectra with those previously reported (Henkin et al., 2012).

6.1.2.3 Synthesis of (*S*)-2-(6-((3-((15-carboxypentadecyl)disulfanyl)propoxy) carbonyloxy)benzo[*d*]thiazol-2-yl)-4,5-dihydrothiazole-4-carboxylic acid (FFA-SS-luc) (**8**)

1. Place 49.9 mg (0.09 mmol, 1 equiv.) of the compound **7** and 13.5 mg (0.09 mmol, 1 equiv.) of D-cysteine hydrochloride under nitrogen in a 25-ml flask equipped with a stir bar.
2. Add to the reaction flask 5 ml of methanol and 5 ml of dichloromethane. These solvents should be purged with nitrogen for 20 min prior to use.
3. Prepare in a separate flask a solution of potassium carbonate (11.8 mg, 0.09 mmol, 1 equiv.) in a mixture of water (2 ml) and methanol (5 ml). Purge this solution with nitrogen for 20 min.
4. Add the prepared solution of potassium carbonate to the reaction flask and allow the reaction mixture to stir for 30 min at room temperature in darkness.
5. Quench the reaction mixture by addition of 1 *M* HCl in water to pH 3–4.
6. Evaporate the solvents *in vacuo*.
7. Purify the residue by flash silica gel chromatography using a mixture of DCM/ethyl acetate 4:1 (v/v) containing 2% (v/v) of acetic acid as an eluent. Determine which fractions contain the product by TLC and evaporate the solvent *in vacuo*.

8. Prove the structure and check the purity of the product after purification by ^1H and ^{13}C NMR using acetone-d^6 as a solvent for NMR sample preparation and compare the spectra with those previously reported (Henkin et al., 2012).

6.1.2.4 Synthesis of (*S*)-2-(6-((3-((15-carboxypentadecyl)disulfanyl)propoxy) carbonyloxy)-benzo[*d*]thiazol-2-yl)-4,5-dihydrothiazole-4-carboxylic acid (FFA-SS-luc) (**8**)

9. Place 49.9 mg (0.09 mmol, 1 equiv.) of the compound **7** and 13.5 mg (0.09 mmol, 1 equiv.) of D-cysteine hydrochloride under nitrogen into a 25-ml flask equipped with a stir bar.
10. Add to the reaction flask 5 ml of methanol and 5 ml of dichloromethane. These solvents should be purged with nitrogen for 20 min prior to use.
11. Prepare in a separate flask a solution of potassium carbonate (11.8 mg, 0.09 mmol, 1 equiv.) in a mixture of water (2 ml) and methanol (5 ml). Purge this solution with nitrogen for 20 min.
12. Add the prepared solution of potassium carbonate to the reaction flask and allow the reaction mixture to stir for 30 min at room temperature in darkness.
13. Quench the reaction mixture by addition of 1 *M* HCl in water to pH 3–4.
14. Evaporate the solvents *in vacuo*.
15. Purify the residue by flash silica gel chromatography using a mixture of DCM/ethyl acetate 4:1 (v/v) containing 2% (v/v) of acetic acid as an eluent. Determine which fractions contain the product by TLC and evaporate the solvent *in vacuo*.
16. Prove the structure and check the purity of the product after purification by ^1H and ^{13}C NMR using acetone-d^6 as a solvent for NMR sample preparation and compare the spectra with those previously reported (Henkin et al., 2012).

6.1.3 Synthesis of Me-FFA-SS-luc control compound

The synthesis of Me-FFA-SS-luc probe is outlined in Scheme 7.2.

Carbonate **6**, which contains thiopyridyl moiety, is converted to the conjugate **10** by reaction with methyl-16-mercaptohexadecanoate **9** in DMF in the presence of triethylamine. In the last step, condensation of **10** with D-cysteine formed *in situ* from D-cysteine hydrochloride and K_2CO_3 results in the desired Me-FFA-SS-luc probe **11**.

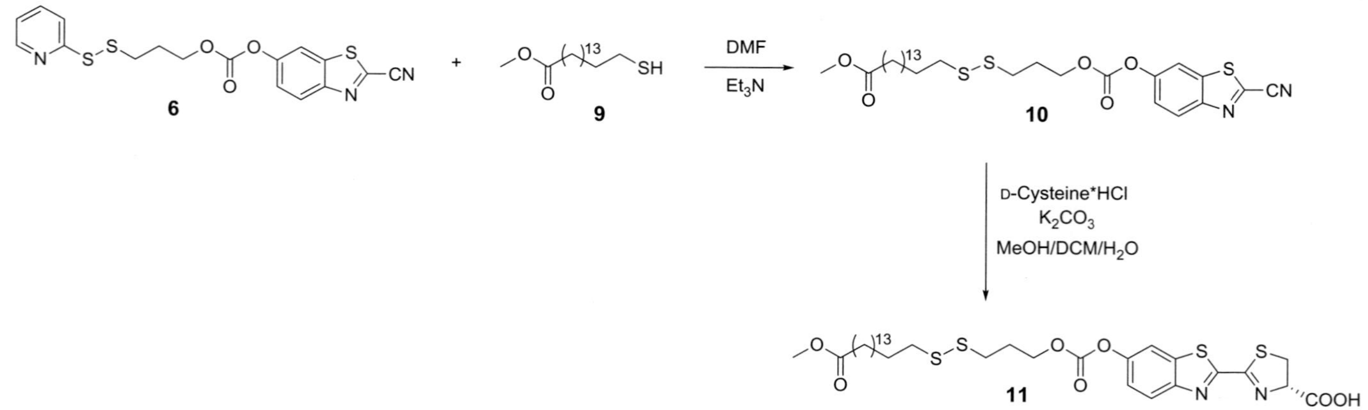

DMF

Et_3N

D-Cysteine*HCl

K_2CO_3

MeOH/DCM/H_2O

Scheme 7.2 Synthesis of Me-FFA-SS-luc probe.

6.1.3.1 Synthesis of methyl 16-((3-((2-cyanobenzo[*d*]thiazol-6-yloxy)carbonyloxy)propyl)disulfanyl)hexadecanoate (**10**)

1. Place 56 mg (0.14 mmol, 1.4 equiv.) of the compound **6** and 30.6 mg (0.1 mmol, 1 equiv.) of methyl-16-mercaptohexadecanoate **9** under nitrogen in an oven-dried 25-ml flask equipped with a stir bar. Add 10 ml of dry DMF and 29 μl (0.21 mmol, 2.1 equiv.) of dry triethylamine to the reaction flask.
2. Allow the reaction mixture to stir at room temperature for 2 h.
3. Evaporate the solvent *in vacuo*.
4. Purify the residue by RP-HPLC, changing the eluent from 40% methanol/60% water to 100% methanol for 45 min and then 100% methanol for 30 min at a flow rate 10 ml/min using a preparative column Microsorb C-18 (21.4 × 250 mm) or equivalent. Determine which fractions contain the product by LC–MS.
5. Evaporate the solvent from fractions containing the product and purify the product again by flash silica gel chromatography using gradient elution with hexane/ethyl acetate 9:1 → 7:1 (v/v). Determine the fractions that contain the product by TLC and evaporate the solvent *in vacuo*.
6. Prove the structure and check the purity of the product after purification by ^{1}H and ^{13}C NMR using $CDCl_3$ as a solvent for NMR sample preparation and compare the spectra with those previously reported (Henkin et al., 2012).

6.1.3.2 Synthesis of (*S*)-2-(6-((3-((16-methoxy-16-oxohexadecyl)disulfanyl)propoxy)carbonyloxy)benzo[*d*]thiazol-2-yl)-4,5-dihydrothiazole-4-carboxylic acid (Me-FFA-SS-luc) (**11**)

1. Place 37.2 mg (0.06 mmol, 1 equiv.) of the compound **10** and 10.7 mg (0.07 mmol, 1.17 equiv.) of D-cysteine hydrochloride under nitrogen in a 25-ml flask equipped with a stir bar.
2. Add to the reaction flask 3.4 ml of methanol and 3.4 ml of dichloromethane. These solvents should be purged with nitrogen for 20 min prior to use.
3. Prepare in a separate flask a solution of potassium carbonate (8.7 mg, 0.06 mmol, 1 equiv.) in a mixture of water (1.4 ml) and methanol (3.4 ml). Purge this solution with nitrogen for 20 min.
4. Add the prepared solution of potassium carbonate to the reaction flask and allow the reaction mixture to stir for 30 min at room temperature in darkness.

5. Quench the reaction mixture by addition of 1 *M* HCl in water to pH 3–4.
6. Evaporate the solvents *in vacuo*.
7. Purify the residue by RP-HPLC, changing the eluent from 40% methanol/60% water to 100% methanol for 45 min and then 100% methanol for 30 min at a flow rate 10 ml/min using a preparative column Microsorb C-18 (21.4 × 250 mm) or equivalent. Determine the fractions that contain the product by LC–MS.
8. Evaporate the solvent from fractions containing the product and purify the product again by RP-HPLC, changing the eluent from 80% methanol/20% water to 100% methanol for 45 min and then 100% methanol for 30 min at a flow rate 10 ml/min using a preparative column Microsorb C-18 (21.4 × 250 mm) or equivalent. Determine which fractions contain the product by LC–MS.
9. Evaporate the solvent from fractions containing the product.
10. Prove the structure and check the purity of the product after purification by ^{1}H and ^{13}C NMR using acetone-d^{6} as a solvent for NMR sample preparation and compare the spectra with those previously reported (Henkin et al., 2012).

6.2. *In vitro* bioluminescent uptake assays with adipocytes

6.2.1 Required materials

3T3-L1 adipocytes stably transfected with the pGL4.51[luc2/CMV/Neo] vector (Promega, Madison, WI) (3T3-L1-luc) and differentiated as described earlier

Black-wall/clear-bottom 96-well plates (Costar, Corning Inc., Corning, NY)

Bioluminescent imager (IVIS Spectrum, Caliper Life Sciences—PerkinElmer, Waltham, MA; or equivalent)

Solutions

Uptake buffer

Dissolve bioluminescent fatty acid in DMSO to yield a 1 m*M* stock solution. Prepare a 0.1% (w/v) solution of fatty acid-free BSA in HBSS and warm to 37 °C. Slowly add bioluminescent fatty acid stock solution to BSA/HBSS to a final concentration of 2–100 μ*M* and use immediately.

1. Install the plate adapter in the imager (see Fig. 7.4) and prewarm the imaging stage to 37 °C.

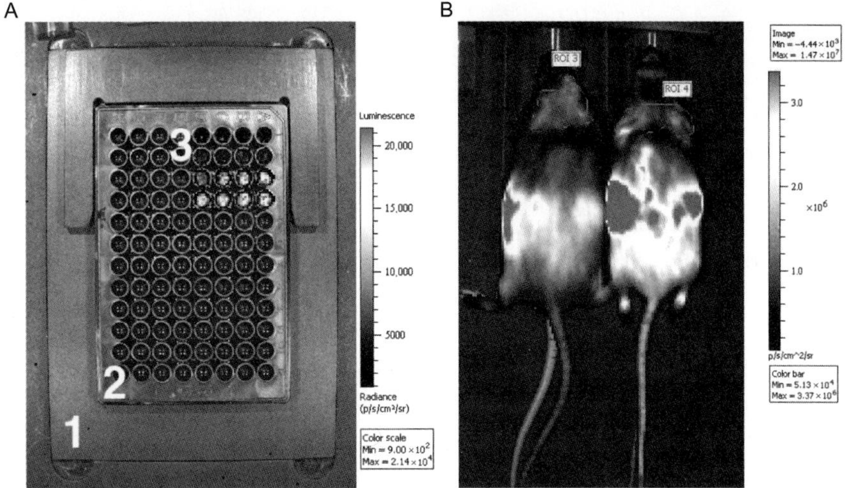

Figure 7.4 Bioluminescent fatty acid uptake assays. (A) Using a plate adapter (1), a 96-well plate (2) containing mature 3T3-L1 adipocytes and undifferentiated fibroblasts (3) was placed into an IVIS Spectrum. Fatty acid uptake rates were monitored following the addition of 100 μl of 100 μ*M* BSA-bound 100 μ*M* FFA-SS-luc. (B) L2G85 mice injected with 100 μl of a 200 μ*M* solution intraperitoneally were imaged using autoexposure settings. Red regions of interest denote the BAT area in both animals. (See color plate.)

2. Seed 5000 cells/well of differentiated 3T3-L1-luc cells into a black-wall/clear-bottom 96-well plate and let adhere for 6–8 h.
3. Rinse once with prewarmed 0.1% (w/v) solution of fatty acid-free BSA in HBSS.
4. Add 100 μl of prewarmed uptake buffer containing 2–100 μ*M* of either FFA-SS-luc or control compounds.
5. Immediately place plate on heated (37 °C) imaging stage and commence image acquisition using either autoexposure or predetermined exposure times, that is, 5 min exposures.
6. Continuously image plates for 60 min.
7. Analyze images using IVIS Living Image software (or equivalent) by aligning a 96-well grid region of interest with the plate. Fatty acid uptake rates can be expressed as photon flux (photons/s) or as radiance (photons/s/cm^2/sr).

7. *IN VIVO* IMAGING OF BAT FATTY ACID UPTAKE RATES

7.1. Materials

Transgenic mice expressing luciferase in BAT such as whole-body luciferase-expressing animals (L2G85 mice, The Jackson Laboratory)
Bioluminescent imager (IVIS Spectrum, Caliper Life Sciences—PerkinElmer, Waltham, MA; or equivalent) with anesthesia setup
Solutions
Injection solution
Dissolve bioluminescent fatty acid in DMSO to yield a 1-m*M* stock solution. Add bioluminescent fatty acid to a prewarmed solution of 0.1% (w/v) solution of fatty acid-free BSA in PBS to yield a final concentration of 200 μ*M*. Prepare sufficient injection solution to allow for 100 μl per animal. Sterile filter using a syringe filter and use immediately.

7.2. Imaging protocol

1. Anesthetize mice with isoflurane/oxygen and inject intraperitoneally with 100 μl of injection solution.
2. Place animals on heated imaging stage with dorsal site facing the camera. Securely place nose cones over animals and maintain anesthesia with isoflurane/oxygen.
3. Adjust imaging stage height to achieve optimal magnification of the interscapular area and commence imaging using either autoexposure or a predetermined fixed exposure time, for example, 3 min.
4. Continue imaging for 30 min. Analyze images using IVIS Living Image software (or equivalent) after identifying a region of interest such as the interscapular BAT of each mouse (see Fig. 7.4B). Fatty acid uptake rates can be expressed as photon flux (photons/s) or as radiance (photons/s/cm^2/sr).

ACKNOWLEDGMENTS

Work that led to the development of these methods was sponsored in part by R01 NIH Grants DK089202 and DK066336 to A. S.

REFERENCES

Anderson, C. M., & Stahl, A. (2013). SLC27 fatty acid transport proteins. *Molecular Aspects of Medicine*, *34*, 516–528.

Berk, P. D., & Stump, D. D. (1999). Mechanisms of cellular uptake of long chain free fatty acids. *Molecular and Cellular Biochemistry*, *192*, 17–31.

Bernlohr, D. A., Coe, N. R., & LiCata, V. J. (1999). Fatty acid trafficking in the adipocyte. *Seminars in Cell and Developmental Biology*, *10*, 43–49.

Blackburn, C., Guan, B., Brown, J., Cullis, C., Condon, S. M., Jenkins, T. J., et al. (2006). Identification and characterization of 4-aryl-3,4-dihydropyrimidin-2(1H)-ones as inhibitors of the fatty acid transporter FATP4. *Bioorganic and Medicinal Chemistry Letters*, *16*, 3504–3509.

Chiu, H. C., Kovacs, A., Ford, D. A., Hsu, F. F., Garcia, R., Herrero, P., et al. (2001). A novel mouse model of lipotoxic cardiomyopathy. *The Journal of Clinical Investigation*, *107*, 813–822.

Coburn, C. T., Knapp, F. F., Jr., Febbraio, M., Beets, A. L., Silverstein, R. L., & Abumrad, N. A. (2000). Defective uptake and utilization of long chain fatty acids in muscle and adipose tissues of CD36 knockout mice. *The Journal of Biological Chemistry*, *275*, 32523–32529.

Cypess, A. M., & Kahn, C. R. (2010). Brown fat as a therapy for obesity and diabetes. *Current Opinion in Endocrinology, Diabetes, and Obesity*, *17*, 143–149.

Dirusso, C. C., Connell, E. J., Faergeman, N. J., Knudsen, J., Hansen, J. K., & Black, P. N. (2000). Murine FATP alleviates growth and biochemical deficiencies of yeast fat1Delta strains. *European Journal of Biochemistry*, *267*, 4422–4433.

DiRusso, C. C., Li, H., Darwis, D., Watkins, P. A., Berger, J., & Black, P. N. (2005). Comparative biochemical studies of the murine fatty acid transport proteins (FATP) expressed in yeast. *The Journal of Biological Chemistry*, *280*, 16829–16837.

Doege, H., Grimm, D., Falcon, A., Tsang, B., Storm, T. A., Xu, H., et al. (2008). Silencing of hepatic fatty acid transporter protein 5 in vivo reverses diet-induced non-alcoholic fatty liver disease and improves hyperglycemia. *The Journal of Biological Chemistry*, *283*, 22186–22192.

Doege, H., & Stahl, A. (2006). Protein-mediated fatty acid uptake: Novel insights from in vivo models. *Physiology (Bethesda)*, *21*, 259–268.

Febbraio, M., Abumrad, N. A., Hajjar, D. P., Sharma, K., Cheng, W., Pearce, S. F., et al. (1999). A null mutation in murine CD36 reveals an important role in fatty acid and lipoprotein metabolism. *The Journal of Biological Chemistry*, *274*, 19055–19062.

Fedorenko, A., Lishko, P. V., & Kirichok, Y. (2012). Mechanism of fatty-acid-dependent UCP1 uncoupling in brown fat mitochondria. *Cell*, *151*, 400–413.

Hagberg, C. E., Falkevall, A., Wang, X., Larsson, E., Huusko, J., Nilsson, I., et al. (2010). Vascular endothelial growth factor B controls endothelial fatty acid uptake. *Nature*, *464*, 917–921.

Henkin, A. H., Cohen, A. S., Dubikovskaya, E. A., Park, H. M., Nikitin, G. F., Auzias, M. G., et al. (2012). Real-time noninvasive imaging of fatty acid uptake in vivo. *ACS Chemical Biology*, *7*, 1884–1891.

Hirano, K., Kuwasako, T., Nakagawa-Toyama, Y., Janabi, M., Yamashita, S., & Matsuzawa, Y. (2003). Pathophysiology of human genetic CD36 deficiency. *Trends in Cardiovascular Medicine*, *13*, 136–141.

Hui, T. Y., & Bernlohr, D. A. (1997). Fatty acid transporters in animal cells. *Frontiers in Bioscience*, *2*, d222–d231.

Isola, L. M., Zhou, S. L., Kiang, C. L., Stump, D. D., Bradbury, M. W., & Berk, P. D. (1995). 3 T3 fibroblasts transfected with a cDNA for mitochondrial aspartate aminotransferase express plasma membrane fatty acid-binding protein and saturable fatty acid uptake. *Proceedings of the National Academy of Sciences of the United States of America*, *92*, 9866–9870.

Kampf, J. P., & Kleinfeld, A. M. (2004). Fatty acid transport in adipocytes monitored by imaging intracellular free fatty acid levels. *The Journal of Biological Chemistry*, *279*, 35775–35780.

Kampf, J. P., Parmley, D., & Kleinfeld, A. M. (2007). Free fatty acid transport across adipocytes is mediated by an unknown membrane protein pump. *American Journal of Physiology Endocrinology and Metabolism*, *293*, E1207–E1214.

Kasurinen, J. (1992). A novel fluorescent fatty acid, 5-methyl-BDY-3-dodecanoic acid, is a potential probe in lipid transport studies by incorporating selectively to lipid classes of BHK cells. *Biochemical and Biophysical Research Communications*, *187*, 1594–1601.

Liao, J., Sportsman, R., Harris, J., & Stahl, A. (2005). Real-time quantification of fatty acid uptake using a novel fluorescence assay. *Journal of Lipid Research*, *46*, 597–602.

Lobo, S., Wiczer, B. M., Smith, A. J., Hall, A. M., & Bernlohr, D. A. (2007). Fatty acid metabolism in adipocytes: Functional analysis of fatty acid transport proteins 1 and 4. *Journal of Lipid Research*, *48*, 609–620.

Ma, S. W., & Foster, D. O. (1986). Uptake of glucose and release of fatty acids and glycerol by rat brown adipose tissue in vivo. *Canadian Journal of Physiology and Pharmacology*, *64*, 609–614.

Samuel, V. T., Petersen, K. F., & Shulman, G. I. (2010). Lipid-induced insulin resistance: Unravelling the mechanism. *Lancet*, *375*, 2267–2277.

Stahl, A., Evans, J. G., Pattel, S., Hirsch, D., & Lodish, H. F. (2002). Insulin causes fatty acid transport protein translocation and enhanced fatty acid uptake in adipocytes. *Developmental Cell*, *2*, 477–488.

Stahl, A., Hirsch, D. J., Gimeno, R. E., Punreddy, S., Ge, P., Watson, N., et al. (1999). Identification of the major intestinal fatty acid transport protein. *Molecular Cell*, *4*, 299–308.

Su, X., & Abumrad, N. A. (2009). Cellular fatty acid uptake: A pathway under construction. *Trends in Endocrinology and Metabolism*, *20*, 72–77.

Wang, H., & Eckel, R. H. (2009). Lipoprotein lipase: From gene to obesity. *American Journal of Physiology Endocrinology and Metabolism*, *297*, E271–E288.

Wu, Q., Kazantzis, M., Doege, H., Ortegon, A. M., Tsang, B., Falcon, A., et al. (2006). Fatty acid transport protein 1 is required for nonshivering thermogenesis in brown adipose tissue. *Diabetes*, *55*, 3229–3237.

Wu, Q., Ortegon, A. M., Tsang, B., Doege, H., Feingold, K. R., & Stahl, A. (2006). FATP1 is an insulin-sensitive fatty acid transporter involved in diet-induced obesity. *Molecular and Cellular Biology*, *26*, 3455–3467.

CHAPTER EIGHT

Measurement of the Unfolded Protein Response to Investigate Its Role in Adipogenesis and Obesity

Jaeseok Han, Randal J. Kaufman[1]

Center for Neuroscience, Aging, and Stem Cell Research, Sanford Burnham Medical Research Institute, La Jolla, California, USA

[1]Corresponding author: e-mail address: rkaufman@sanfordburnham.org

Contents

Abstract

The endoplasmic reticulum (ER) is the cellular organelle responsible for the folding of proteins destined for secretion and the intramembrane system of the cell, biosynthesis of lipids, and storage of calcium for regulated release. Extracellular stimuli and changes in intracellular homeostasis can alter the protein-folding environment of the ER and cause the accumulation of misfolded or unfolded proteins, a stress condition called ER stress. To resolve protein misfolding, cells have evolved a collection of adaptive signaling pathways, called the unfolded protein response (UPR). It is now recognized that ER stress contributes to many pathophysiological conditions. Increasing lines of evidence suggest that obesity/insulin resistance and subsequent type 2 diabetes are associated with ER stress and UPR activation in adipose tissue. However, whether and/or how ER stress and the UPR contribute to the pathogenesis of metabolic syndrome and obesity is not entirely clear. In this section, we describe how the UPR may contribute to the pathology of obesity, methods to measure UPR induction, and approaches to investigate the role of the UPR during adipocyte differentiation and in mature adipose tissue.

Methods in Enzymology, Volume 538
ISSN 0076-6879
http://dx.doi.org/10.1016/B978-0-12-800280-3.00008-6

1. INTRODUCTION

The endoplasmic reticulum (ER) is the cellular organelle responsible for the folding and assembly of proteins destined for cellular membranes and secretion, lipid and sterol biosynthesis, and calcium storage (Back & Kaufman, 2012). As these functions are essential for cell physiology, the environment of the ER is optimized for these processes. However, numerous insults, including nutrient depletion, hypoxia, lipid excess, mutation in protein chaperones or their client proteins, increased synthesis of secretory proteins, DNA damage, bacterial/virus infection, and inflammation, perturb ER homeostasis (Walter & Ron, 2011). Under these conditions, misfolded or unfolded proteins accumulate in the ER lumen, which is detrimental to cell function. To resolve this stressful condition, cells evolved a complex of signaling pathways collectively called the unfolded protein response (UPR). UPR induction is initiated by the dissociation of the protein chaperone BiP/GRP78 from the three basic UPR sensors: double-stranded RNA-dependent protein kinase-like ER kinase (PERK), inositol-requiring enzyme 1α (IRE1α), and activating transcription factor 6α (Walter & Ron, 2011; Wang & Kaufman, 2012). Activated PERK phosphorylates the alpha subunit of eukaryotic translation initiation factor 2 (eIF2α) at Ser51 leading to rapid and transient attenuation of protein synthesis (Harding et al., 2000; Scheuner et al., 2001). This allows time for cells to resolve the ER stress by reducing the amount of newly synthesized proteins entering the ER lumen. Paradoxically, under these conditions, translation of *Atf4* mRNA is selectively enhanced, which induces transcription of the C/EBP homologous protein (CHOP) (Han, Back, et al., 2013; Harding et al., 2003; Marciniak et al., 2004; Scheuner et al., 2001). Activated IRE1α elicits an endoribonuclease function that initiates unconventional splicing of *Xbp1* mRNA (Calfon et al., 2002; Shen et al., 2001), to produce a functional bZIP transcription factor, XBP1, that induces expression of genes encoding ER protein chaperones, lipid biosynthetic enzymes, and components of ER-associated protein degradation (ERAD) (Zhang & Kaufman, 2008). Upon ER stress, ATF6α traffics to the Golgi apparatus where it is cleaved by the processing enzymes S1P and S2P, to liberate an N-terminal fragment that migrates into the nucleus to induce genes encoding ER protein chaperones and ERAD functions (Haze, Yoshida, Yanagi, Yura, & Mori, 1999; Shen, Chen, Hendershot, & Prywes, 2002; Wu et al., 2007).

In addition to storing energy as triacylglycerols, adipose tissue serves an important endocrine function. Adipocytes secret large amounts of adipocyte-generated factors, including adipokines, cytokines, and complement components (Kershaw & Flier, 2004). Consequently, during adipogenesis, the ER increases its capacity for protein secretion. Recent studies in rodent models suggest that ER stress causes insulin resistance in adipose tissues. Phosphorylation of PERK was observed in the fat tissues from murine obese models (Ozcan et al., 2004, 2008). In addition, there were enhanced IRE1α activation, *Xbp1* mRNA splicing, and JNK phosphorylation in fat tissues from obese, insulin-resistant human subjects (Boden et al., 2008). There were significantly decreased levels of spliced *Xbp1* mRNA, BiP, and reduced phosphorylation of PERK and JNK in the adipose tissues of subjects in response to gastric bypass therapy for weight loss (Gregor et al., 2009). These results suggest a role of the UPR in adipocyte pathogenesis under ER stress conditions. Therefore, it is important to understand how ER stress and subsequent induction of UPR play roles in the development of obesity and their impact on adipose tissue. In this chapter, we will describe methods to measure UPR activation and how to investigate its role in adipocyte differentiation and physiology.

2. METHODS TO MEASURE UPR INDUCTION IN ADIPOSE TISSUES

Here, we describe optimal conditions and procedures to detect activation of three major UPR pathways during adipocyte differentiation. We use preadipocytes or mature adipocytes differentiated from 3T3-L1 cells or primary mouse embryonic fibroblasts (MEFs) *in vitro* and adipose tissues from mice *in vivo*. For *in vitro* induction of the UPR, we use pharmacological stresses, such as tunicamycin (Cat # 654380, EMD Millipore, Billerica, MA, United States), an *N*-glycosylation inhibitor, or thapsigargin (Cat # T9033, Sigma-Aldrich, St. Louis, MO, United States), a SERCA inhibitor, or physiological insults such as palmitate (0.3 m*M*, Cat # P9767, Sigma-Aldrich) or oxygen deprivation (1% oxygen).

2.1. Measuring the activation of PERK-eIF2α pathway

The eIF2α catalyzes the first regulated step of translation initiation, promoting the binding of the met-tRNA to 40S ribosomal subunits (Kimball, 1999). Upon ER stress, activated PERK phosphorylates and inhibits the activity of eIF2α, resulting in the inhibition of global translation.

Paradoxically, *Atf4* mRNA is preferentially translated into functional ATF4 protein under this condition. Subsequently, ATF4 induces transcription of downstream target genes, which include *Chop* and *Gadd34*. Since GADD34 is a protein phosphatase specific for eIF2α, phosphorylation of eIF2α by PERK is transient after GADD34 induction by ATF4 and CHOP (Novoa, Zeng, Harding, & Ron, 2001).

2.1.1 Measurement of PERK activation

Since autophosphorylation of PERK is the initial step for transmitting the UPR signal into the eIF2α subpathway, measurement of phosphorylated PERK is a critical determinant for the induction of this subpathway. Expression and phosphorylation of PERK can be analyzed by an upward mobility shift upon sodium dodecyl sulfate polyacrylamide gel electrophoresis (SDS–PAGE) containing 50 μ*M* Phos-tag™ and Western blot analysis using anti-PERK antibody. The conditions and procedures to detect levels of total and phosphorylated PERK proteins are as follows: (1) protein lysates are prepared from differentiated adipocytes treated with ER stress-inducing reagents or from frozen adipose tissues (~100 mg) in appropriate volume of NP-40 lysis buffer (1% NP-40; 50 m*M* Tris–HCl, pH 7.5; 150 m*M* NaCl; 0.1% SDS; 5 m*M* EDTA, pH 8; and 1% deoxycholate, supplemented with protease and phosphatase inhibitors (Cat # 78441, Thermo Scientific, Rockford, IL, United States)); (2) protein concentrations of the lysates are determined by the Bradford assay using the commercial kit (Cat # 500-0205, Bio-Rad Laboratories, Hercules, CA, United States); (3) appropriate amounts of 2 × SDS–PAGE sample buffer (130 m*M* Tris–HCl, pH 6.8, 20% glycerol, 4.6% SDS, and 0.02% bromophenol blue, supplemented with 2% DTT) are added to the protein lysate solution followed by heating at 95 °C for 5 min; (4) approximately 20–40 μg of denatured proteins is separated on 10% Tris–glycine SDS–PAGE containing 50 μ*M* of Phos-tag™ (Cat # AAL-107, Wako Chemicals USA, Richmond, VA, United States) and transferred to a 0.45 mm PVDF membrane (Cat # RPN1416F, GE Healthcare Life Science, Uppsala, Sweden); (5) to detect levels of PERK protein, the blots are incubated with a rabbit anti-PERK antibody (Cat # 5683, Cell Signaling Technology, Danvers, MA, United States) at a 1:1000 dilution in 0.1% Tween 20–PBS buffer containing 5% bovine serum albumin (BSA) overnight at 4 °C; (6) after the primary antibody incubation, the blots are washed and then incubated with horseradish peroxidase (HRP)-conjugated anti-rabbit antibody (Cat # NA9340, GE Healthcare Life Science) at a 1:3000 dilution for 2 h at room temperature; and (7) membrane-bound antibodies are detected by an enhanced chemiluminescence (ECL) detection reagent (Cat # 32106, Thermo Scientific). The size of detected mouse PERK protein is approximately 140 kDa, and size of phosphorylated PERK protein is shifted upward slightly.

2.1.2 Measurement of eIF2α phosphorylation

The status of eIF2α phosphorylation is a key determinant for transmitting the stress signals to the protein synthesis machinery. The conditions and procedures to detect levels of total and phosphorylated eIF2α proteins are as follows: (1) protein lysates are prepared from adipocytes in appropriate volume of NP-40 lysis buffer; (2) after determination of protein concentration, appropriate amounts of 2 × SDS–PAGE sample buffer are added into the protein lysate solution followed by heating at 95 °C for 5 min; (3) approximately 20–40 μg of denatured proteins is separated by 10% Tris–glycine SDS–PAGE and transferred to a 0.45 mm PVDF membrane; (4) to detect levels of total or phosphorylated eIF2α, the blots are incubated with a rabbit anti-eIF2α (Cat # AHO1182, Invitrogen, Carlsbad, CA, United States) or antiphosphorylated eIF2α antibody (Cat # 44782, Invitrogen) at a 1:1000 dilution in 0.1% Tween 20–PBS buffer containing 5% BSA overnight at 4 °C; (5) after the primary antibody incubation, the blots are washed and then incubated with HRP-conjugated anti-rabbit antibody at a 1:3000 dilution for 2 h at room temperature; and (6) membrane-bound antibodies are detected by an ECL detection reagent. The size of detected mouse eIF2α protein is approximately 37 kDa.

2.1.3 *Measuring* Atf4, Chop, *and* Gadd34 *expression*

Under global translational attenuation status caused by ER stress-mediated eIF2α phosphorylation, translation of *Atf4* mRNA is preferentially favored, with subsequent induction of *Chop* as well as other genes involved in amino acid metabolism and the oxidative stress response (Harding et al., 2003). Recently, it was reported that CHOP and ATF4 interact to cooperatively regulate genes involved in UPR pathways, as well as protein synthesis (Han, Back, et al., 2013; Krokowski et al., 2013). Therefore, the expressions of ATF4 and CHOP proteins are critical indicators for the activation of PERK-eIF2α UPR pathway. The procedure to measure protein levels of ATF4 and CHOP in adipocytes is as follows: (1) protein lysates from adipocytes are prepared as described earlier; (2) after determination of protein concentration, appropriate amounts of 2 × SDS–PAGE sample buffer are added into the protein lysate solutions followed by heating at 95 °C for

5 min; (3) approximately 20–40 μg of denatured proteins is separated by 10% Tris–glycine SDS–PAGE and transferred to a 0.45 mm PVDF membrane; (4) the blots are incubated with a rabbit anti-ATF4 (Cat # sc200, Santa Cruz Biotechnology, Dallas, TX, United States) or anti-CHOP (Cat # sc793, Santa Cruz Biotechnology) antibody at a 1:200 dilution in 0.1% Tween 20–PBS buffer containing 5% BSA overnight at 4 °C; (5) after the primary antibody incubation, the blots are washed and then incubated with HRP-conjugated anti-rabbit antibody at a 1:3000 dilution for 2 h at room temperature; and (6) membrane-bound antibodies are detected by an ECL detection reagent. The size of detected mouse ATF4 and CHOP proteins is approximately 50 and 29 kDa, respectively.

For quantitative real-time RT-PCR analysis of murine *Aft4* and *Chop* mRNAs, a pair of real-time PCR primers is designed: *Atf4*, 5′-ATG GCC GGC TAT GGA TGA T CGA-3′ and 5′-AGT CAA ACT CTT TCA GAT CCA TT-3′ and *Chop*, 5′-CTG CCT TTC ACC TTG GAG AC-3′ and 5′-CGT TTC CTG GGG ATG AGA TA-3′. As an internal control for normalization, a primer pair for β-actin mRNA is used: 5′-GAT CTG GCA CCA CAC CTT CT-3′ and 5′-GGG GTG TTG AAG GTC TCA AA-3′. Procedures to determine the mRNA levels of ATF4 and CHOP in the adipocytes are as follows: (1) Total RNAs are prepared from adipocytes using RNeasy mini kit (Cat # 74104, Qiagen, Germantown, MD, United States); (2) 300 n*M* of forward and reverse primers, 10 ng cDNAs prepared from total RNA, 5 μl SYBR Green Supermix (Cat # 172-5124, Bio-Rad Laboratories), and RNase-free water are mixed for 10 μl reaction volume; (3) the reaction is performed in CFX384 Touch™ real-time PCR detection system (Bio-Rad Laboratories). The PCR parameters for step 1 are 95 °C for 5 min and for step 2 are 95 °C for 15 s and 60 °C for 30 s followed by detection. Step 2 is repeated for 45 cycles; (4) fold changes of mRNA levels are determined after normalization to internal control β-actin mRNA levels using the 2-ΔΔC(t) method (Livak & Schmittgen, 2001).

2.2. Measuring the activation of IRE1α pathway

2.2.1 Phosphorylation of IRE1α

Activation of IRE1α also requires homodimerization and autophosphorylation, like PERK. Therefore, phosphorylation of IRE1α is a key proximal indicator of UPR induction for the IRE1α pathway. The amount of total and phosphorylated IRE1α can be analyzed by an upward mobility shift upon SDS–PAGE containing 50 μ*M* Phos-tag, followed by Western blot analysis using anti-IRE1α antibody (Cat # IMG-30755, Imgenex, San Diego, CA, United States). The conditions and procedures to detect levels of total and phosphorylated IRE1α proteins are as follows: (1) lysates are prepared from adipocytes in an appropriate volume of NP-40 lysis buffer; (2) after determination of protein concentration, appropriate amounts of 2 × SDS–PAGE sample buffer are added into the protein lysate solutions followed by heating at 95 °C for 5 min; (3) approximately 20–40 μg of denatured proteins is separated by 10% Tris–glycine SDS–PAGE containing 50 μ*M* of Phos-tag™ and transferred to a 0.45 mm PVDF membrane; (4) the blots are incubated with a rabbit anti-IRE1α antibody at a 1:1000 dilution in 0.1% Tween 20–PBS buffer containing 5% BSA overnight at 4 °C; (5) after the primary antibody incubation, the blots are washed and then incubated with HRP-conjugated anti-rabbit antibody at a 1:3000 dilution for 2 h at room temperature; and (6) membrane-bound antibodies are detected by an ECL detection reagent. The size of detected murine IRE1α protein is approximately 120 kDa.

2.2.2 Xbp1 *mRNA splicing and expression of spliced form of XBP1 protein*

After activation under ER stress conditions, IRE1α activates its highly sequence-specific endoribonuclease (RNase) activity to initiate unconventional splicing of *Xbp1* mRNA to remove a 26-nucleotide intron. The splicing of *Xbp1* mRNA creates a translational frameshift to introduce a new C-terminus, encoding a potent transcriptional activation domain and producing a functional transcription factor that induces the expression of a large number of genes involved in ER protein folding, trafficking, and ERAD. However, unspliced *Xbp1* mRNA encodes a nonfunctional protein that is rapidly degraded by the proteasome (Cencic et al., 2009). It was suggested that XBP1 protein encoded from unspliced *Xbp1* mRNA inhibits the function of XBP1 encoded from spliced *Xbp1* mRNA (Yoshida, Oku, Suzuki, & Mori, 2006). Thus, quantitative analyses of *Xbp1* mRNA splicing and the expression levels of XBP1 protein encoded from spliced *Xbp1* mRNA are critical in measuring the activation status of the IRE1α-mediated UPR subpathway.

For quantitative real-time RT-PCR analysis of murine total and spliced *Xbp1* mRNA, a pair of real-time PCR primers are designed as follows: spliced form of *Xbp1*, 5′-GAG TCC GCA GCA GGT G-3′ and 5′-GTG TCA GAG TCC ATG GGA-3′, and total *Xbp1*, 5′-AAG AAC ACG CTT GGG AAT GG-3′ and 5′-ACT CCC CTT GGC CTC CAC-3′. Procedures to measure the mRNA levels of *Xbp1* in adipocytes are as follows: (1) Total RNAs are prepared from adipocytes using RNeasy mini kit. (2) cDNAs are prepared; 300 n*M* forward and reverse primers, 100 ng cDNAs prepared from total RNA, 5 μl SYBR Green Supermix, and RNase-free water are mixed for 10 μl reaction volume; (3) the reaction is performed in CFX384 Touch™ real-time PCR detection system. The PCR parameters are, for step 1, 95 °C for 5 min; the PCR parameters are, for step 2, 95 °C for 15 s and 60 °C for 30 s followed by detection. Step 2 is repeated for 45 cycles; (4) fold changes of mRNA levels are determined after normalization to internal control β-actin mRNA levels using the 2-ΔΔC(t) method.

2.3. Measuring the activation of ATF6α pathway

ATF6α is a bZIP transcription factor localized to the ER membrane (Haze et al., 1999). Upon ER stress and the accumulation of misfolded proteins in the ER lumen, ATF6α traffics to the Golgi complex, where it is cleaved by site 1 and site 2 proteases. The resulting N-terminal fragment of ATF6α translocates to nucleus to induce gene expression through ER stress elements in the promoter regions (Yoshida et al., 2000). For quantitative real-time RT-PCR analysis of murine *Atf6α* mRNA, a pair of real-time PCR primers is designed: *Atf6α*, 5′-CTT CCT CCA GTT GCT CCA TC-3′ and 5′-CAA CTC CTC AGG AAC GTG CT-3′. The procedure to measure mRNA levels of *Atf6α* in the adipocytes is as follows: (1) Total RNAs are prepared from adipocytes or adipose tissues; (2) cDNAs are prepared; 300 n*M* forward and reverse primers, 100 ng cDNAs prepared from total RNA, 5 μl SYBR Green Supermix (Bio-Rad), and RNase-free water are mixed for 10 μl reaction volume; (3) the reaction is performed in the CFX384 Touch™ real-time PCR detection system (Bio-Rad). The PCR parameters are, for step 1, 95 °C for 5 min; the PCR parameters are, for step 2, 95 °C for 15 s and 60 °C for 30 s followed by detection. Step 2 is repeated for 45 cycles; (4) fold changes of mRNA levels are determined after normalization to internal control β-actin mRNA levels using the 2-ΔΔC (t) method.

3. METHODS TO INVESTIGATE ROLE OF UPR IN METABOLIC PHENOTYPES OF ADIPOCYTES

3.1. Exogenous stimulus to induce ER stress

3.1.1 Pharmacological induction of ER stress

Tunicamycin (Cat # 654380, EMD Millipore) or thapsigargin (Cat # T9033, Sigma-Aldrich) disturbs ER homeostasis and causes the accumulation of misfolded or unfolded proteins in the ER. The advantage of this approach is to obtain strong induction of UPR in short time period. However, the induction of the UPR under these conditions is either transient or toxic. It is notable that pharmacological induction of the UPR is relatively stronger and more exaggerated compared to physiologically induced ER stress conditions. In addition, since pharmacological treatment causes death at higher dosages and under long incubation time, it is necessary to optimize the conditions that cause ER stress but not cell death. In preadipocytes during differentiation, 50 ng/ml of tunicamycin induces the UPR without cell death (Han, Murthy, et al., 2013).

3.1.2 Physiological induction of ER stress

It was reported that adipose tissue in obese mice exhibits hypoxia that causes ER stress and UPR induction (Hosogai et al., 2007), indicating that hypoxia is a physiological ER stress-inducing factor *in vivo*. In our studies, 12 h hypoxia induces eIF2α phosphorylation and subsequent induction of CHOP, indicating that hypoxia is a potential ER stress-inducing stimulus (Han, Murthy, et al., 2013). Increased serum lipid levels are associated with obesity in humans and animal models, suggesting a pathological role of lipids in the regulation of adipocyte functions (Szkudelski, 2007). Interestingly, free fatty acids (FFAs) induce expression of CHOP and BiP protein, increase *Xbp1* mRNA splicing, and activate JNK in 3T3-L1 and rat primary adipocytes (Guo, Wong, Xie, Lei, & Luo, 2007; Nguyen et al., 2005), suggesting FFAs might exert adverse effects on adipose tissues through ER stress. Among FFAs, palmitate or other saturated FFAs are known to activate the UPR *in vitro*. The range of concentration of palmitate to induce the UPR is 0.2–1 m*M* for preadipocytes.

3.1.3 Chemical chaperone treatment to relieve ER stress

Chemical chaperones such as 4-phenylbutyrate (PBA) or tauroursodeoxycholic acid (TUDCA) are known to relieve ER stress conditions in liver and adipose tissues (Basseri, Lhotak, Sharma, & Austin, 2009; Ozcan et al., 2006). Therefore, chemical chaperones provide a good tool to investigate whether certain pathological conditions of interest originate through ER stress and are mediated by the UPR. PBA (Cat # SPB005, Scandinavian Formulas, Sellersville, PA, United States) can be delivered to mice at 500 mg/kg body weight/day through oral gavage for 3 weeks (Cao et al., 2013). TUDCA (Cat # 580549, EMD Chemicals, 500 mg/kg body weight) also can be administered through oral gavage daily for 3 weeks (Cao et al., 2013). These substances can also be administered through intraperitoneal injection.

3.2. Genetic approaches to investigate the role of UPR in adipocytes

3.2.1 Loss of function study using knockout models

To investigate the role of each UPR subpathway, it is possible to knock out genes of interest. We recently published that phosphorylation of eIF2α and subsequent induction of CHOP expression inhibits adipocyte differentiation (Han, Murthy, et al., 2013). In that report, we used primary MEFs from *Chop*$^{+/+}$ and *Chop*$^{-/-}$ mice. In addition, we used primary MEFs from mouse embryos with two wild-type *Eif2α* alleles (*Eif2α*$^{S/S}$) or with two mutant eIF2α alleles that have alanine substitutions at Ser51 to prevent phosphorylation (*Eif2α*$^{A/A}$). Since the *Eif2α*$^{A/A}$ mice are perinatal lethal, mice were engineered to constitutively express a wild-type floxed *Eif2α* transgene under a ubiquitous promoter (Fig. 8.1A) (Back et al., 2009). Expression of the wild-type eIF2α prevents lethality of the homozygous Ser51Ala eIF2α mutation and can be deleted by Cre recombinase in a temporal and/or tissue-specific manner to elicit the homozygous Ser51Ala eIF2α mutant phenotype. The procedure to prepare primary MEFs was described previously (Back et al., 2009). The procedures for the differentiation of primary MEFs to mature adipocytes are as follows: (1) primary MEFs are maintained in DMEM supplemented with 10% calf serum, nonessential amino acids (Cat # 11140, Invitrogen), MEM amino acids (Cat # 11130, Invitrogen), and penicillin/streptomycin (Cat # 15140, Invitrogen); (2) 2 days after confluence, cell differentiation into adipocytes is induced with DMEM containing 10 μg/ml insulin (Cat # I-5500, Sigma-Aldrich), 0.5 m*M* 3-isobutyl-1-methylxanthine (Cat # I-7018, Sigma-Aldrich), 1 μ*M* dexamethasone (Cat # D-4902, Sigma-Aldrich), and 50 n*M* rosiglitazone (Cat # 71740, Cayman Chemical, Ann Arbor, MI, United States) for 4 days; (3) DMEM supplemented with 10% FBS and 10 μg/ml insulin are replaced for another 3 days; and (4) after differentiation, mature adipocytes are fed every other day with DMEM containing 10% FBS. It is noteworthy that the concentration of insulin is 10 times greater for the differentiation of primary MEFs compared to 3T3-L1 cells.

When adipocyte differentiation is affected by knocking out genes of interest, alternative approach is to make primary MEFs from mice with a

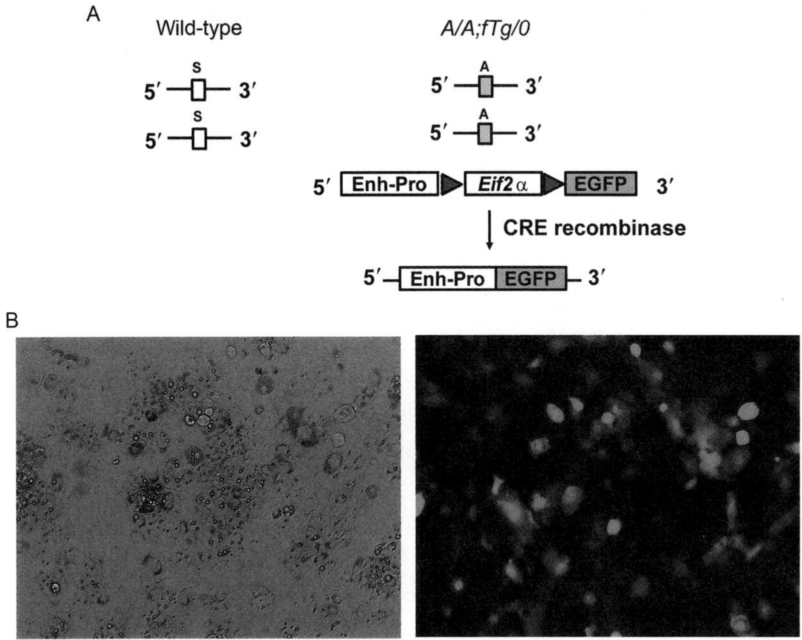

Figure 8.1 Deletion of *Eif2α* transgene by CRE expression. (A) Diagram illustrating the genotypes of wild type and *A/A;fTg/0*. *A/A;fTg/0* MEFs harbor two Ser51Ala mutant *Eif2α* alleles (*A/A*) as well as a floxed wild-type *Eif2α* transgene (*fTg*) driven by the CMV enhancer and chicken β-actin promoter. LoxP sequences (red arrowheads) allow excision of floxed wild-type *Eif2α* transgene and turn on EGFP expression. S indicates endogenous wild-type *Eif2α* allele and A indicates the Ser51Ala mutant allele of *Eif2α*. (B) Primary *A/A;fTg/0* MEFs are differentiated into mature adipocytes as described in the text. On day 10 after differentiation, adenoviruses expressing CRE recombinase gene under CMV promoter are infected at MOI 100. After deletion of the wild-type *Eif2α* transgene (*fTg*), EGFP expression is expressed. DIC (differential interference contrast, left panel) and GFP fluorescence (right panel) images are taken after 3 days of adenovirus infection. Magnification: 100×. (See color plate.)

conditional knockout gene, which can be subsequently deleted by introducing CRE activity after differentiation is complete. Procedures to differentiate primary MEFs from conditional knockout mice are the same as described for primary MEFs. Cre expression can be mediated by transfection of CRE-expressing plasmids (pBS185 CMV-Cre, Cat # 11916, Addgene, Cambridge, MA, United States) or infection of viruses expressing CRE. Adenovirus expressing CRE can be infected at MOI 100 in mature adipocytes to excise gene of interest flanked by LoxP sites (Fig. 8.1B).

3.2.2 Gain-of-function studies using inducible systems

It is difficult to attribute an effect to any specific UPR subpathway since all three UPR subpathways are activated under conditions of ER stress. In addition, since some components of the UPR are essential for normal cell function, constitutively altering their expression or activity may negatively affect the physiology of the cells and complicate interpretation of results. To avoid these problems and to investigate the role of individual UPR subpathways, it is possible to use inducible systems to remove or to force expression of certain components of the UPR at a desired time.

3.2.2.1 Tetracycline-mediated inducible system

To induce gene expression at a designated time, the tetracycline (Tet)-mediated inducible system is widely used. Most genes involved in UPR pathways are suitable for this approach. Those genes include *Chop*, *Atf4*, N-terminal fragment of *Atf6α*, *Ire1α*, and the spliced form of *Xbp1*. Although IRE1α activity is regulated at the level of phosphorylation, it was shown that overexpression of IRE1α can autophosphorylate and activate *Xbp1* mRNA splicing (Fig. 8.2A) (Han, Murthy, et al., 2013). In addition, overexpression of IRE1α with mutation in the RNase domain acts as dominant-negative inhibitor of *Xbp1* mRNA splicing (Fig. 8.2B). The procedure to generate 3T3-L1 preadipocytes expressing the gene of interest under Tet-inducible promoter is as follows: (1) cDNA from gene of interest is cloned to appropriate restriction enzyme sites of pLVX-Tight-Puro vector (Cat # 632162, Clontech Laboratories, Inc., Mountain View, CA, United States); (2) the construct is transfected to 293T cells (Cat # 632180, Clontech Laboratories, Inc.) according to manufacturer's instruction to generate lentivirus; (3) lentiviruses (10^5 PFU) containing gene of interest under the Tet-inducible promoter are infected into Tet-inducible 3T3-L1 cells in the presence of 8 μg/ml of polybrene (Cat # H-9268, Sigma-Aldrich); (4) the infected cells are incubated in media containing 2 μg/ml of puromycin (Cat # P-8833, Sigma-Aldrich) for 2 weeks; and (5) after selection in the presence of puromycin, cell clones are isolated and tested for Tet inducibility by treating doxycycline at 1 μg/ml, followed by quantitative real-time PCR or Western blot analysis for the gene of interest. Using this system, it is possible to prevent expression of the gene of interest during differentiation and then induce it in mature adipocytes to study the effect on adipocyte physiology.

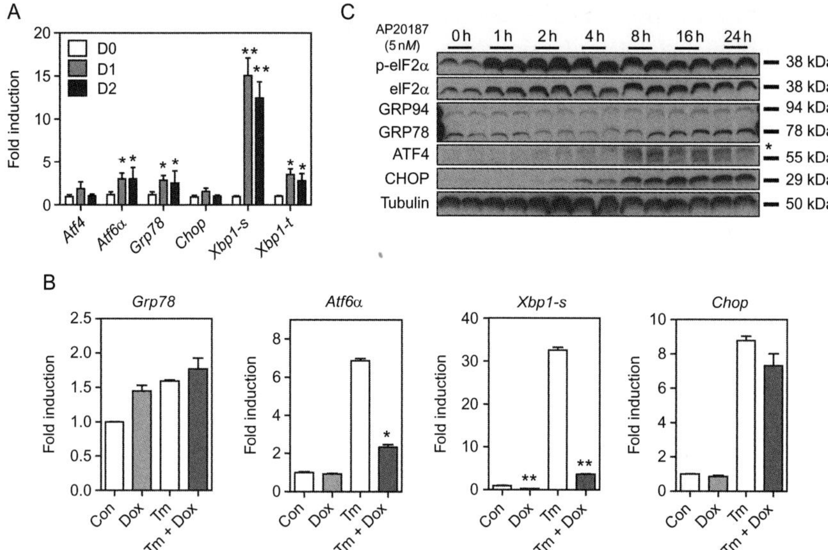

Figure 8.2 Inducible system for IRE1α and eIF2α phosphorylation. (A) Gene expression profile for 3T3-L1 cells expressing wild-type IRE1α under the control of a tetracycline-inducible promoter. Total RNAs were collected on day 0 (D0), day 1 (D1), and day 2 (D3) after doxycycline (5 μg/ml) treatment, followed by quantitative real-time PCR. Data are presented as means ± SEM of three independent experiments with triplicates. (B) Gene expression profile for stable 3T3-L1 cells expressing dominant-negative RNase mutant IRE1α (K907A) under the control of a tetracycline-inducible promoter in the absence or presence of Tm-induced ER stress. Data are presented as means ± SEM of three independent experiments with triplicates. (C) Protein expression profile in a stable cell line that expresses chimeric Fv2E-PERK. Cell lysates were collected at the indicated times after AP20187 treatment (5 n*M*) for Western blot analysis. * indicates nonspecific band. *From Han, Murthy, et al. (2013) with permission.* (See color plate.)

3.2.2.2 AP20187-mediated dimerization system

The activity of eIF2α is mainly dependent on phosphorylation status regulated by eIF2α kinases, including PERK, and not the eIF2α expression level. Therefore, induced expression of eIF2α protein by Tet induction is not appropriate to study the role of eIF2α kinases. To control the phosphorylation of eIF2α, the kinase domain of PERK can be fused to a protein module, FK506-binding protein, that binds a small dimerizer molecule, AP20187 (Cat # 635060, Clontech) (Lu et al., 2004). This artificial eIF2α kinase construct, Fv2E-PERK, is transfected into 3T3-L1 cells and cells are stably selected by puromycin treatment as described earlier. Upon AP20187 treatment (5μg/ml), the kinase is activated to obtain conditional phosphorylation of eIF2α and induction of downstream genes, including ATF4 and CHOP (Fig. 8.2C).

ACKNOWLEDGMENT

Grant support is from NIH HL057346, HL052173, DK042394, DK088227, and the Crohn's and Colitis Foundation of America (R. J. K.).

REFERENCES

Back, S. H., & Kaufman, R. J. (2012). Endoplasmic reticulum stress and type 2 diabetes. *Annual Review of Biochemistry*, *81*, 767–793. http://dx.doi.org/10.1146/annurev-biochem-072909-095555.

Back, S. H., Scheuner, D., Han, J., Song, B., Ribick, M., Wang, J., et al. (2009). Translation attenuation through eIF2alpha phosphorylation prevents oxidative stress and maintains the differentiated state in beta cells. *Cell Metabolism*, *10*(1), 13–26. http://dx.doi.org/10.1016/j.cmet.2009.06.002.

Basseri, S., Lhotak, S., Sharma, A. M., & Austin, R. C. (2009). The chemical chaperone 4-phenylbutyrate inhibits adipogenesis by modulating the unfolded protein response. *Journal of Lipid Research*, *50*(12), 2486–2501. http://dx.doi.org/10.1194/jlr.M900216-JLR200.

Boden, G., Duan, X., Homko, C., Molina, E. J., Song, W., Perez, O., et al. (2008). Increase in endoplasmic reticulum stress-related proteins and genes in adipose tissue of obese, insulin-resistant individuals. *Diabetes*, *57*(9), 2438–2444. http://dx.doi.org/10.2337/db08-0604.

Calfon, M., Zeng, H., Urano, F., Till, J. H., Hubbard, S. R., Harding, H. P., et al. (2002). IRE1 couples endoplasmic reticulum load to secretory capacity by processing the XBP-1 mRNA. *Nature*, *415*(6867), 92–96. http://dx.doi.org/10.1038/415092a.

Cao, S. S., Zimmermann, E. M., Chuang, B. M., Song, B., Nwokoye, A., Wilkinson, J. E., et al. (2013). The unfolded protein response and chemical chaperones reduce protein misfolding and colitis in mice. *Gastroenterology*, *144*(5), 989–1000. http://dx.doi.org/10.1053/j.gastro.2013.01.023, e1006.

Cencic, R., Carrier, M., Galicia-Vazquez, G., Bordeleau, M. E., Sukarieh, R., Bourdeau, A., et al. (2009). Antitumor activity and mechanism of action of the cyclopenta[b]benzofuran, silvestrol. *PLoS One*, *4*(4), e5223. http://dx.doi.org/10.1371/journal.pone.0005223.

Gregor, M. F., Yang, L., Fabbrini, E., Mohammed, B. S., Eagon, J. C., Hotamisligil, G. S., et al. (2009). Endoplasmic reticulum stress is reduced in tissues of obese subjects after weight loss. *Diabetes*, *58*(3), 693–700. http://dx.doi.org/10.2337/db08-1220.

Guo, W., Wong, S., Xie, W., Lei, T., & Luo, Z. (2007). Palmitate modulates intracellular signaling, induces endoplasmic reticulum stress, and causes apoptosis in mouse 3T3-L1 and rat primary preadipocytes. *American Journal of Physiology. Endocrinology and Metabolism*, *293*(2), E576–E586. http://dx.doi.org/10.1152/ajpendo.00523.2006.

Han, J., Back, S. H., Hur, J., Lin, Y. H., Gildersleeve, R., Shan, J., et al. (2013). ER-stress-induced transcriptional regulation increases protein synthesis leading to cell death. *Nature Cell Biology*, *15*(5), 481–490. http://dx.doi.org/10.1038/ncb2738.

Han, J., Murthy, R., Wood, B., Song, B., Wang, S., Sun, B., et al. (2013). ER stress signalling through eIF2alpha and CHOP, but not IRE1alpha, attenuates adipogenesis in mice. *Diabetologia*, *56*(4), 911–924. http://dx.doi.org/10.1007/s00125-012-2809-5.

Harding, H. P., Novoa, I., Zhang, Y., Zeng, H., Wek, R., Schapira, M., et al. (2000). Regulated translation initiation controls stress-induced gene expression in mammalian cells. *Molecular Cell*, *6*(5), 1099–1108.

Harding, H. P., Zhang, Y., Zeng, H., Novoa, I., Lu, P. D., Calfon, M., et al. (2003). An integrated stress response regulates amino acid metabolism and resistance to oxidative stress. *Molecular Cell*, *11*(3), 619–633.

Haze, K., Yoshida, H., Yanagi, H., Yura, T., & Mori, K. (1999). Mammalian transcription factor ATF6 is synthesized as a transmembrane protein and activated by proteolysis in response to endoplasmic reticulum stress. *Molecular Biology of the Cell*, *10*(11), 3787–3799.

Hosogai, N., Fukuhara, A., Oshima, K., Miyata, Y., Tanaka, S., Segawa, K., et al. (2007). Adipose tissue hypoxia in obesity and its impact on adipocytokine dysregulation. *Diabetes*, *56*(4), 901–911. http://dx.doi.org/10.2337/db06-0911.

Kershaw, E. E., & Flier, J. S. (2004). Adipose tissue as an endocrine organ. *The Journal of Clinical Endocrinology and Metabolism*, *89*(6), 2548–2556. http://dx.doi.org/10.1210/jc.2004-0395.

Kimball, S. R. (1999). Eukaryotic initiation factor eIF2. *The International Journal of Biochemistry & Cell Biology*, *31*(1), 25–29, S1357-2725(98)00128-9 [pii].

Krokowski, D., Han, J., Saikia, M., Majumder, M., Yuan, C. L., Guan, B. J., et al. (2013). A self-defeating anabolic program leads to beta-cell apoptosis in endoplasmic reticulum stress-induced diabetes via regulation of amino acid flux. *The Journal of Biological Chemistry*, *288*(24), 17202–17213. http://dx.doi.org/10.1074/jbc.M113.466920.

Livak, K. J., & Schmittgen, T. D. (2001). Analysis of relative gene expression data using real-time quantitative PCR and the 2(-Delta Delta C(T)) method. *Methods*, *25*(4), 402–408. http://dx.doi.org/10.1006/meth.2001.1262.

Lu, P. D., Jousse, C., Marciniak, S. J., Zhang, Y., Novoa, I., Scheuner, D., et al. (2004). Cytoprotection by pre-emptive conditional phosphorylation of translation initiation factor 2. *The EMBO Journal*, *23*(1), 169–179. http://dx.doi.org/10.1038/sj.emboj.7600030.

Marciniak, S. J., Yun, C. Y., Oyadomari, S., Novoa, I., Zhang, Y., Jungreis, R., et al. (2004). CHOP induces death by promoting protein synthesis and oxidation in the stressed endoplasmic reticulum. *Genes & Development*, *18*(24), 3066–3077. http://dx.doi.org/10.1101/gad.1250704.

Nguyen, M. T., Satoh, H., Favelyukis, S., Babendure, J. L., Imamura, T., Sbodio, J. I., et al. (2005). JNK and tumor necrosis factor-alpha mediate free fatty acid-induced insulin resistance in 3T3-L1 adipocytes. *The Journal of Biological Chemistry*, *280*(42), 35361–35371. http://dx.doi.org/10.1074/jbc.M504611200.

Novoa, I., Zeng, H., Harding, H. P., & Ron, D. (2001). Feedback inhibition of the unfolded protein response by GADD34-mediated dephosphorylation of eIF2alpha. *The Journal of Cell Biology*, *153*(5), 1011–1022.

Ozcan, U., Cao, Q., Yilmaz, E., Lee, A. H., Iwakoshi, N. N., Ozdelen, E., et al. (2004). Endoplasmic reticulum stress links obesity, insulin action, and type 2 diabetes. *Science*, *306*(5695), 457–461. http://dx.doi.org/10.1126/science.1103160.

Ozcan, U., Ozcan, L., Yilmaz, E., Duvel, K., Sahin, M., Manning, B. D., et al. (2008). Loss of the tuberous sclerosis complex tumor suppressors triggers the unfolded protein response to regulate insulin signaling and apoptosis. *Molecular Cell*, *29*(5), 541–551. http://dx.doi.org/10.1016/j.molcel.2007.12.023.

Ozcan, U., Yilmaz, E., Ozcan, L., Furuhashi, M., Vaillancourt, E., Smith, R. O., et al. (2006). Chemical chaperones reduce ER stress and restore glucose homeostasis in a mouse model of type 2 diabetes. *Science*, *313*(5790), 1137–1140. http://dx.doi.org/10.1126/science.1128294.

Scheuner, D., Song, B., McEwen, E., Liu, C., Laybutt, R., Gillespie, P., et al. (2001). Translational control is required for the unfolded protein response and in vivo glucose homeostasis. *Molecular Cell*, 7(6), 1165–1176, S1097-2765(01)00265-9 [pii].

Shen, J., Chen, X., Hendershot, L., & Prywes, R. (2002). ER stress regulation of ATF6 localization by dissociation of BiP/GRP78 binding and unmasking of Golgi localization signals. *Developmental Cell*, *3*(1), 99–111.

Shen, X., Ellis, R. E., Lee, K., Liu, C. Y., Yang, K., Solomon, A., et al. (2001). Complementary signaling pathways regulate the unfolded protein response and are required for C. elegans development. *Cell*, *107*(7), 893–903.

Szkudelski, T. (2007). Intracellular mediators in regulation of leptin secretion from adipocytes. *Physiological Research*, *56*(5), 503–512.

Walter, P., & Ron, D. (2011). The unfolded protein response: From stress pathway to homeostatic regulation. *Science*, *334*(6059), 1081–1086. http://dx.doi.org/10.1126/science.1209038.

Wang, S., & Kaufman, R. J. (2012). The impact of the unfolded protein response on human disease. *The Journal of Cell Biology*, *197*(7), 857–867. http://dx.doi.org/10.1083/jcb.201110131.

Wu, J., Rutkowski, D. T., Dubois, M., Swathirajan, J., Saunders, T., Wang, J., et al. (2007). ATF6alpha optimizes long-term endoplasmic reticulum function to protect cells from chronic stress. *Developmental Cell*, *13*(3), 351–364.

Yoshida, H., Okada, T., Haze, K., Yanagi, H., Yura, T., Negishi, M., et al. (2000). ATF6 activated by proteolysis binds in the presence of NF-Y (CBF) directly to the cis-acting element responsible for the mammalian unfolded protein response. *Molecular and Cellular Biology*, *20*(18), 6755–6767.

Yoshida, H., Oku, M., Suzuki, M., & Mori, K. (2006). pXBP1(U) encoded in XBP1 pre-mRNA negatively regulates unfolded protein response activator pXBP1(S) in mammalian ER stress response. *The Journal of Cell Biology*, *172*(4), 565–575. http://dx.doi.org/10.1083/jcb.200508145.

Zhang, K., & Kaufman, R. J. (2008). From endoplasmic-reticulum stress to the inflammatory response. *Nature*, *454*(7203), 455–462. http://dx.doi.org/10.1038/nature07203.

CHAPTER NINE

Application of Activity-Based Protein Profiling to Study Enzyme Function in Adipocytes

Andrea Galmozzi, Eduardo Dominguez, Benjamin F. Cravatt, Enrique Saez[1]

Department of Chemical Physiology, The Skaggs Institute for Chemical Biology, The Scripps Research Institute, La Jolla, California, USA

[1]Corresponding author: e-mail address: esaez@scripps.edu

Contents

Abstract

Activity-based protein profiling (ABPP) is a chemical proteomics approach that utilizes small-molecule probes to determine the functional state of enzymes directly in native systems. ABPP probes selectively label active enzymes, but not their inactive forms, facilitating the characterization of changes in enzyme activity that occur without alterations in protein levels. ABPP can be a tool superior to conventional gene expression and proteomic profiling methods to discover new enzymes active in adipocytes and to detect differences in the activity of characterized enzymes that may be associated with disorders of adipose tissue function. ABPP probes have been developed that react selectively with most members of specific enzyme classes. Here, using as an example the serine hydrolase family that includes many enzymes with critical roles in adipocyte physiology, we describe methods to apply ABPP analysis to the study of adipocyte enzymatic pathways.

Methods in Enzymology, Volume 538
ISSN 0076-6879
http://dx.doi.org/10.1016/B978-0-12-800280-3.00009-8

1. INTRODUCTION

One of the greatest challenges of the postgenomic era is to identify and assign function to the large number of potential new enzymes that genome sequencing projects have unearthed. Adipocytes are cells that depend on multiple enzymatic pathways to carry out their function: to store and release energy as needed (white adipocytes) or to dissipate energy to generate heat (brown adipocytes). Although many of the enzymes involved in these processes are well known, gene expression and proteomic profiling studies have revealed that numerous poorly characterized enzymes are present in adipocytes. Recent advances in chemical proteomic methods have provided tools to establish the level of enzyme activity in cells and tissues. Application of these approaches to the study of adipogenesis and fat cell physiology is likely to provide significant insight into the enzymatic pathways active in adipocytes and those that may be abnormally regulated in disease. Among the tools developed by chemical biologists to analyze enzyme activity, activity-based protein profiling (ABPP) stands out as a particularly powerful method to discover new enzymes involved in normal adipocyte physiology, as well as to uncover differences in activity of well-characterized enzymes in pathological states involving adipose depots. ABPP is a chemical proteomics approach that utilizes small-molecule probes to determine the functional state of enzymes directly in native systems (Cravatt, Wright, & Kozarich, 2008; Heal, Dang, & Tate, 2011). An ABPP probe contains at least two key features: (1) a *reactive group* that binds and covalently modifies the active sites of a large number of enzymes that share conserved mechanistic and/or structural features and (2) a *reporter tag*, such as a fluorophore or biotin, to enable detection, enrichment, and identification of probe-labeled enzymes by gel electrophoresis and in-gel fluorescence scanning (gel-based ABPP; Jessani, Liu, Humphrey, & Cravatt, 2002) or liquid chromatography–mass spectrometry (LC–MS, e.g., multidimensional protein identification technology, MudPIT; ABPP–MudPIT; Jessani et al., 2005; Fig. 9.1). ABPP probes have been generated for more than a dozen enzyme classes, including probes that react specifically with serine hydrolases (SHs) (Liu, Patricelli, & Cravatt, 1999), cysteine proteases (Greenbaum, Medzihradszky, Burlingame, & Bogyo, 2000), kinases (Patricelli et al., 2007), and glycosidases (Hekmat, Kim, Williams, He, & Withers, 2005; Table 9.1). These probes selectively label active enzymes, but not their inactive forms, facilitating the characterization of changes in enzyme activity that occur without alterations in protein levels (Jessani et al., 2004; Kidd, Liu, & Cravatt, 2001).

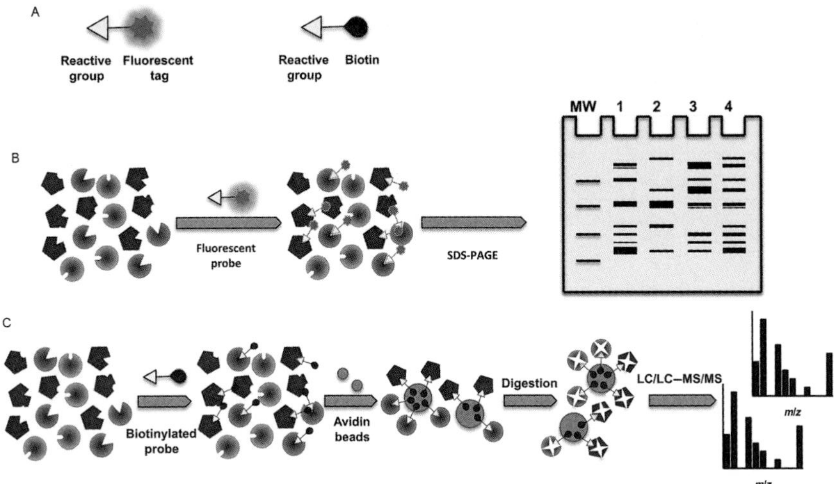

Figure 9.1 Basics of activity-based protein profiling (ABPP). (A) Schematic of ABPP probes showing their basic features: (1) a reactive head group that targets a specific enzyme class and (2) a reporter tag. (B) Gel-based ABPP. A fluorophosphonate-reactive group can be coupled to a fluorophore tag (e.g., rhodamine) to covalently label and detect active serine hydrolases by SDS-PAGE and in-gel fluorescence scanning. (C) ABPP–MudPIT. Biotin-conjugated ABPP probes can be used to label, enrich, and identify by mass spectrometry active serine hydrolases (SH). (See color plate.)

A major application of ABPP technology is target discovery. Two or more proteomes are analyzed by ABPP in parallel to identify enzymes with different levels of activity. If the samples analyzed reflect different biological conditions (e.g., physiological vs. pathological), then the enzyme(s) showing differential activity can be hypothesized to be involved in the phenotype in question. The applications of ABPP methods to the study of adipocyte physiology are multiple. For example, ABPP may be used to profile and identify enzyme activities that vary between subcutaneous and visceral white adipose depots and that may explain the opposite relationship that these depots show to the development of insulin resistance and metabolic syndrome. Or it may be used to monitor enzyme activities that are altered in adipose tissue of diabetics relative to healthy individuals. ABPP can also be used to identify enzymes whose activity is modified upon exposure to chemical or environmental stimuli, as in the response to cold of brown adipose tissue. Because ABPP reports changes in enzyme activity, not gene or protein levels, it can be more powerful than traditional gene or proteomic profiling approaches in revealing those enzymes that play central molecular roles in the phenotypic

Table 9.1 ABPP probes are available for a variety of enzyme classes

Probe	**Enzyme class**	**References**
Fluorophosphonate	Serine hydrolases	Liu et al. (1999)
2-Deoxy-2-fluoro glycoside	Exo- and endoglycosidases	Vocadlo and Bertozzi (2004) and Hekmat et al. (2005)
Acyloxymethyl ketone	Cysteine proteases	Kato et al. (2005)
E-64 based	Papain-like cysteine proteases	Bogyo, Verhelst, Bellingard-Dubouchaud, Toba, and Greenbaum (2000)
Photoreactive benzophenone-hydroxamate	Metalloproteases	Saghatelian, Jessani, Joseph, Humphrey, and Cravatt (2004); Chan, Chattopadhaya, Panicker, Huang, and Yao (2004)
Photoreactive benzophenone-hydroxyl-ethylene	Aspartyl proteases	Li et al. (2000)
Quinolimine methide	Proteases	Zhu, Girish, Chattopadhaya, and Yao (2004)
Quinone methide	Tyrosine phosphatases–glycosidases	Lo et al. (2002) and Tsai et al. (2002)
Sulfonate ester	Dehydrogenases–glutathione *S*-transferases–sugar kinases–epoxide hydrolases–transglutaminases	Adam, Sorensen, and Cravatt (2002)
Vinyl sulfone	Proteasome	Bogyo (2005); Verdoes et al. (2006)
Vinyl sulfone	Ubiquitin-specific proteases	Borodovsky et al. (2005)
Wortmannin analogues	Lipid and protein kinases	Yee, Fas, Stohlmeyer, Wandless, and Cimprich (2005); Liu, Shreder, Gai, Corral, Ferris, and Rosenblum (2005)
α-Bromobenzylphosphonate	Tyrosine phosphatases	Kumar et al. (2004)
Acyl-phosphate	Kinases/ATPases	Patricelli et al. (2007)

differences under comparison. An additional advantage of ABPP relative to other approaches is that it can detect changes in activity of very low-abundance enzymes in highly complex samples.

When performed in a *competitive mode*, ABPP can also be used to screen for enzyme inhibitors or to identify the target(s) to which a small-molecule inhibitor binds (Bachovchin et al., 2010; Li, Blankman, & Cravatt, 2007) directly in native cell and tissue proteomes (Bachovchin & Cravatt, 2012). In competitive ABPP, proteomes are first incubated with the small molecule of interest (e.g., a phenotypic hit that targets an unknown enzyme) and then labeled with a broad ABPP probe that reacts with most enzymes in the class under study (Fig. 9.2). Competitive ABPP can be performed using irreversible or reversible inhibitors, although the identification of reversible inhibitors requires a kinetically controlled condition in which labeling with the ABPP probe has not reached saturation. In the case of activity-based

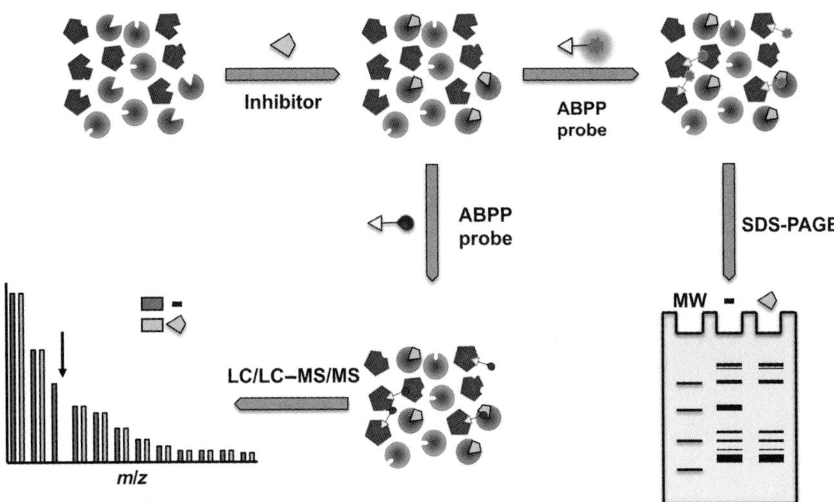

Figure 9.2 Competitive ABPP for enzyme and inhibitor discovery. In a competitive activity-based protein profiling experiment, cells, animals, or prepared proteomes are first treated with either an inhibitor or a vehicle and subsequently labeled with a broad ABPP probe. Gel-based ABPP or ABPP–MudPIT is then used to either visualize or identify active enzymes (i.e., labeled by the ABPP probe). An enzyme that is the target of an inhibitor will show reduced signals in the inhibitor-treated samples relative to vehicle controls. (See color plate.)

irreversible enzyme inhibitors, the enzymes they target will not be subsequently labeled with the broad ABPP probe, and loss of probe labeling will reveal the identity of the inhibitor's target(s). Enzyme activities present in the vehicle, but not the inhibitor-treated sample, represent the molecular target of the compound in question.

To illustrate the application of ABPP methods to the study of adipocyte enzymes, we describe here several basic protocols that use activity-based probes that specifically label enzymes of the SH family. SHs constitute one of the largest and most diverse enzyme classes in mammals, where they perform crucial roles in many biological processes (Bachovchin & Cravatt, 2012; Long & Cravatt, 2011). There are ~240 human SHs, and these enzymes include lipases, peptidases/proteases, and (thio)esterases, all of which hydrolyze their substrates through a conserved mechanism involving an active-site serine nucleophile. A substantial fraction of human SHs (>50%) remain unannotated, with no described function or identified physiological substrates, and an even greater number (>80%) lack selective inhibitors to aid in their characterization (Bachovchin & Cravatt, 2012). SHs play a particularly prominent role in adipocytes, where they regulate lipolysis (hormone-sensitive lipase, adipose triglyceride lipase, monoacylglycerol lipase, and PNPLA4) and lipid uptake (lipoprotein lipase) and synthesis (fatty acid synthase). Adipocytes also express many poorly characterized SHs that may play equally important roles in their cellular physiology (Soukas, Socci, Saatkamp, Novelli, & Friedman, 2001).

ABPP can prove to be a powerful tool to discern the function of these unannotated SHs in adipocyte biology and to establish which may serve as therapeutic targets for metabolic disorders. ABPP probes bearing a fluorophosphonate (FP) electrophile that shows broad reactivity with SHs and negligible interaction with other enzymes have been generated (Kidd et al., 2001; Liu et al., 1999). These FP-based probes have been shown to label >80% of the 115+ predicted mouse metabolic SHs in tissue/cell proteomes (Bachovchin et al., 2010). An FP probe conjugated to a fluorophore (e.g., FP-rhodamine) can be used to reveal differences in SH activity by *gel-based ABPP*, while an FP-biotin probe can be used to enrich and identify the specific SHs in question using *ABPP–MudPIT*. An example of data generated with these broad FP-based ABPP probes is shown in Fig. 9.3.

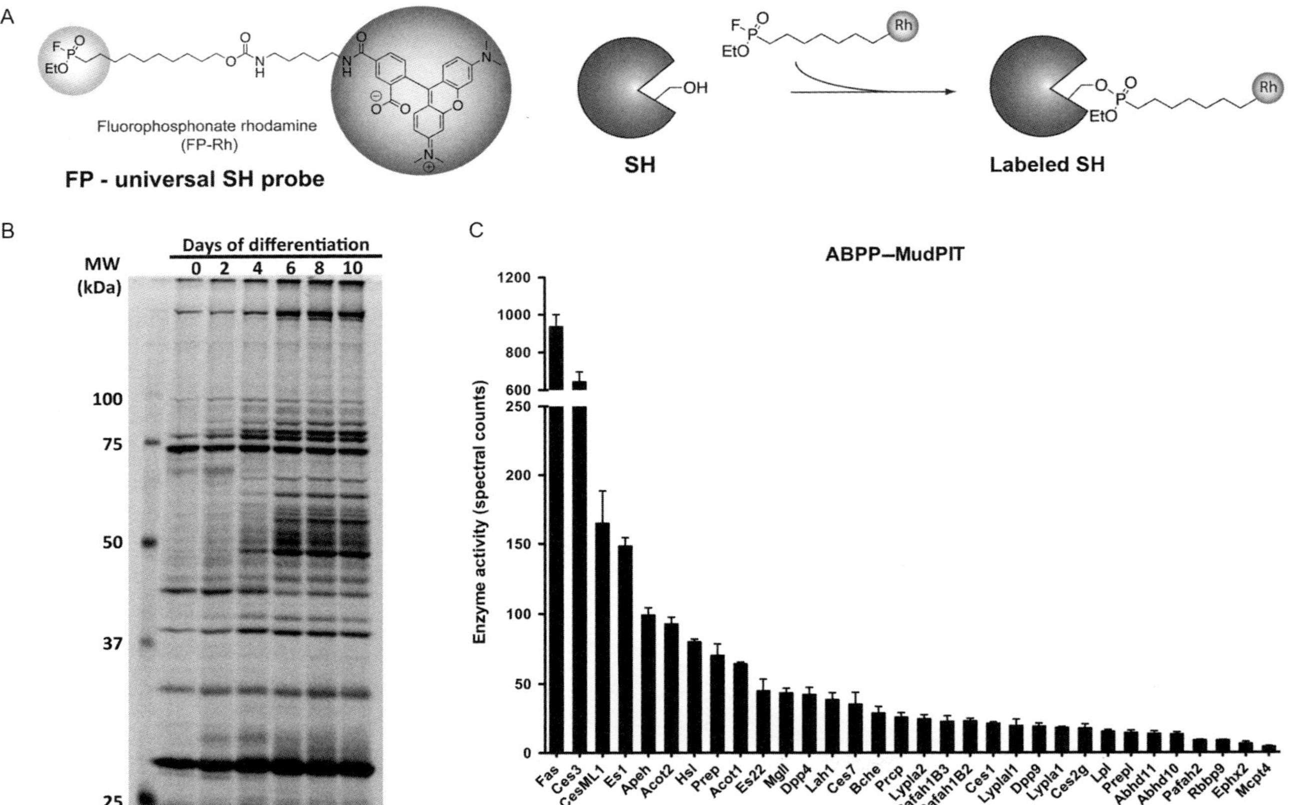

Figure 9.3 Sample ABPP data obtained with FP-based probes. (A) Reporter-tagged fluorophosphonate ABPP probes that react covalently only with the active form of serine hydrolases can be used to (B) profile the pattern of serine hydrolase activity by gel-based ABPP during differentiation of 10T1/2 adipocytes or (C) identify active serine hydrolases in mouse white adipose tissue using ABPP–MudPIT. (See color plate.)

2. TECHNICAL ASPECTS

2.1. Preparation of proteomes for ABPP analysis

2.1.1 Total proteome extracts

To preserve enzyme activity, only nondenaturing lysis buffers should be used in ABPP analysis. Mechanical disruption of cells in PBS is the best procedure to maintain proteome integrity. Protease inhibitors may be added to the PBS solution, but they are not recommended, as they may inhibit the activity of some enzymes and interfere with labeling of enzymes with the broad activity probe. To maintain proteome integrity and minimize protein degradation, all solutions and collecting tubes must be precooled and kept on ice, and all procedures must be carried out at 4 °C.

2.1.2 Materials

Cell culture or tissue sample
Ice-cold phosphate-buffered saline (PBS) without calcium or magnesium (e.g., Corning catalog number 21-031-CV)
Dounce homogenizer
TissueLyser (Qiagen) or similar mechanical tissue disruption equipment (e.g., Polytron homogenizer)
Beads (for use with TissueLyser)
Probe sonicator
Ultracentrifuge (required only if fractionation of soluble and membrane-bound proteomes is desired)

1. In the case of cells, wash and scrape the cells of interest from the culture plate using PBS and centrifuge to pellet them. Start with enough cells to yield sufficient proteome for gel-based ABPP (50 μg/reaction) or ABPP–MudPIT (1 mg/reaction) as desired; bear in mind the low protein yield of adipocytes. Resuspend the cell pellet in four volumes of ice-cold PBS and homogenize on ice by performing 10–15 strokes with a Dounce homogenizer or by sonicating five times for 5 s at 50% power. To prepare proteomes from tissues (a reasonable starting amount is 30–100 mg of adipose tissue), add four volumes of PBS to freshly isolated (preferred) or frozen tissue, and homogenize using a bead-based TissueLyser for 3 min at maximum shaking frequency. Repeat the procedure until the solution is homogeneous. Once the tissue is fully homogenized, sonicate the solution five times for 5 s at 30% power. If a TissueLyser (or similar instrument) is not available, alternative methods

of mechanical tissue disruption are also appropriate (e.g., manual Polytron homogenizer) as long as the protein solution is not unduly heated.

2. Centrifuge proteome 30 min at 14,000 × *g* at 4 °C and collect the supernatant. Be careful to avoid the fat cake that is present on top of the solution.
3. Quantify protein concentration and adjust to desired concentration (1–2 mg/mL).
4. Aliquot samples and store at −80 °C, or proceed with ABPP probe labeling if desired.
5. Prior to ABPP, thaw samples on ice and briefly allow them to warm to room temperature.

2.1.3 Fractionation of soluble and membrane-associated proteomes

To capture the greatest number of enzyme activities in ABPP experiments, and to simplify interpretation of enzyme activity profiles in gel-based ABPP or mass spectrometry (i.e., ABPP–MudPIT) analyses, it is often useful to separate the soluble from the membrane-bound proteome of cells/tissues and to analyze them separately:

1. Mechanically homogenize cells or tissues as in step 1 mentioned earlier.
2. Centrifuge for 5 min at 1000 × *g* at 4 °C to eliminate nuclei and unbroken cells or small pieces of tissue.
3. Transfer supernatant to a clean ultracentrifuge tube, being careful to avoid the fat cake. Centrifuge for 1 h at 100,000 × *g* at 4 °C to pellet the membrane-associated fraction and separate it from the soluble proteome.
4. Transfer supernatant to a clean tube. Gently wash the pellet once with ice-cold PBS. Discard wash and add 1–2 volumes of cold PBS to the pellet. Sonicate pellet three times at 30% power to resuspend membrane-associated proteome.
5. Quantify concentration of membrane-associated and soluble proteomes and adjust to desired concentration (1–2 mg/mL).
6. Aliquot samples and store at −80 °C, or proceed with ABPP probe labeling if desired.

2.2. Gel-based ABPP

The following protocol is the standard procedure for labeling enzymes of the SH family in proteomes prepared from cells or tissues (i.e., *in vitro* ABPP labeling) using FP-based ABPP probes (e.g., FP-rhodamine) and visualizing differences in enzyme activity levels in SDS-PAGE gels. This basic protocol can be modified as necessary and used to profile the activity levels of enzymes in other families for which appropriate activity-based probes are available (see, e.g., in Table 9.1). An unlabeled control sample should be used for comparison to the experimental samples to confirm efficiency of the labeling reaction. Each enzyme class may have different ABPP analysis requirements. For example, SHs are often modified at the posttranslational level (e.g., subject to glycosylation). This can result in altered migration patterns in electrophoresis gels and in the presence of multiple bands corresponding to the same enzyme. To reduce the number of bands visualized in gels and simplify interpretation of data, an optional deglycosylation step can be performed after the labeling procedure. An additional control sample labeled with the ABPP probe but not subjected to deglycosylation is required in this case (to verify efficiency of the procedure).

2.2.1 Materials

Fluorophore-conjugated FP ABPP probe (e.g., FP-rhodamine; FP-TAMRA, Thermo Fisher catalog number 88318)
Proteome (cell/tissue lysate prepared as described earlier)
Cold PBS without calcium or magnesium pH 7.4
50 × FP fluorophore-conjugated ABPP probe (50 μ*M* stock in DMSO; can be stored at −20 °C for over a year)
DMSO
10% Nonidet P-40 (NP-40) (w/v) solution (New England Biolabs, catalog number B2704S)
10 × Glycoprotein denaturing buffer (New England Biolabs, catalog number B1704S)
Reaction buffer G7 (New England Biolabs, catalog number B3704S)
PNGase F (New England Biolabs, catalog number P0704S)
4 × LDS sample buffer (Life Technologies, catalog number NP0007)

2.2.2 Sample labeling

1. For each experimental and control sample, aliquot 49 μL of 1 mg/mL cell/tissue proteome into microcentrifuge tubes.
2. To the experimental samples, add 1 μL of 50 × ABPP probe stock (50 μ*M*) for a final concentration of 1 μ*M* FP probe. To the control sample, add 1 μL of DMSO.

3. Vortex samples and allow the reaction to proceed for 1 h at RT. The labeling reaction can also be carried out at 37 °C, which can result in better visualization of slow-labeling enzymes.

   Optional: Deglycosylation step. Go to step 8 if deglycosylation is not desired.
4. Add 5 μL of 10 × glycoprotein denaturing buffer.
5. Heat samples for 10 min at 95 °C.
6. Add 5 μL of 10% NP-40 solution, 5 μL of reaction buffer G7, and 0.5 μL of PNGase F to each sample. Do not add PNGase F to the labeled control sample.
7. Vortex samples and incubate for 1 h at 37 °C.
8. Add 20 μL of 4 × LDS sample buffer to quench the reaction.
9. Heat samples for 5 min at 95 °C. At this point, samples can be stored at −20 °C overnight.
10. Run 40 μL of labeled sample on a 10% SDS-PAGE gel (250 V constant for a large gel). To avoid overheating due to the high voltage, electrophoresis may be performed at 4 °C or cooling plates added to the outside of the gel box.
11. Visualize enzyme activities using an in-gel fluorescence flatbed scanner (e.g., Hitachi FMBIO IIe). Inverse grayscale images are typically most useful to analyze and present data.

2.3. ABPP–MudPIT

This technique allows the identification by mass spectrometry of specific enzyme activities, in these case SHs. It is highly recommended that all mass spectrometry experiments be run in four or more biological replicates.

2.3.1 Materials

Biotin-conjugated FP ABPP probe (e.g., FP-biotin; FP-desthiobiotin Thermo Fisher catalog number 88317). Make a 1-m*M* stock solution in DMSO.

Proteome (cell/tissue lysate) prepared as described in Section 2.1

10% Triton X-100 (e.g., Sigma-Aldrich, catalog number T-9284) solution in PBS

PBS without calcium or magnesium, pH 7.4

PD-10 separation columns (GE Healthcare, catalog number 17-0851-0)

10% SDS (e.g., Fluka, catalog number BP166-500) solution in water

Avidin beads (avidin–agarose beads from egg white, Sigma-Aldrich, catalog number A9207)

Water, LC–MS grade (e.g., Fluka, catalog number 39253)
Urea (e.g., Fluka, catalog number U15-500)
Tris (2-carboxyethyl) phosphine hydrochloride (TCEP) (Sigma-Aldrich, catalog number C4706)
Iodoacetamide (e.g., Sigma-Aldrich, catalog number I6125)
Trypsin (Promega, catalog number V5111)
Formic acid (e.g., Fluka, catalog number O6440)

2.3.2 Sample labeling

1. For each reaction, label 1 mg of proteome (as a 1 mg/mL solution) with FP-biotin at 5 μ*M* final concentration (5 μL of 1 m*M* stock solution in DMSO) for 2 h at room temperature.
2. Solubilize proteome by adding 100 μL of Triton X-100 (10% stock solution in PBS) for a final concentration of approximately 1%. The final volume of the reaction should now be 1.1 mL.
3. Rotate at 4 °C for 1 h.
4. Equilibrate PD-10 column with 25 mL of PBS.
5. To eliminate the excess of ABPP probe, add the 1.1 mL labeling reaction to the column and rinse the tubes where it was carried out twice with 700 μL of PBS, adding each wash to the column. The final volume of the reaction should now be 2.5 mL.
6. Elute the labeled proteome with 3.5 mL of PBS and collect the fraction. The final volume of the reaction should now be 6 mL.
7. Add 0.3 mL of SDS (10% stock solution in water) for a final concentration of about 0.5%.
8. Heat samples at 90 °C for 8 min.
9. Cool samples to room temperature.
10. Add PBS to reach a final volume of 8.5 mL.
11. When working with avidin beads, use cut pipette tips to ensure integrity of beads. Wash avidin beads (100 μL of slurry per sample; 100 μL contains 50 μL of glycerol and 50 μL of beads) three times with 10 mL of PBS. Centrifuge 1 min at 400 × *g* between washes. Resuspend the washed beads pellet in one volume of PBS.
12. Mix labeled proteome and washed beads (100 μL/sample) in a 15-mL conical tube.
13. Rotate at room temperature for 1 h.
14. Centrifuge labeled proteome/bead mixture at 400 × *g* for 3 min.
15. Discard supernatant.

From this point on, use only mass spectrometry-grade water or its equivalent for all solutions.

16. Wash pelleted beads twice with 8 mL of 1% SDS. For each wash, rotate tubes at room temperature for 5 min, prior to pelleting beads (400 × *g* for 3 min). This same washing procedure should be used for all subsequent washes described later.
17. Wash beads twice with 8 mL of 6 *M* urea.
18. Wash beads three times with 8 mL of PBS.
19. After the final wash, bring beads up to 1 mL in PBS and transfer to screw top microcentrifuge tubes.
20. Spin in tabletop microcentrifuge at 400 × *g* for 3 min.
21. Discard buffer and add 150 μL of 8 *M* urea to the pellet.
22. Add 10 μL of 100 m*M* TCEP for a final concentration of approximately 5 m*M*.
23. Add 5 μL of freshly made 500 m*M* iodoacetamide (~12 m*M* final concentration) to each sample and rotate at room temperature in the dark for 30 min.
24. Spin on tabletop microcentrifuge at 400 × *g* for 3 min. Discard all but 50 μL of supernatant and add 250 μL of PBS to the pellet in order to reduce the concentration of urea to less than 2 *M* prior to tryptic digestion.
25. Add 4 μL of 0.5 μg/μL trypsin to each sample and incubate at 37 °C overnight.
26. Spin in tabletop microcentrifuge at 400 × *g* for 4 min.
27. Transfer peptide supernatant to mass spectrometry-grade tubes. Be careful not to aspirate beads that can clog the HPLC columns during mass spectrometry analysis.
28. Add 0.1 mL of PBS to the beads, gently shake, and centrifuge at 1600 × *g* for 4 min. Add supernatant to that obtained in the previous step. Volume should now be ~400 μL.
29. Add formic acid to a final concentration of 5%.
30. Store samples at −80 °C or apply to equilibrated capillary column for mass spectrometry analysis.

Protocols to carry out mass spectrometry analysis can vary widely. For an example of a MudPIT protocol, the reader is referred to Kline and Wu (2009).

2.4. Competitive ABPP

As mentioned earlier, competitive ABPP can be performed using irreversible or reversible inhibitors, but the clear identification of reversible inhibitors requires a kinetically controlled condition in which labeling with the ABPP probe has not reached saturation. The protocol described here is tailored for irreversible enzyme inhibitors, but incubation with the ABPP probe can be shortened to enable the isolation of reversible inhibitors.

1. For each compound to be tested, aliquot 50 μg in 49 μL (for gel-based ABPP) or 1 mg in 1 mL (for ABPP–MudPIT) of proteome into microcentrifuge tubes.
2. To each sample, add 0.5 μL of each compound to be evaluated for gel-based ABPP or 5 μL for ABPP–MudPIT. Compound stock solutions should be made to reflect the final concentration desired in the competitive ABPP analysis (e.g., if the compound is to be tested at a 10 μ*M* dose, stock should be 1 m*M*). Add 0.5 μL of DMSO to the control sample for gel-based ABPP or 5 μL for ABPP–MudPIT.
3. Vortex samples and incubate for 1 h at room temperature.
4. Label samples by adding 0.5 μL of ABPP probe stock for gel-based ABPP or 5 μL for ABPP–MudPIT, keeping in mind that for gel-based ABPP, a final probe concentration of 1 μ*M* should be used, while for ABPP–MudPIT, the final probe concentration should be 5 μ*M*.

From this point on, the labeling reactions should be treated just as described in the protocols mentioned earlier for sample labeling, picking up at the step immediately after ABPP probe addition.

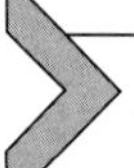

3. DISCUSSION

3.1. Critical parameters and troubleshooting

Protein degradation during the preparation of the proteomes may affect the quality of ABPP analysis. The use of protease inhibitors will help avoid this problem, but it may affect the ability of ABPP probes to label certain enzyme classes. This is a particular problem for SHs, some of which are themselves proteases (e.g., DPP-IV). If warranted, protease inhibitors should be avoided and all proteome preparation steps carried out at 4 °C. Use of mass spectrometry-grade water is critical in ABPP–MudPIT analysis to reduce background signals. To maximize tryptic digestion, the TCEP solution must be prepared fresh every time (TCEP powder should be stored at 4 °C) in order to obtain complete protein denaturation. Similarly, to prevent disulfide bond regeneration, the iodoacetamide solution must be fresh and not be exposed to light. Significant nonspecific protein binding is frequently observed with streptavidin beads (as used in ABPP–MudPIT). Until the investigator is familiar enough with the procedure to recognize common nonspecific proteomic analysis artifacts, two controls are suggested. First, it is important to include a control sample that has not been labeled with the ABPP probe but that has been processed in parallel with labeled samples. Proteins identified by MudPIT in both datasets represent proteins that bind nonspecifically to the streptavidin beads and should be filtered out in the analysis. Second, to obtain the complete peptide spectrum of the proteomes in question (i.e., the input condition), a control tryptic sample obtained by digesting the intact proteome prior to any enrichment should be performed. Peptides identified only in the streptavidin-enriched samples but absent in the "input" tryptic digest sample should be discarded as false-positives. It is important to keep in mind for all ABPP experiments that once the labeling reaction has started, the procedure must be carried out to completion until a freezing step is indicated. Investigators are encouraged to plan ahead, particularly while performing ABPP–MudPIT experiments with multiple samples. Although ABPP probes label previously frozen proteomes well, greater labeling efficiency may be observed in fresh lysates; whenever possible, fresh proteomes are preferred for ABPP analysis. Finally, the lipid content of adipocytes results in the generation of a fat cake layer during cell/tissue homogenization that should be avoided while harvesting supernatants. Lipid contamination of proteomes may interfere with labeling by ABPP probes.

In addition to the basic ABPP techniques described earlier, there are several embodiments of ABPP that merit brief mention.

3.2. *In situ* ABPP

Because the majority of ABPP probes have limited cell permeability due to their bulky reporter tag, labeling of enzymes with tagged ABPP probes is performed in homogenized proteomes (*in vitro* ABPP as described in all protocols earlier), not intact cells or tissues. This limits the ability of *in vitro* ABPP to report on the activity of all endogenous enzymes, as some are likely to degrade during the proteome preparation steps. Moreover, some enzymes may work as part of complexes that need to remain intact to be active. To enable activity-based labeling of enzymes directly in living cells and organisms, cell-permeable ABPP probes have been developed in which the reporter tag is substituted with a small, latent chemical handle (an alkyne or an azide), which allows cell permeability. An orthogonally functionalized reporter tag is then attached to the probe using click chemistry methods (Cu(I)-catalyzed stepwise azide–alkyne cycloaddition of Huisgen)

(Kolb & Sharpless, 2003). Following the click reaction that couples a reporter tag to the ABPP probe, labeled proteins can then be analyzed by gel-based ABPP or ABPP–MudPIT. Interested readers are referred to Speers and Cravatt (2009) for protocols for *in situ* ABPP.

3.3. ABPP–SILAC

Stable isotope labeling by amino acids in cell culture (SILAC) enables *in vivo* incorporation of a label into proteins for mass spectrometry-based quantitative proteomic analysis (Ong et al., 2002). SILAC relies on the metabolic incorporation of a "light" or "heavy" form of an amino acid (e.g., labeled with deuterium, ^{13}C, or ^{15}N) into all newly synthesized proteins. SILAC involves growing cells in isotopically "light" and "heavy" amino acid media for several population doublings (typically 5) to allow full incorporation of the heavy amino acid into the proteome. Because the "light" and "heavy" forms of the amino acids used are virtually identical, "heavy" cells behave similar to cells grown in "light" media. Incorporation of isotopically labeled

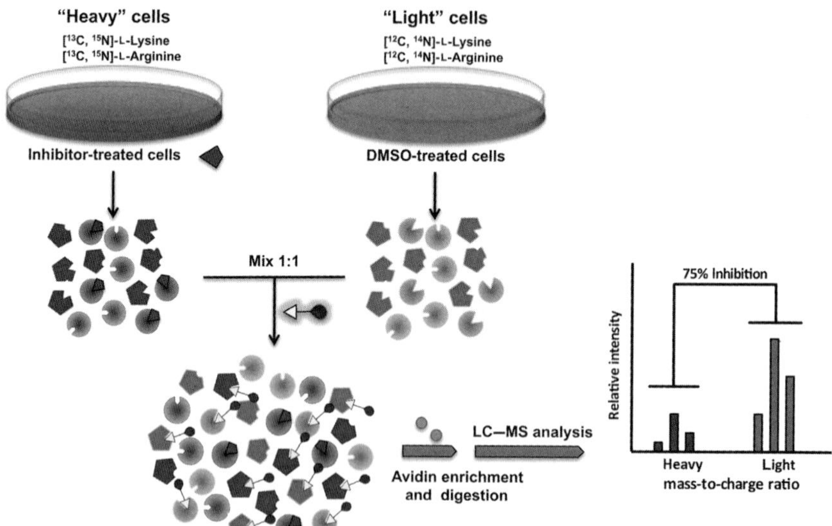

Figure 9.4 Competitive ABPP–SILAC to identify the target of enzyme inhibitors. Cells are grown in different media with isotopically stable amino acids. Proteomes prepared from compound-treated "heavy" and vehicle-treated "light" cells are labeled with a biotin-conjugated ABPP probe and mixed in a 1:1 ratio. Following avidin enrichment and on-bead digestion, samples are analyzed by mass spectrometry and representative peptides belonging to treated or control cells are identified by mass–charge shift. (See color plate.)

amino acids into proteins results in a mass shift of the corresponding peptides that can be detected by a mass spectrometer. When "light" and "heavy" samples are combined, the ratio of peak intensities in the mass spectrum reflects the relative protein abundance (Fig. 9.4). Because quantitative differences between samples can be established with greater confidence that with other methods, SILAC has been integrated with ABPP analysis (Adibekian et al., 2011). A representative competitive ABPP–SILAC experiment to determine the target of a SH inhibitor would involve growing cells in isotopically "light" and "heavy" amino acid media, which would then be treated with inhibitor or DMSO, respectively, for a given time (typically 1–4 h), lysed, combined, and treated with an FP-biotin probe. FP-labeled SHs would then be enriched by avidin chromatography and identified and quantified by LC–MS based on analysis of MS2 spectra and MS1 profiles using a mass spectrometer. ABPP–SILAC can also be used in combination with *in situ* ABPP (Adibekian et al., 2012), thus maximizing the range of applications and quantitative power of ABPP.

REFERENCES

Adam, G. C., Sorensen, E. J., & Cravatt, B. F. (2002). Proteomic profiling of mechanistically distinct enzyme classes using a common chemotype. *Nature Biotechnology*, *20*, 805–809.

Adibekian, A., Martin, B. R., Chang, J. W., Hsu, K. L., Tsuboi, K., Bachovchin, D. A., et al. (2012). Confirming target engagement for reversible inhibitors in vivo by kinetically tuned activity-based probes. *Journal of the American Chemical Society*, *134*(25), 10345–10348.

Adibekian, A., Martin, B. R., Wang, C., Hsu, K. L., Bachovchin, D. A., Niessen, S., et al. (2011). Click-generated triazole ureas as ultrapotent in vivo-active serine hydrolase inhibitors. *Nature Chemical Biology*, *7*(7), 469–478.

Bachovchin, D. A., & Cravatt, B. F. (2012). The pharmacological landscape and therapeutic potential of serine hydrolases. *Nature Reviews Drug Discovery*, *11*(1), 52–68.

Bachovchin, D. A., Ji, T., Li, W., Simon, G. M., Blankman, J. L., Adibekian, A., et al. (2010). Superfamily-wide portrait of serine hydrolase inhibition achieved by library-versus-library screening. *Proceedings of the National Academy of Sciences of the United States of America*, *107*(49), 20941–20946.

Bogyo, M., Verhelst, S., Bellingard-Dubouchaud, V., Toba, S., & Greenbaum, D. (2000). Selective targeting of lysosomal cysteine proteases with radiolabeled electrophilic substrate analogs. *Chemistry & Biology*, *7*, 27–38.

Bogyo, M. (2005). Screening for selective small molecule inhibitors of the proteasome using activity-based probes. *Methods in Enzymology*, *399*, 609–622.

Borodovsky, A., Ovaa, H., Meester, W. J., Venanzi, E. S., Bogyo, M. S., Hekking, B. G., et al. (2005). Small-molecule inhibitors and probes for ubiquitin- and ubiquitin-like-specific proteases. *ChemBioChem*, *6*, 287–291.

Chan, E. W., Chattopadhaya, S., Panicker, R. C., Huang, X., & Yao, S. Q. (2004). Developing photoactive affinity probes for proteomic profiling: Hydroxamate-based probes for metalloproteases. *Journal of the American Chemical Society*, *126*(44), 14435–14446.

Cravatt, B. F., Wright, A. T., & Kozarich, J. W. (2008). Activity-based protein profiling: From enzyme chemistry to proteomic chemistry. *Annual Review of Biochemistry*, *77*, 383–414.

Greenbaum, D., Medzihradszky, K. F., Burlingame, A., & Bogyo, M. (2000). Epoxide electrophiles as activity-dependent cysteine protease profiling and discovery tools. *Chemistry & Biology*, *7*, 569–581.

Heal, W. P., Dang, T. H., & Tate, E. W. (2011). Activity-based probes: Discovering new biology and new drug targets. *Chemical Society Reviews*, *40*(1), 246–257.

Hekmat, O., Kim, Y. W., Williams, S. J., He, S., & Withers, S. G. (2005). Active-site peptide "fingerprinting" of glycosidases in complex mixtures by mass spectrometry. Discovery of a novel retaining beta-1,4-glycanase in Cellulomonas fimi. *The Journal of Biological Chemistry*, *280*(42), 35126–35135.

Jessani, N., Humphrey, M., McDonald, W. H., Niessen, S., Masuda, K., Gangadharan, B., et al. (2004). Carcinoma and stromal enzyme activity profiles associated with breast tumor growth in vivo. *Proceedings of the National Academy of Sciences of the United States of America*, *101*(38), 13756–13761.

Jessani, N., Liu, Y., Humphrey, M., & Cravatt, B. F. (2002). Enzyme activity profiles of the secreted and membrane proteome that depict cancer invasiveness. *Proceedings of the National Academy of Sciences of the United States of America*, *99*, 10335–10340.

Jessani, N., Niessen, S., Wei, B. Q., Nicolau, M., Humphrey, M., Ji, Y., et al. (2005). A streamlined platform for high-content functional proteomics of primary human specimens. *Nature Methods*, *2*(9), 691–697.

Kato, D., Boatright, K. M., Berger, A. B., Nazif, T., Blum, G., Ryan, C., et al. (2005). Activity-based probes that target diverse cysteine protease families. *Nature Chemical Biology*, *1*(1), 33–38.

Kidd, D., Liu, Y., & Cravatt, B. F. (2001). Profiling serine hydrolase activities in complex proteomes. *Biochemistry*, *40*(13), 4005–4015.

Kline, K. G., & Wu, C. C. (2009). MudPIT analysis: Application to human heart tissue. *Methods in Molecular Biology*, *528*, 281–293.

Kolb, H. C., & Sharpless, K. B. (2003). The growing impact of click chemistry on drug discovery. *Drug Discovery Today*, *8*(24), 1128–1137.

Kumar, S., Zhou, B., Liang, F., Wang, W. Q., Huang, Z., & Zhang, Z. Y. (2004). Activity-based probes for protein tyrosine phosphatases. *Proceedings of the National Academy of Sciences of the United States of America*, *101*, 7943–7948.

Li, W., Blankman, J. L., & Cravatt, B. F. (2007). A functional proteomic strategy to discover inhibitors for uncharacterized hydrolases. *Journal of the American Chemical Society*, *129*(31), 9594–9595.

Li, Y. M., Xu, M., Lai, M. T., Huang, Q., Castro, J. L., DiMuzio- Mower, J., et al. (2000). Photoactivated gamma-secretase inhibitors directed to the active site covalently label presenilin 1. *Nature*, *405*, 689–694.

Liu, Y., Patricelli, M. P., & Cravatt, B. F. (1999). Activity-based protein profiling: The serine hydrolases. *Proceedings of the National Academy of Sciences of the United States of America*, *96*(26), 14694–14699.

Liu, Y., Shreder, K. R., Gai, W., Corral, S., Ferris, D. K., & Rosenblum, J. S. (2005). Wortmannin, a widely used phosphoinositide 3-kinase inhibitor, also potently inhibits mammalian polo-like kinase. *Chemistry & Biology*, *12*(1), 99–107.

Lo, L. C., Pang, T. L., Kuo, C. H., Chiang, Y. L., Wang, H. Y., & Lin, J. J. (2002). Design and synthesis of class-selective activity probes for protein tyrosine phosphatases. *Journal of Proteome Research*, *1*, 35–40.

Long, J. Z., & Cravatt, B. F. (2011). The metabolic serine hydrolases and their functions in mammalian physiology and disease. *Chemistry Review*, *111*(10), 6022–6063.

Ong, S. E., Blagoev, B., Kratchmarova, I., Kristensen, D. B., Steen, H., Pandey, A., et al. (2002). Stable isotope labeling by amino acids in cell culture, SILAC, as a simple and accurate approach to expression proteomics. *Molecular and Cellular Proteomics*, *1*(5), 376–386.

Patricelli, M. P., Szardenings, A. K., Liyanage, M., Nomanbhoy, T. K., Wu, M., Weissig, H., et al. (2007). Functional interrogation of the kinome using nucleotide acyl phosphates. *Biochemistry*, *46*(2), 350–358.

Saghatelian, A., Jessani, N., Joseph, A., Humphrey, M., & Cravatt, B. F. (2004). Activity-based probes for the proteomic profiling of metalloproteases. *Proceedings of the National Academy of Sciences of the United States*, *101*(27), 10000–10005.

Soukas, A., Socci, N. D., Saatkamp, B. D., Novelli, S., & Friedman, J. M. (2001). Distinct transcriptional profiles of adipogenesis in vivo and in vitro. *Journal of Biological Chemistry*, *276*(36), 34167–34174.

Speers, A. E., & Cravatt, B. F. (2009). Activity-based protein profiling (ABPP) and click chemistry (CC)-ABPP by MudPIT mass spectrometry. *Current Protocols in Chemical Biology*, *1*, 29–41.

Tsai, C. S., Li, Y. K., Lo, L. C., Tsai, C. S., Li, Y. K., & Lo, L. C. (2002). Design and synthesis of activity probes for glycosidases. *Organic Letters*, *4*(3607), 3610.

Verdoes, M., Florea, B. I., Menendez-Benito, V., Maynard, C. J., Witte, M. D., van der Linden, W. A., et al. (2006). A fluorescent broad-spectrum proteasome inhibitor for labeling proteasomes in vitro and in vivo. *Chemistry & Biology*, *13*(11), 1217–1226.

Vocadlo, D. J., & Bertozzi, C. R. (2004). A strategy for functional proteomic analysis of glycosidase activity from cell lysates. *Angewandte Chemie, International Edition*, *43*, 5338–5342.

Yee, M. C., Fas, S. C., Stohlmeyer, M. M., Wandless, T. J., & Cimprich, K. A. (2005). A cell-permeable, activity-based probe for protein and lipid kinases. *The Journal of Biological Chemistry*, *280*, 2905–2909.

Zhu, Q., Girish, A., Chattopadhaya, S., & Yao, S. Q. (2004). Developing novel activity-based fluorescent probes that target different classes of proteases. *Chemical Communications*, *13*, 1512–1513.

CHAPTER TEN

Measurement of Lipolysis

Martina Schweiger[1], Thomas O. Eichmann, Ulrike Taschler, Robert Zimmermann, Rudolf Zechner, Achim Lass[1]

Institute of Molecular Biosciences, University of Graz, Graz, Austria

[1]Corresponding authors: e-mail address: tina.schweiger@uni-graz.at; achim.lass@uni-graz.at

Contents

Abstract

Lipolysis is defined as the hydrolytic cleavage of ester bonds in triglycerides (TGs), resulting in the generation of fatty acids (FAs) and glycerol. The two major TG pools in the body of vertebrates comprise intracellular TGs and plasma/nutritional TGs. Accordingly, this leads to the discrimination between intracellular and intravascular/gastrointestinal lipolysis, respectively. This chapter focuses exclusively on intracellular lipolysis, referred to as lipolysis herein. The lipolytic cleavage of TGs occurs in essentially all cells and tissues of the body. In all of them, the resulting FAs are utilized endogenously for energy production or biosynthetic pathways with one exception, white adipose tissue (WAT). WAT releases FAs and glycerol to supply nonadipose tissues at times of nutrient deprivation. The fundamental role of lipolysis in lipid and energy homeostasis requires the accurate measurement of lipase activities and lipolytic rates. The recent discovery of new enzymes and regulators that mediate the hydrolysis of TG has made these measurements more complex. Here, we describe detailed methodology for how to measure lipolysis and specific enzymes' activities in cells, organs, and their respective extracts.

Methods in Enzymology, Volume 538

ISSN 0076-6879

http://dx.doi.org/10.1016/B978-0-12-800280-3.00010-4

1. INTRODUCTION

In times of abundant nutrients, virtually all eukaryotic organisms are able to store excess energy in form of triglycerides (TGs) that can be reutilized in times of nutrient deprivation or high-energy demand. In all cell types, TGs are stored in lipid droplets (LDs) composed of variable amounts of neutral lipids, phospholipids, and LD-associated proteins. Vertebrates, in particular, have developed white adipose tissue (WAT) as a highly specialized tissue for the deposition and remobilization of TG. In WAT, a tight balance between lipolysis and lipogenesis controls homeostasis and body composition. Dysregulation of this balance leads to either obesity or lipodystrophy. In nonadipose tissues, TG storage is normally much more short term with a small number of LDs and a high LD turnover. Nevertheless, lipolysis is also essential in nonadipose tissues for instant provision of fatty acids (FAs) upon demand, particularly in highly oxidative tissues such as the brown adipose tissue, liver, cardiac and skeletal muscles, or macrophages. In these tissues, FAs not only are efficient energy substrates but also represent basic building blocks for the synthesis of membrane lipids or intracellular signaling molecules. In addition, FAs have also been recognized as signaling molecules *per se*, which can bind to and thereby activate members of the nuclear receptor family of transcription factors (Kliewer et al., 1997). Defective lipolysis in nonadipose tissues impairs their normal function, leading to excessive TG accumulation and neutral lipid storage disease. Conversely, overabundance of FAs due to unrestrained lipolysis results in lipotoxicity, which is characterized by ER stress, mitochondrial dysfunction, and cell death. Elevated cellular FA concentrations, independent of whether they originate from exogenous lipolysis in WAT or endogenous lipolysis in nonadipose tissues, cause prevalent metabolic disorders such as insulin resistance, type II diabetes, liver steatosis, or atherosclerosis. Additionally, dysregulated lipolysis is also observed in other pathological conditions such as in cancer-induced cachexia.

1.1. Current model of lipolysis

Lipolysis requires specific enzymes commonly called lipases for each step of ester bond hydrolysis in TGs. To date, three major lipases have been identified: (i) Adipose triglyceride lipase (ATGL) performs the first step of TG hydrolysis, generating diglycerides (DGs) and FAs (Zimmermann et al., 2004). Its activity is tightly regulated by two accessory proteins: comparative gene identification-58 (CGI-58) coactivates the hydrolase activity of ATGL and G0/G1 switch gene 2 (G0S2) inactivates the hydrolase activity of ATGL (Lass et al., 2006; Yamaguchi, 2010; Yamaguchi & Osumi, 2009; Yang et al., 2010). (ii) Hormone-sensitive lipase (HSL) performs the second step and hydrolyzes DGs, generating monoglycerides (MGs) and FAs. In contrast to ATGL, HSL exhibits a broader substrate specificity, also hydrolyzing TGs, cholesteryl esters, MGs, and retinyl esters in addition to DGs. (iii) Monoglyceride lipase (MGL) is selective for MGs and generates glycerol and the third FA (Vaughan, Berger, & Steinberg, 1964). Thus, lipolysis constitutes a coordinated three-step process catalyzed by three different enzymes, which degrade TG into glycerol and FAs (Fig. 10.1). The relative importance of these three enzymes, however, highly depends on their tissue-specific expression pattern and the metabolic state. In WAT, ATGL, HSL, and MGL are responsible for more than 90% of its lipolytic activity (Schweiger et al., 2006). This may be different in other tissues such as skeletal muscle or cardiac muscle, where other, possibly unknown enzymes may also contribute to intracellular TG hydrolase activity (Haemmerle et al., 2006).

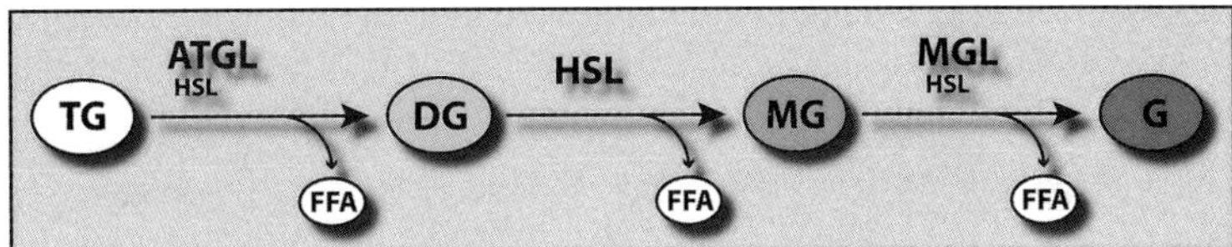

Figure 10.1 Coordinated breakdown of triglycerides in the course of lipolysis. (See color plate.)

1.2. Regulation of adipose tissue lipolysis

As mentioned earlier, WAT provides FAs to tissues with high-energy demand at times of fasting. This necessitates that FAs and glycerol derived from WAT lipolysis enter the bloodstream, which makes WAT lipolysis the main determinant of plasma FA concentrations. Numerous physiological conditions, hormones, and cytokines regulate WAT lipolysis on multiple levels. The most important regulatory steps to control enzyme activity include (i) regulation of lipase gene expression, (ii) translation and stability of lipases and accessory proteins, (iii) control of intracellular localization of lipases, (iv) posttranslational modifications (e.g., phosphorylation), and (v) regulation of lipase activity via coregulators (e.g., CGI-58, G0S2, and perilipins).

Regulation of each of these steps is tissue-specific and often varies among different organisms. For example, while mRNA expression of ATGL is upregulated in response to fasting in murine WAT (reviewed in Zechner, Kienesberger, Haemmerle, Zimmermann, & Lass, 2009), it is reduced in human WAT. Nevertheless, this reduction of ATGL mRNA in fasted human WAT is associated with increased ATGL protein concentrations (Nielsen et al., 2011) and, presumably, increased lipase activity. Similarly, despite drastically elevated lipolysis, ATGL and HSL mRNA expression is unaltered upon endurance training (Huijsman et al., 2009) or even decreased by the lipolysis-inducing (lipolytic) hormones isoproterenol and TNF-α (Kralisch et al., 2005). This indicates that lipolysis is predominantly regulated at the posttranscriptional and posttranslational level (Fig. 10.2). Accordingly, measurements of lipase mRNA levels are inadequate markers for the estimation of actual lipase activities.

To date, most work related to regulatory mechanisms controlling lipolysis focused on WAT. Oxidative tissues share many of the lipases and

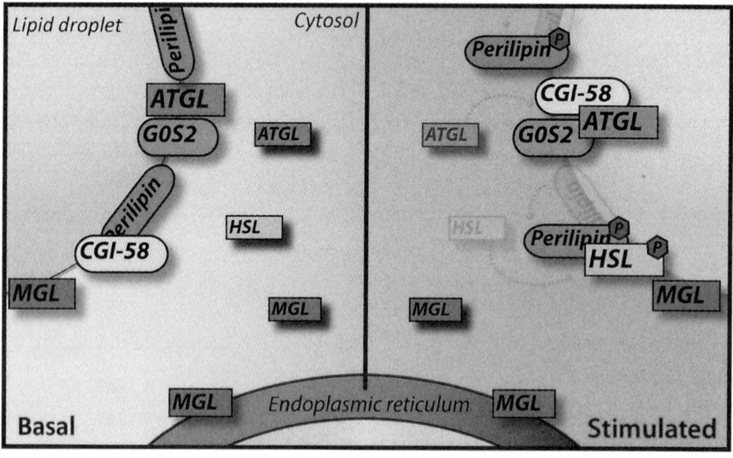

Figure 10.2 Intracellular localization of lipases and regulatory proteins in the basal/non-lipolytic-stimulated state and upon lipolytic stimulation. In the basal/non-lipolytic-stimulated state, CGI-58 is bound to perilipin-1 and unavailable for ATGL binding/activation. HSL resides in the cytosol. β-Adrenergic stimulation leads to the phosphorylation of HSL and perilipin-1, which causes HSL to translocate onto the LD. ATGL also translocates from the cytosol to the LD surface. In both lipolytic states, G0S2 is localized on LD, bound to ATGL. In addition, phosphorylated perilipin-1 releases CGI-58, which is now available for ATGL binding and activation. MGL is localized on the LD, the cytosol, and the ER independent of the metabolic state of the cell. (See color plate.)

accessory proteins with adipose tissue, specific differences exist regarding expression of hormone receptors, signaling pathways, and LD composition and structure. These differences require better characterization. In WAT, the most important mechanism regulating lipolysis involve the activation of ATGL by CGI-58 and the protein kinase A (PKA) - mediated phosphorylation of HSL and perilipin-1. In the nonstimulated/basal state, CGI-58 is bound to perilipin-1 and unavailable for ATGL binding/activation. At the same time, HSL resides in the cytosol. Upon hormonal (e.g., β-adrenergic) stimulation, PKA phosphorylates both HSL and perilipin-1 on multiple serines, which causes HSL translocation to the LD (Clifford, Londos, Kraemer, Vernon, & Yeaman, 2000; Egan et al., 1992). Phosphorylated perilipin-1 also releases CGI-58, which is now available for ATGL binding and activation (Granneman et al., 2006). Following hormone stimulation, ATGL also translocates from the cytosol to the LD surface. However, the underlying mechanism is not clear (Bezaire et al., 2009; Yang et al., 2010) (Fig. 10.2). G0S2 efficiently inhibits ATGL activity in *in vitro* lipase assays. Moreover, G0S2-deficient mice exhibit increased adipose tissue lipolysis, improved insulin sensitivity, and are resistant to diet induced liver steatosis (Zhang et al., 2013). Notably, G0S2 appears to be involved in the activation of lipolysis in adipocytes by TNF-α. Experiments in 3T3-L1 adipocytes showed that TNF-α decreases G0S2 mRNA and protein levels, leading to enhanced lipolysis (Yang, Zhang, Heckmann, Lu, & Liu, 2011).

1.3. Strategies to assess lipolysis

1.3.1 Non-activity-based methods

1.3.1.1 Quantitation of cell and tissue mRNA levels of lipases and regulatory proteins

Due to the convenience and simplicity of the involved methodology, the lipolytic activities in cells or tissues are often estimated by quantification of mRNA levels of lipases and regulatory proteins. It is assumed that changes in mRNA levels will lead to respective changes in the protein concentrations, which will finally translate into changes in the lipolytic activity. This conclusion, as mentioned before, is mostly unwarranted. Accordingly, determination of mRNA levels is only meaningful when the transcription, mRNA processing, or mRNA degradation of lipolytic genes is in question.

1.3.1.2 Quantitation of protein concentrations of lipases and regulatory proteins

To date, a number of very specific antibodies for lipolytic enzymes and regulatory proteins are available for various species including murine and human proteins. These tools have substantially broadened the assessment of protein levels of lipases and regulatory proteins, determination of which is necessary if translation from RNA, protein stability, or protein degradation is in question. Since lipase activities are regulated at the posttranscriptional level, a direct correlation between protein expression levels and activities usually does not exist.

1.3.1.3 Quantitation of HSL and perilipin phosphorylation state

The phosphorylation state of perilipin-1 and HSL is a good measure to investigate hormone-stimulated lipolysis. More precisely, the phosphorylation of HSL and perilipin is a direct readout for PKA activity, since these proteins are known substrates for PKA. Thus, this method is well suited when PKA activity has to be determined. However, PKA-mediated phosphorylation of HSL and perilipin does not necessarily correlate with lipolytic activities, since lipolysis is a complex orchestration between different lipolytic enzymes and regulatory proteins.

1.3.2 Activity-based methods

1.3.2.1 *In vitro* TG hydrolase activity

To directly measure lipolytic activities of tissues or cells, the *in vitro* TG hydrolase activity assay is widely used. This assay is relatively easy to perform and meaningful in that it provides a good measure for the lipolytic capacity in a given sample. It is based on an artificial, radioactive isotope-labeled triolein substrate where the lipolytic activity of the sample leads to a release of radiolabeled FAs. Hence, it is important that samples do not contain large amounts of endogenous TG, since these would also be hydrolyzed but as they are not labeled they would not be detected in the assay. Thus, endogenous TG lowers the rate of hydrolysis in the assay by diluting the artificial substrate. Usually, a 20,000 × *g* infranatant of tissue homogenates is used because LDs will float on top and can be easily removed. Yet, by removing the LDs from the sample, also, a portion of lipases and regulatory proteins is removed and lost for the measurement. As shown in Fig. 10.2, the majority of ATGL and HSL translocates from the cytosol to the LD surface upon β-adrenergic stimulation. Thus, under these conditions, the major lipolytic activity of the sample is not accessible for TG hydrolase activity measurement.

1.3.2.2 LD-associated activity

To circumvent the limitations of the *in vitro* TG hydrolase assay, an analysis of the LD-associated lipolytic capacity can be performed. For this, LDs have to be isolated and purified by differential centrifugation (density gradient). Then, these LDs are incubated for "self-digestion" and released FAs are determined by an enzymatic test, which is a direct readout for the lipolytic capacity. Generally, LDs can be isolated from either cultured cells or tissues. The advantage of using cultured cells for LD isolation is that they can be transfected with a gene of interest (i.e., an LD-associated protein), whose role in lipolysis can then be analyzed. Furthermore, the TG moiety of LDs of cultured cells can also be radiolabeled, which enhances sensitivity and accuracy of the assay. LDs, isolated from tissues of genetically modified animals, can be used as substrate to more precisely determine the effect of the genetic manipulation on lipolysis. Furthermore, during "self-digestion" of LDs, the exogenous addition of proteins of interest (either purified or in lysates) allows investigators to assess their respective effect on lipolysis in a much-closer-to-reality system since the substrate, the isolated LDs, is composed of a biological complex mixture of lipid species and already contains numerous proteins, depending on cell type or tissue. Although this method is convenient, isolation of LDs is time-consuming and inactivation or mislocalization of proteins may occur.

1.3.2.3 Lipolysis of cultured adipocytes and organ explants

The most direct assessment of lipolysis is the measurement of the lipolytic products released by adipocytes or fat explants. This method is based on the measurement of lipolytic products (FAs and glycerol), which are released into the incubation media. In this scenario, two lipolytic states can be differentiated, the so-called basal and β-adrenergic-stimulated states. It is important to add bovine serum albumin (BSA) as FA acceptor to the incubation media since the extracellular accumulation of free FAs will promote reuptake of FAs and reesterification of glycerolipids. Despite an excess of FA acceptor in the media, not all FAs and glycerols generated during lipolysis are released to the media but used for intracellular needs. Thereby, FAs are readily activated by acyl-CoA synthetases (ACS) to FA-CoA, which serve as precursors for MG, DG, TG, and phospholipid (PL) synthesis. Although to a lesser extent, glycerol can also be recycled by glycerol kinase, glycerolphosphate-acyltransferase, acylglycerolphosphate-acyltransferase, and lipin to DGs (Fig. 10.3). Thus, the molar ratio between FA and glycerol released into the media is usually closer to 1:1 under basal/non-lipolytic-stimulated conditions and only reaches a ratio of 3:1 under lipolytic-stimulated conditions (i.e., by β-adrenergic stimulation) or by the inhibition of ACS (i.e., by triacsin C; Paar et al., 2012).

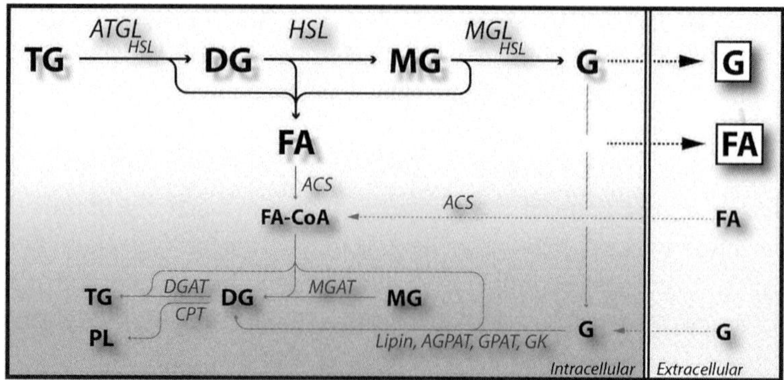

Figure 10.3 The fate of free fatty acids (FA) and glycerol (G) generated during intracellular lipolysis. FAs generated by the activity of ATGL, HSL, and MGL are either released into the circulation or activated by acyl-CoA synthetases (ACS) to FA-CoA. These FA-CoAs serve as building blocks for the synthesis of MG, DG, TG, and phospholipids (PL), by the activities of MGAT (monoacylglycerol-acyltransferase), DGAT (diacylglycerol-acyltransferase), or CPT (cholinephosphotransferase). Glycerol (G) is recycled by GK (glycerol kinase), GPAT (glycerolphosphate-acyltransferase), AGPAT (acylglycerolphosphate-acyltransferase), and lipin to DGs. (See color plate.)

2. METHODOLOGICAL DESCRIPTION OF ACTIVITY-BASED ASSAYS

2.1. Measurement of *in vitro* TG hydrolase activity

2.1.1 Validity of the in vitro *TG hydrolase assay to determine ATGL activity*

The measurement of TG hydrolase activity in adipose tissue was originally optimized for HSL (Holm, Olivecrona, & Ottosson, 2001). It was suggested that HSL hydrolyzes TG and DG to form MG and FA. MGL then hydrolyzes MG. At that time, TG hydrolysis was found to be highest at neutral pH, using a substrate of triolein emulsified with phospholipids. Not surprisingly, ATGL is also maximally active under the same conditions. In the presence of CGI-58, the optimal pH for ATGL-mediated TG hydrolysis is pH 7.0 (Fig. 10.4). An increase to pH 8.0 or decrease to pH 6.0 results in a loss of nearly half of the activity. Using triolein as substrate, the acylglycerol hydrolase activity present in adipose tissue lysates derives mostly (>90%) from ATGL, HSL, and MGL. To differentiate between TG hydrolase activities mediated by ATGL or HSL, a specific inhibitor for HSL can be used (Schweiger et al., 2006). Inhibition of HSL activity in WAT reduces the acylglycerol

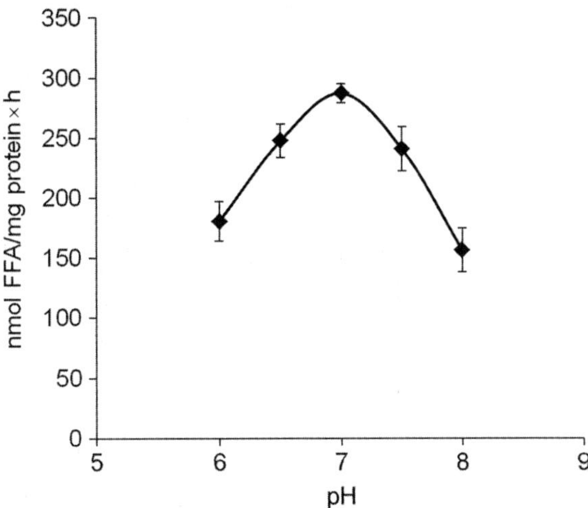

Figure 10.4 pH dependence of CGI-58-coactivated ATGL hydrolysis of TG. COS-7 cell lysates containing recombinant CGI-58 and ATGL were incubated with 0.32 m*M* triolein as substrate and ^{3}H-triolein as tracer in 100 m*M* potassium phosphate buffer of various pH (6–8). Substrates were emulsified in phospholipids (PC/PI, 3:1, w/w) containing 5% BSA (FA-free). After incubation, FAs were extracted and radioactivity was determined by scintillation counting. Data are means ± standard deviation of three replicates.

activity by approximately 60%, which is expected since DG accumulates, and DG has two out of three FAs still esterified to the glycerol backbone. The absence of ATGL activity reduces the acylglycerol hydrolase activity of WAT by approximately 80%, and no intermediary degradation products accumulate (Haemmerle et al., 2006).

For the determination of TG hydrolase activity, triolein is widely used as substrate. Detergents are often used for the preparation of emulsions, including phospholipids (PC/PI), cholate, Triton X-100, and glycerol or a combination thereof. BSA (FA-free) has to be added as FA acceptor since the solubility of FAs is limited and accumulation of FAs will inhibit rate of hydrolysis. Figure 10.5 depicts the effect of different substrate preparations on rate of hydrolysis of HSL, ATGL, and ATGL coactivated by CGI-58. HSL exhibits highest activity when phospholipids (PC/PI) are used for substrate preparation. When the substrate is prepared using Triton X-100 (0.025%) or glycerol (2%), HSL loses approximately 50% of the activity. In the presence of cholate (20 m*M*), HSL activity is reduced by ~80% (Fig. 10.5A). For ATGL (Fig. 10.5B), the highest activity was obtained with PC/PI- and Triton X-100-stabilized substrates. Preparation of substrates either with cholate or glycerol reduced activities to 35% and 8%,

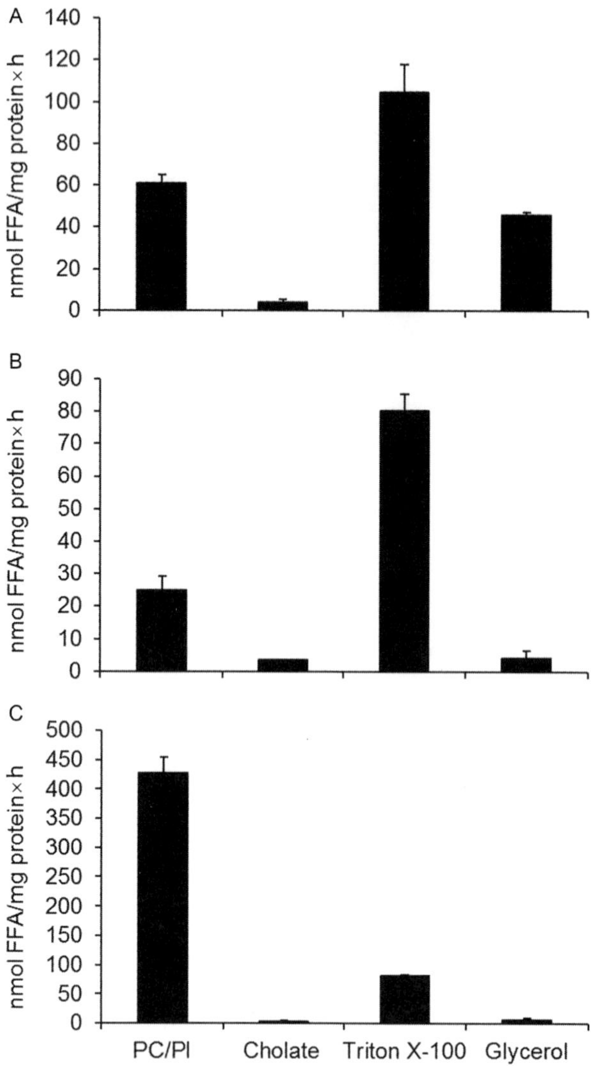

Figure 10.5 TG hydrolase activities of HSL and ATGL using different detergents for substrate preparation. COS-7 cell lysates containing recombinant HSL (A), ATGL (B), or ATGL and CGI-58 (C) were subjected to TG hydrolase assay using 0.32 m*M* triolein as substrate and ^{3}H-triolein as tracer. As control, lysates containing recombinant β-galactosidase (LacZ) were used. Substrates were prepared in 100 m*M* potassium phosphate buffer, pH 7.0, containing 5% BSA (FA-free). For stabilization 45 μ*M* PC/PI (3:1, w/w), 20 m*M* cholate, 2% glycerol, or 0.025% Triton X-100 were used. After incubation, FAs were extracted and radioactivity was determined by scintillation counting. Data are means ± standard deviation of three replicates.

respectively. This suggests that for measurement of ATGL activity, PC/PI and Triton X-100 are suitable. However, this is not the case for CGI-58-coactivated ATGL activity. CGI-58 coactivation of ATGL is only seen using a PC/PI-emulsified substrate (Fig. 10.5C), implicating that the interaction of ATGL and CGI-58 is highly affected by the nature of a detergent.

2.1.2 Required materials

- Solution A: 0.25 *M* sucrose, 1 m*M* EDTA, 1 m*M* DTT, pH 7.0, 1 μg/ml pepstatin, 2 μg/ml antipain, and 20 μg/ml leupeptin as protease inhibitors (Pi's).
- Prepare 20 mg/ml (~25 m*M*) phosphatidylcholine (PC) from egg yolk and phosphatidylinositol (PI) from soybean (Sigma-Aldrich, St Louis, MO, cat. no. P-3556 and cat. no. P-0639, respectively) with a ratio of 3:1 in chloroform and store at −20 °C.
- Prepare a 100 mg/ml (113 m*M*) solution of triolein (trioleoylglycerol, TO; Sigma-Aldrich, cat. no. T-7140) in toluene and store at −20 °C.
- Prepare a 0.5 μCi/μl solution of TO [9,10(N)-3H] (PerkinElmer, Waltham, MA, cat. no. NET431L005MC) in toluene and store in aliquots of 2 ml at −20 °C.
- 0.1 *M* potassium phosphate buffer (KPB; pH 7.0) is prepared freshly.
- Prepare a 20% solution of BSA (FA-free; Sigma-Aldrich, cat. no. A6003) in 0.1 *M* KPB (pH 7.0) and store in aliquots at −20 °C.
- Assay tubes (12 ml, round bottom, polypropylene; Greiner Bio-One GmbH, cat. no. 160201).
- Extraction solution I: Methanol/chloroform/*n*-heptane with a ratio of 10:9:7 (v/v/v) is stored light-protected at 4 °C. Prewarm solution before use.
- Extraction solution II: 0.1 *M* potassium carbonate, pH 10.5 adjusted with saturated boric acid.
- Liquid scintillation cocktail (Roth, Karlsruhe, Germany, cat. no. 0016.3, or equivalent).
- Sonicator Virsonic 475 (Virtis, Gardiner, NJ) or equivalent.
- Liquid scintillation counter (Packard or equivalent).

2.1.3 Procedure

2.1.3.1 Preparation of cellular or tissue extracts

Homogenize tissue samples on ice in solution A using an UltraTurrax (IKA, Janke & Kunkel, Germany) and centrifuge at 1,000 × *g*, 4 °C for 10 min to remove nondisrupted tissue and nuclei. Thereafter, transfer the supernatant to a new tube and centrifuge at 20,000 × *g*, 4 °C for 30 min. Carefully collect the infranatant, leaving the fat layer behind. TG hydrolase activity is determined on the 20,000 × *g* infranatant. For preparation of cell extracts, harvest cultured cells and disrupt cells in buffer A by sonication on ice. Thereafter, centrifuge cell lysates at 1,000 × *g*, 4 °C for 10 min to pellet intact cells and nuclei. For cells that do not contain large amounts of TG (e.g., COS-7 cells), the 1,000 × *g* supernatant is used to measure TG hydrolase activity. Be aware that ATGL, CGI-58, HSL, and MGL may localize on LDs of cells and tissues. Thus, when assaying TG hydrolytic activity of the 20,000 × *g* infranatant, a part of the total lipolytic activity might be lost in the fat cake.

2.1.3.2 Measurement of TG hydrolase activity

According to Holm and Østerlund with some modifications (Holm et al., 2001):

1. Add 1.67 m*M* TO, 10 μCi ^{3}H-TO/ml, and 190 μ*M* PC/PI in an assay tube resistant to organic solvent. For nonadipose tissue samples, a 0.32 m*M* TO and 45 μ*M* PC/PI substrate with same amounts of radiolabel can be used, which will increase the sensitivity of the assay. Bring the solvents to complete dryness under a stream of nitrogen.
2. Add initially not more than 2 ml of 0.1 *M* KPB, pH 7.0. Sonicate (Virsonic 475, Virtis, Gardiner, NJ) for 30 s with an output power of 20% on ice. Wait for 30 s and repeat sonication two more times. Be careful that sonicator tip does not touch the walls or the bottom during sonication and avoid foam formation. Ideally, sonicator tip should be on the upper third of the tube.
3. Add remaining KPB as required and sonicate for 15 s on ice.
4. Add FA-free BSA to give a final concentration of 5%.
5. Measure aliquot of the substrate to determine specific substrate activity (should be around 1 × 10^{6}cpm/100 μl).
6. Prepare samples on ice. For each determination, use 100 μl samples (tissue homogenates or cell lysates) in solution A. As blank, use 100 μl solution A. *Important*: Adapt the amount of protein in the assay experimentally to obtain rates in the linear range of the assay. Prepare at least triplicates for samples and blank. Pipette samples and blanks onto the bottom of assay tubes.
7. Mix 100 μl samples with 100 μl of substrate by pipetting substrate onto the sample and incubate for 1 h in a water bath at 37 °C under constant shaking.

8. Terminate the reaction by the addition of 3.25 ml extraction solution I. Add 1.05 ml extraction solution II and mix vigorously for 5 s using a vortex.
9. Centrifuge samples at 1,000 × *g* for 10 min. Transfer 200 μl of the upper aqueous phase to a scintillation vial containing 2 ml of scintillation cocktail and analyze radioactivity.

2.1.4 Calculation

Rates of TG/DG/MG or acylglycerol hydrolase activities are presented as amount (nmol) of released FAs per hour and milligram protein. For the calculation, the partition coefficient of 1.9 for the extraction of FAs into the water phase (22 °C), meaning that 71.5% of FA is recovered into the water phase, is used. Rates are be calculated using the equation as follows:

$$\frac{(\text{cpm sample} - \text{cpm blank}) \times (V_1/V_2)}{(\text{cpm substrate/nmol FA})^1 \times \text{mg protein} \times 0.715 \times t} = \text{nmol FA/mg protein/h}$$

where V_1 is the total volume of upper water phase (2.45 ml); V_2, the volume measured by liquid scintillation (0.2 ml); and t, incubation time (h); [1]501 nmol FA/100 μl for 1.67 m*M* TG substrate; 90 nmol FA/100 μl for 0.3 m*M* TG substrate; 60 nmol FA/100 μl for 0.3 m*M* DG substrate; 400 nmol/100 μl for 4 m*M* MG substrate.

2.2. Measurement of *in vitro* DG hydrolase activity

2.2.1 Required materials

- Prepare a 10 mg/ml DG solution (select isomer as required; Sigma-Aldrich, cat. no. D3627, *sn*-1,3 DG; cat. no. 42494, *sn*-1,2 DG; cat. no. D8394, *rac*-1,2/2,3 DG; cat. no. D8894, *rac*-1,2/2,3/1,3 DG) in chloroform and store at −20 °C.
- Prepare a 0.5 μCi/μl radiolabeled DG isomer solution (American Radiolabeled Chemicals, Saint Louis, United States, *rac*-DG [9,10(N)-3*H*], cat. no. ART0643) in chloroform and store in aliquots of 2 ml at −20 °C.

2.2.2 Procedure

1. Add 0.3 m*M* DG, 10 μCi radiolabeled DG/ml, and 50 μ*M* PC/PI in an assay tube resistant to organic solvents (polypropylene test tubes) and bring to complete dryness under a stream of nitrogen.

2. Further steps are identical as described in steps 2–9 of Section 2.1.3.2 for measurement of TG hydrolase activity and Section 2.1.4 for calculation of lipase activity.

2.3. Measurement of *in vitro* MG hydrolase activity

2.3.1 Required materials

- Prepare a 10 mg/ml MG solution (select isomer as required; Sigma-Aldrich, cat. no. M7765, 1-oleoyl-*rac*-glycerol; M2787 2-oleoyl-glycerol) in chloroform and store at −20 °C.
- Prepare a 0.5 μCi/μl radiolabeled MG isomer solution (American Radiolabeled Chemicals, e.g., 2-mono oleoyl [9,10-3H(N)] glycerol, cat. no. ART1158) in chloroform and store in aliquots of 2 ml at −20 °C.

2.3.2 Procedure

1. Add 4 m*M* MG, 2.5 μCi radiolabeled MG/ml, and 1 m*M* PC/PI in an assay tube resistant to organic solvents (polypropylene test tubes) and bring to complete dryness under a stream of nitrogen.
2. Steps 2–9 are identical as described in Section 2.1.3.2 for measurement of TG hydrolase activity with the modification that the incubation time of the assay is 15 min.
3. For calculation, see Section 2.1.4.

2.4. Measurement of ATGL-, HSL-, and/or MGL-specific activity using inhibitors

To determine the TG hydrolase activity of ATGL in tissue samples or cells, HSL activity can be inhibited by specific inhibitors such as NNC0076-0000-0079 (*N*-methyl-phenyl carbamoyl triazole, Novo Nordisk, Denmark) (Ebdrup, Sørensen, Olsen, & Jacobsen, 2004; Schweiger et al., 2006) or 4-isopropyl-3-methyl-2-{1-[3-(*S*)-methyl-piperidin-1-yl]-methanoyl}-2*H*-isoxazol-5-1 (BAY; Lowe et al., 2004). Alternatively, murine ATGL can be specifically inhibited by Atglistatin (3-(4′-(Dimethylamino)-[1,1′-biphenyl]-3-yl)-N,N-dimethylurea) (Mayer et al., 2013). To determine MGL-independent MG hydrolase activity, the MGL-specific inhibitor JZL 184 (Cayman Chemicals, Michigan, United States, cat. no. 13158) can be used (Long, Nomura, & Cravatt, 2009). Stock solution (20 m*M*) of respective inhibitors is prepared in DMSO and stored at −80 °C. The inhibitors will be added to the samples prior to the addition of substrate. Full inhibition of ATGL, HSL, and MGL activity is obtained at 50 μ*M*, 10 μ*M*, and 100 n*M*, respectively. Rates obtained in the presence of inhibitors as compared to that in the absence will give a good approximation for ATGL and HSL activity, respectively.

2.5. Measurement of LD-associated lipolysis of cultured cells and tissues

For the analysis of LD-associated lipolytic capacity, LDs of either cultured cells or tissues can be used. The advantage of using LDs from cultured cells is that cells can be transfected with a gene of interest, whose function on LDs can then be analyzed. Furthermore, the TG moiety of cellular LDs can also be radiolabeled, which enhances sensitivity and accuracy of the assay. To investigate lipolytic activity of tissues (e.g., from genetically modified animals, fed/fasted animals, or animals on specific diets), LDs can also be isolated from tissue samples and used as substrate for "self-digestion" experiments.

2.5.1 Isolation of LDs from cultured cells and tissues

2.5.1.1 Required materials

- Prepare 4 m*M* oleic acid (OA) complexed to 1.35 m*M* BSA by the dropwise addition of prewarmed (37 °C) OA (8 m*M* in water) to prewarmed BSA (FA-free, 2.7 m*M* in Dulbecco-PBS, DPBS) under shaking conditions.
- Prepare ^{3}H-9,10-OA complexed to BSA (1 μCi/μl; 1.4 Ci/mmol) by dissolving 1,000 μCi ^{3}H-9,10-OA (5 μCi/μl; cat. no. ART 0198, American Radiolabeled Chemicals) in 50 μl 20 m*M* NaOH for 15 min at 37 °C under shaking conditions. Thereafter, add 950 μl of 1% BSA (in water) and incubate for 5 min while shaking. Ideally, this solution has 1×10^{6} cpm/μl as determined by liquid scintillation counting.
- DPBS.
- Solution A: 0.25 *M* sucrose, 1 m*M* EDTA, 1 m*M* DTT, pH 7.0, 1 μg/ml pepstatin, 2 μg/ml antipain, and 20 μg/ml leupeptin as Pi's.
- Overlay buffer: 50 m*M* KPB (pH 7.4), 100 m*M* KCl, 1 m*M* EDTA, 1 μg/ml pepstatin, 2 μg/ml antipain, and 20 μg/ml leupeptin as Pi's.
- Siliconizing Fluid (AquaSil; cat. no. TS-42799, Thermo Fisher Scientific, Rockford, United States).
- Swing out rotor (SW 41, Beckman or equivalent), corresponding tubes (Herolab or equivalent), and ultracentrifuge (Beckman or equivalent).
- TG reagent (cat. no. TR 22421, Thermo Fisher Scientific) and glycerol standard solution (cat. no. F6428, Sigma-Aldrich).
- BCA reagent and BSA standard (cat. no. 23227, Thermo Fisher Scientific).

2.5.1.2 Procedure

(A) *To isolate radiolabeled LDs from cultured cells:*

1. For labeling of TG stores, seed cells at a density of 50,000 cells/cm^2. Incubate semiconfluent cells for 24 h in the presence of 0.4 m*M* OA complexed to BSA, containing 0.75 mCi ^{3}H-9,10-OA/mmol as tracer.
2. Intensively wash cells with DPBS, suspend in solution A, and disrupt by sonication. Then, proceed with step 3 later.

(B) *To isolate LDs from tissue samples:*

1. Excise tissue (white and brown adipose tissue or liver), wash intensively with DPBS, and place on ice. For isolation of LDs, frozen tissue is not recommended.
2. Homogenize tissue in solution A on ice using an UltraTurrax (IKA, Janke & Kunkel, Germany) and centrifuge at 1,000 × *g*, 4 °C for 10 min to remove nondisrupted tissue and nuclei.
3. Transfer cell/tissue lysates to siliconized tubes, overlay with overlay buffer, and centrifuge 2 h, 100,000 × *g*, 4 °C.
4. Collect floated LDs from the top of the tubes and resuspend in overlay buffer by brief sonication.
5. Determine the TG content of LDs using TG reagent.
6. Perform protein determination using BCA reagent. As lipids interfere with protein measurement, incubate 20 μl of the LD fraction with 200 μl BCA reagent. Then, add 0.5% Triton X-100, mix, and incubate for 10 min at RT. After centrifugation at 20,000 × *g* for 10 min, read the absorbance of the underlying solution at 562 nm (typically the protein content of LDs is 0.5 mg protein/mmol TG).

2.5.2 Measurement of lipolysis of isolated LDs from cultured cells

2.5.2.1 Required materials

- Isolated radiolabeled LDs.
- 0.1 *M* KPB (pH 7.0).
- Pi (1,000 × stock): 1 mg/ml pepstatin, 2 mg/ml antipain, and 20 mg/ml leupeptin.
- 20% FA-free BSA (Sigma-Aldrich) dissolved in 0.1 *M* KPB (pH 7.0).
- Extraction solution I: Methanol/chloroform/*n*-heptane with a ratio of 10:9:7 (v/v/v) is stored light-protected at 4 °C. Prewarm solution before use.
- Extraction solution II: 0.1 *M* potassium carbonate, 0.1 *M* boric acid, pH 10.5.
- Liquid scintillation cocktail (Roth or equivalent).
- Liquid scintillation counter (Packard or equivalent).

2.5.2.2 Procedure

1. Prepare samples on ice. For each determination, prepare 100 μl sample (~0.2 m*M* TG from isolated radiolabeled LD; see 2.5.1). As blank, use 100 μl overlay buffer. Pipette samples and blanks directly onto the bottom of the assay tubes.
2. Add 100 μl of 0.1 *M* KPB, containing 1 × Pi and 1% BSA (FA-free) to samples and blank to give a final volume of 200 μl. To study the effect of recombinant proteins (e.g. lipases or LD proteins) on hydrolysis of LDs, add purified enzymes or cell lysates containing proteins of interest.
3. Incubate samples and blanks for 1 h at 37 °C in a water bath under constant shaking.
4. Terminate the reaction by the addition of 3.25 ml extraction solution I. Add 1.05 ml of extraction solution II and mix vigorously for 5 s using a vortex.
5. Centrifuge samples at 1,000 × *g* for 10 min. Then, transfer 200 μl of the upper aqueous phase to a scintillation vial containing 2 ml of scintillation cocktail and analyze radioactivity.

2.5.2.3 Calculation

LD-associated lipolytic activity is calculated as the amount of released FA (nmol) per TG (μmol) or per protein (mg) amount of the LD and per hour. The partition coefficient for FAs into the water phase is 1.9 (22 °C), meaning that 71.5% of OA is recovered in the water phase. For the calculation, the following equation can be used:

$$\frac{(\text{cpm sample} - \text{cpm blank}) \times (V_1/V_2)}{(\text{cpm substrate/nmol FA})^1 \times \text{mg protein} \times 0.715 \times t} = \text{nmol FA/mg protein/h}$$

where V_1 is the total volume of the upper water phase (2.45 ml); V_2, volume measured by liquid scintillation (0.2 ml); and t, incubation time (h); [1] 60 nmol FA/100 μl for 0.2 m*M* TG substrate.

2.5.3 Lipolysis of isolated tissue LDs

2.5.3.1 Required materials

- Isolated LDs.
- 0.1 *M* KPB (pH 7.0).
- Pi (1,000 × stock): 1 mg/ml pepstatin, 2 mg/ml antipain, and 20 mg/ml leupeptin.

- 20% BSA (FA-free; Sigma-Aldrich) in 0.1 *M* KPB (pH 7.0).
- NEFA kit and standard solution (cat. no. 999-34691; 995-3479; 991-34891; 993-35191; 276-76491, Wako Chemicals, Neuss, Germany).
- 10% Triton X-100.

2.5.3.2 Procedure

1. Incubate 1 m*M* TG of isolated LDs with 1% BSA in 0.1 *M* KPB containing 1 × Pi in a final volume of 200 μl for 1 h at 37 °C in a water bath under constant shaking. In order to evaluate the capacity of ATGL or HSL to hydrolyze natural LDs, add purified enzymes or cell lysates overexpressing the enzyme of interest.
2. Solubilize LDs by the addition of 20 μl 10% Triton X-100 (final conc. 1%) to the samples, vortex, and incubate for 10 min at RT.
3. Centrifuge samples for 30 min at 20,000 × *g* and carefully transfer the underlying solution, without any lipids in a new tube.
4. Determine FA content in the underlying solution using NEFA kit.

2.5.3.3 Calculation

To calculate net FA release of the tissue LDs, FA levels are also measured from the LD substrate prior to incubation (blank). After subtraction of the blank, lipolysis of tissue LDs is calculated as nmol FA per TG (μmol) or per protein (mg) amount of the LDs.

2.6. Measurement of lipolysis in cultured adipocytes and organ explants

The most direct assessment of lipolysis is the measurement of the lipolytic products, released by adipocytes or fat explants. This method is based on the measurement of the lipolytic products, FAs, and glycerol in the incubation media. Two lipolytic states can be differentiated, the so-called basal and β-adrenergic-stimulated states. Stimulation of lipolysis leads to a rapid increase in FA and glycerol release within 15 min, with a quasi linear rate during the first hours which levels off thereafter (see Fig. 10.6). Since the rate of stimulated lipolysis is highest and nearly linear between the first and second hour of β-adrenergic stimulation, the measurement of the stimulated state requires a preincubation period. Usually, cells or fat explants are preincubated for 1 h with isoproterenol or forskolin; then, the incubation media is replaced and the released FAs and glycerol of the forthcoming 1 h period are determined.

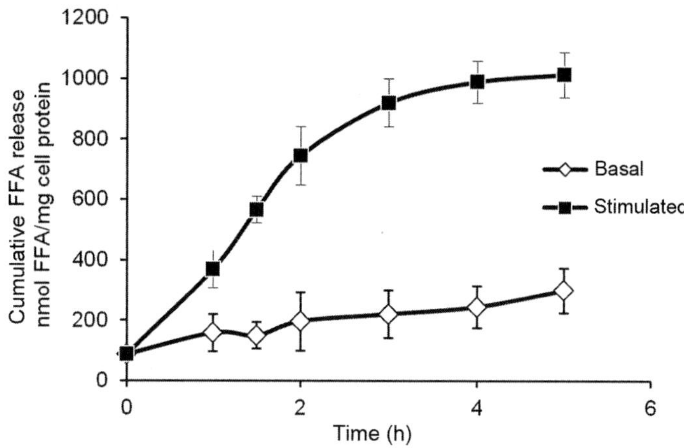

Figure 10.6 Time course of basal and stimulated lipolysis of adipocytes. Differentiated 3T3-L1 adipocytes were incubated in DMEM containing 2% BSA (FA-free) in the absence (basal) and presence (stimulated) of isoproterenol (10 μ*M*). After various times, the FA content of incubation media was determined and normalized to cellular protein content. Data are means ± standard deviation of three replicates.

2.6.1 Measurement of basal and stimulated lipolysis in cultured adipocytes

2.6.1.1 Required material

- Differentiated adipocytes in 12-well or 24-well plates.
- Dulbecco's modified Eagle's medium (DMEM, 1 g/l glucose; GIBCO, Life Technologies, Carlsbad, CA).
- 20% BSA (FA-free) stock solution in PBS (Sigma-Aldrich), sterile.
- DPBS, sterile.
- NEFA kit and standard solution (cat. no. 999-34691; 995-3479; 991-34891; 993-35191; 276-76491,Wako Chemicals, Neuss, Germany)
- Free glycerol reagent and glycerol standard solution (cat. no. F6428; Sigma-Aldrich).
- Extraction solution: Chloroform/methanol (2:1, v/v), 1% glacial acetic acid.
- Lysis solution: NaOH/SDS (0.3 N/0.1%).
- BCA reagent and BSA standard (cat. no. 23227, Thermo Fisher Scientific).
- Forskolin or isoproterenol, 10 m*M* stock solutions in DMSO, stored in aliquots at −20 °C (Sigma-Aldrich).
- Triacsin C, 5 m*M* stock solution in DMSO, stored in aliquots at −20 °C (Triacsin C from *Streptomyces* sp.; cat. no. T4540, Sigma-Aldrich).

2.6.1.2 Procedure

1. For measurement of stimulated lipolysis, preincubate differentiated adipocytes (i.e., 3T3-L1) in DMEM, containing 2% BSA (FA-free) and

forskolin or isoproterenol in the presence or absence of 5 μ*M* triacsin C for 60 min at 37 °C, 5% CO_2, and 95% humidified atmosphere. For measurement of basal lipolysis, preincubation is not required.

2. Replace the medium by an identical fresh medium containing 2% FA-free BSA and incubate for 1 h (=period of measurement). Save media for measurements.
3. Wash cells 2 × with DPBS.
4. Lyse cells with lysis solution (200–500 μl/well) by shaking the plates for 3 h at room temperature.
5. Determine protein content of the cell lysate using BCA reagent and BSA as standard.
6. Determine free FA and glycerol content of the medium using NEFA kit and free glycerol reagent and the appropriate standard solutions, respectively.

2.6.1.3 Calculation

Lipolysis of adipocytes is calculated as nmol FA and/or nmol glycerol per mg cell protein and hour.

2.6.1.4 Limitations

To avoid reesterification of FA and glycerol, it is recommended to incubate cells in the presence of 5 μ*M* triacsin C to inhibit acyl-CoA synthetases. At high lipolytic rates, the capacity of BSA (FA-free) as FA acceptor may become limiting after prolonged incubation time (Fig. 10.6). This will lead to decreased apparent FA release rate and increased reesterification (Paar et al., 2012).

2.6.2 Measurement of basal and stimulated lipolysis of WAT organ explants

2.6.2.1 Required materials

- Dulbecco's modified Eagle's medium (DMEM, 1 g/l glucose; GIBCO, Life Technologies, Carlsbad, CA).
- DPBS, sterile.
- Forceps and scissors for organ explant preparations.
- 20% BSA (FA-free; Sigma-Aldrich) stock solution in PBS, sterile.
- 96-well plates, sterile.
- NEFA kit and standard solution (Wako chemicals).
- Free glycerol reagent and glycerol standard solution (Sigma-Aldrich).
- Extraction solution: Chloroform/methanol (2:1, v/v), 1% glacial acetic acid.
- Lysis solution: NaOH/SDS (0.3 N/0.1%).

- BCA reagent and BSA standard (Pierce).
- Forskolin or isoproterenol, 10 m*M* stock solutions in DMSO, stored in aliquots at −20 °C (Sigma-Aldrich).
- Triacsin C, 5 m*M* stock solution in DMSO, stored in aliquots at −20 °C (Sigma-Aldrich).

2.6.2.2 Procedure

1. Surgically remove gonadal adipose tissue, wash in DPBS, and incubate in prewarmed (37 °C) DMEM till use (perform measurements the same day).
2. For measurement of stimulated lipolysis, cut tissue pieces (~20 mg) and preincubate them in 200 μl DMEM containing 2% BSA (FA-free) and 10 μ*M* forskolin/isoproterenol and in the presence or absence of 5 μ*M* triacsin C in 96-well plates at 37 °C, 5% CO_2, and 95% humidified atmosphere for 60 min. For measurement of basal lipolysis, preincubation is not required.
3. Transfer fat explants into 200 μl of identical, fresh medium and incubate for further 60 min (=period of measurement) at 37 °C, 5% CO_2, and 95% humidified atmosphere. Save incubation media.
4. Transfer fat explants in 1 ml extraction solution and incubate for 60 min at 37 °C under vigorous shaking on a thermomixer.
5. Transfer fat explants in 500 μl lysis solution and incubate overnight at 55 °C under vigorous shaking on a thermomixer.
6. Determine protein content of the fat explant lysates using BCA reagent and BSA as standard.
7. Determine FA and glycerol content of the incubation media using NEFA kit and free glycerol reagent and appropriate standard solutions, respectively.

2.6.2.3 Calculation

Lipolysis of adipose tissue organ cultures is calculated as nmol FA and/or nmol glycerol per mg/protein and hour.

2.6.2.4 Limitations

To avoid reesterification of FA and glycerol, it is recommended to incubate fat explants in the presence of 5 μ*M* triacsin C to inhibit acyl-CoA synthetases. At high lipolytic rates, the capacity of BSA (FA-free) as FA acceptor may become limiting after prolonged incubation times. This will lead to decreased apparent FA release rate and increased reesterification (Paar et al., 2012).

3. CONCLUSION

During the last decade, the perception of lipolysis has dramatically changed. Research steadily uncovers the complexity of the lipidome, proteome, and interactome, the regulation of the lipolytic process, and its interconnection with other cellular processes such as lipid and energy metabolism. It has now been recognized that the measurement of mRNA or protein expression of key players in lipolysis is often not enough to determine the lipolytic state of cells and tissues. Not a single biochemical estimation but the combination of *in vitro, in situ,* and *in vivo* measurements is required to assess lipolytic capacity and the lipolytic state and to unravel the role of lipolysis in energy metabolism, cellular signaling, the maintenance of normophysiology, and the development of pathologies.

ACKNOWLEDGMENTS

This work was supported by the Grants SFB Lipotox and the Wittgenstein Award Z136 (Ru. Ze.), the grants P24294, P18434, and TRP4 (Ro. Zi.), and the Grant P25193 (A. L.), which are funded by the Austrian Science Fund (FWF).

REFERENCES

Bezaire, V., Mairal, A., Ribet, C., Lefort, C., Girousse, A., Jocken, J., et al. (2009). Contribution of adipose triglyceride lipase and hormone-sensitive lipase to lipolysis in hMADS adipocytes. *The Journal of Biological Chemistry, 284*, 18282–18291.

Clifford, G. M., Londos, C., Kraemer, F. B., Vernon, R. G., & Yeaman, S. J. (2000). Translocation of hormone-sensitive lipase and perilipin upon lipolytic stimulation of rat adipocytes. *The Journal of Biological Chemistry, 275*, 5011–5015.

Ebdrup, S., Sørensen, L. G., Olsen, O. H., & Jacobsen, P. (2004). Synthesis and structure-activity relationship for a novel class of potent and selective carbamoyl-triazole based inhibitors of hormone sensitive lipase. *Journal of Medicinal Chemistry, 47*, 400–410.

Egan, J. J., Greenberg, A. S., Chang, M. K., Wek, S. A., Moos, M. C. J., & Londos, C. (1992). Mechanism of hormone-stimulated lipolysis in adipocytes: Translocation of hormone-sensitive lipase to the lipid storage droplet. *Proceedings of the National Academy of Sciences of the United States of America, 89*, 8537–8541.

Granneman, J. G., Moore, H. P., Granneman, R. L., Greenberg, A. S., Obin, M. S., & Zhu, Z. (2006). Analysis of lipolytic protein trafficking and interactions in adipocytes. *Journal of Biological Chemistry, 282*, 5726–5735.

Haemmerle, G., Lass, A., Zimmermann, R., Gorkiewicz, G., Meyer, C., Rozman, J., et al. (2006). Defective lipolysis and altered energy metabolism in mice lacking adipose triglyceride lipase. *Science, 312*, 734–737.

Holm, C., Olivecrona, G., & Ottosson, M. (2001). Assays of lipolytic enzymes. *Methods in Molecular Biology (Clifton, N.J.), 155*, 97–119.

Huijsman, E., Van de Par, C., Economou, C., Van der Poel, C., Lynch, G. S., Schoiswohl, G., et al. (2009). Adipose triacylglycerol lipase deletion alters whole body energy metabolism and impairs exercise performance in mice. *American Journal of Physiology Endocrinology and Metabolism, 297*, E505–E513.

Kliewer, S. A., Sundseth, S. S., Jones, S. A., Brown, P. J., Wisely, G. B., Koble, C. S., et al. (1997). Fatty acids and eicosanoids regulate gene expression through direct interactions with peroxisome proliferator-activated receptors alpha and gamma. *Proceedings of the National Academy of Sciences of the United States of America*, *94*, 4318–4323.

Kralisch, S., Klein, J., Lossner, U., Bluher, M., Paschke, R., Stumvoll, M., et al. (2005). Isoproterenol, TNFalpha, and insulin downregulate adipose triglyceride lipase in 3T3-L1 adipocytes. *Molecular and Cellular Endocrinology*, *240*, 43–49.

Lass, A., Zimmermann, R., Haemmerle, G., Riederer, M., Schoiswohl, G., Schweiger, M., et al. (2006). Adipose triglyceride lipase-mediated lipolysis of cellular fat stores is activated by CGI-58 and defective in Chanarin-Dorfman syndrome. *Cell Metabolism*, *3*, 309–319.

Long, J. Z., Nomura, D. K., & Cravatt, B. F. (2009). Characterization of monoacylglycerol lipase inhibition reveals differences in central and peripheral endocannabinoid metabolism. *Chemistry & Biology*, *16*, 744–753.

Lowe, D. B., Magnuson, S., Qi, N., Campbell, A.-M., Cook, J., Hong, Z., et al. (2004). In vitro SAR of (5-(2H)-isoxazolonyl) ureas, potent inhibitors of hormone-sensitive lipase. *Bioorganic & Medicinal Chemistry Letters*, *14*, 3155–3159.

Mayer, N., Schweiger, M., Romauch, M., Grabner, G. F., Eichmann, T. O., Fuchs, E., et al. (2013). Development of small molecule inhibitors targeting adipose triglyceride lipase. *Nature Chemical Biology*, *9*, 785–787.

Nielsen, T. S., Vendelbo, M. H., Jessen, N., Pedersen, S. B., Jørgensen, J. O., Lund, S., et al. (2011). Fasting, but not exercise, increases adipose triglyceride lipase (ATGL) protein and reduces G(0)/G(1) switch gene 2 (G0S2) protein and mRNA content in human adipose tissue. *The Journal of Clinical Endocrinology and Metabolism*, *96*, E1293–E1297.

Paar, M., Jüngst, C., Steiner, N. a., Magnes, C., Sinner, F., Kolb, D., et al. (2012). Remodeling of lipid droplets during lipolysis and growth in adipocytes. *The Journal of Biological Chemistry*, *287*, 11164–11173.

Schweiger, M., Schreiber, R., Haemmerle, G., Lass, A., Fledelius, C., Jacobsen, P., et al. (2006). Adipose triglyceride lipase and hormone-sensitive lipase are the major enzymes in adipose tissue triacylglycerol catabolism. *Journal of Biological Chemistry*, *281*, 40236–40241.

Vaughan, M., Berger, J. E., & Steinberg, D. (1964). Hormone-sensitive lipase and monoglyceride lipase activities in adipose tissue. *The Journal of Biological Chemistry*, *239*, 401–409.

Yamaguchi, T. (2010). Crucial role of CGI-58/alpha/beta hydrolase domain-containing protein 5 in lipid metabolism. *Biological & Pharmaceutical Bulletin*, *33*, 342–345.

Yamaguchi, T., & Osumi, T. (2009). Chanarin-Dorfman syndrome: Deficiency in CGI-58, a lipid droplet-bound coactivator of lipase. *Biochimica et Biophysica Acta*, *1791*, 519–523.

Yang, Xingyuan, Lu, X., Lombes, M., Rha, G. B., Chi, Y.-I. I., Guerin, T. M., et al. (2010). The G(0)/G(1) switch gene 2 regulates adipose lipolysis through association with adipose triglyceride lipase. *Cell Metabolism*, *11*, 194–205.

Yang, X., Zhang, X., Heckmann, B. L., Lu, X., & Liu, J. (2011). Relative contribution of adipose triglyceride lipase and hormone-sensitive lipase to tumor necrosis factor-alpha (TNF-alpha)-induced lipolysis in adipocytes. *The Journal of Biological Chemistry*, *286*, 40477–40485.

Zhang, X., Xie, X., Heckmann, B. L., Saarinen, A. M., Czyzyk, T. A., & Liu, J. (2013). Target disruption of G0/G1 switch gene 2 enhances adipose lipolysis, alters hepatic energy balance, and alleviates high fat diet-induced liver steatosis. *Diabetes*, [Epub ahead of print].

Zechner, R., Kienesberger, P. C., Haemmerle, G., Zimmermann, R., & Lass, A. (2009). Adipose triglyceride lipase and the lipolytic catabolism of cellular fat stores. *Journal of Lipid Research*, *50*, 3–21.

Zimmermann, R., Strauss, J. G., Haemmerle, G., Schoiswohl, G., Birner-Gruenberger, R., Riederer, M., et al. (2004). Fat mobilization in adipose tissue is promoted by adipose triglyceride lipase. *Science*, *306*, 1383–1386.

CHAPTER ELEVEN

Measurement of Lipolysis Products Secreted by 3T3-L1 Adipocytes Using Microfluidics

Colleen E. Dugan, Robert T. Kennedy[1]

Department of Chemistry, University of Michigan, Ann Arbor, Michigan, USA

[1]Corresponding author: e-mail address: rtkenn@umich.edu

Contents

Abstract

Glass microfluidic devices have been fabricated to monitor the secretion of glycerol or fatty acids from cultured murine 3T3-L1 adipocytes. In the current studies, adipocytes are perfused in a reversibly sealed cell chamber, and secreted products are analyzed by enzyme assay on either a single- or dual-chip device. The analysis of glycerol employed the use of a dual-chip system, which used separate chips for cell perfusion and sample analysis. An improved single-chip device integrated the cell perfusion chamber and analysis component on one platform. The performance of this device was demonstrated by the analysis of fatty acids but could also be applied to analysis of glycerol or other chemicals. The single-chip system required fewer cells and lower flow rates and provided improved temporal response. In both systems, cells were perfused with buffer to monitor basal response followed by lipolysis stimulation with the β-adrenergic agonist isoproterenol. Measured basal glycerol concentration from 50,000 cells was 28 μ*M*, and when stimulated, a spike threefold higher than basal concentration was detected followed by a continuous release 40% above basal levels. Fatty acid basal concentration was 24 μ*M*, measured from 6200 cells, and isoproterenol stimulation resulted in a constant elevated concentration sevenfold higher than basal levels.

Methods in Enzymology, Volume 538
ISSN 0076-6879
http://dx.doi.org/10.1016/B978-0-12-800280-3.00011-6

1. INTRODUCTION

Since the discovery that adipose tissue functions as endocrine tissue that releases hormones to regulate energy metabolism, research delving into its function and interactions with other tissues and organs has become increasingly important. Gaining a deeper understanding of adipose tissue physiology has become more crucial with the growing population of individuals with obesity-related disorders like type 2 diabetes (Rosen & Spiegelman, 2006). The production and secretion of hormones and other chemicals by adipocytes in response to circulation chemicals can be an indication of such disorders (Duncan, Ahmadian, Jaworski, Sarkadi-Nagy, & Sul, 2007). Lipolysis products, nonesterified fatty acids (NEFA) and glycerol, are examples of chemicals secreted from adipocytes that play an important role in physiological homeostasis. Fatty acids provide necessary energy to surrounding tissues, but when circulating at elevated levels, as observed in individuals with increased adiposity, they can lead to insulin resistance or cardiovascular disease (Samuel & Shulman, 2012; Wyne, 2003). To understand these processes, it is important to have a reliable, accurate method of measuring lipolysis products secreted by adipocytes under different conditions.

The conventional method for monitoring chemical uptake or secretion by adipocytes generally involves using large numbers of cells and off-line assay analysis (Millward et al., 2010). Microfluidics is an alternative analysis technique that has gained popularity for cell biology applications because of its ability to allow monitoring of cellular dynamics in near real time. Additionally, microfluidics can offer a cellular environment that more closely reflects the *in vivo* type of conditions, media can be replenished or recirculated, and shear stress can be controlled (El-Ali, Sorger, & Jensen, 2006). Furthermore, it is possible to integrate and automate chemical analysis on the chip by using microfabrication. Cell culture, analysis, and detection can all be performed on the same platform. Many groups have demonstrated the versatility and scope of microfluidic cell biology applications: mimicking lung cell expansion and function (Huh et al., 2010), encapsulating multiple cell lines in hydrogel microbeads to monitor cell–cell interactions (Tumarkin et al., 2011), differentiating adipocytes on-chip using a gradient flow (Lai, Sims, Jeon, & Lee, 2012), and monitoring secretion (Roper, Shackman, Dahlgren, & Kennedy, 2003).

Depending on the desired application, a variety of microfluidic substrates can be utilized. Devices made from plastics like polydimethylsiloxane (PDMS) or polymethyl methacrylate are frequently used because of their low cost and ease of fabrication. The exposed surfaces of these plastic devices are generally hydrophobic and can absorb or become fouled by certain biomolecules or organic chemicals (van Midwoud, Janse, Merema, Groothuis, & Verpoorte, 2012). Surface modifications can be performed to change the chemistry of the channel walls. Alternatively, glass microfluidics can be used. Glass offers a greater range of chemical compatibility (for use with fatty acids and enzymes) and is used in the studies described herein.

Cellular secretions can be monitored online in a variety of ways with microfluidics. For example, glucagon and insulin secretion from islets were measured by immunoassay (Dishinger, Reid, & Kennedy, 2009; Shackman, Reid, Dugan, & Kennedy, 2012), ATP release from red blood cells was measured by a bioluminescent assay (Wan, Ristenpart, & Stone, 2008), and nitric oxide secretion from macrophages was measured by enzyme assay (Goto, Sato, Murakami, Tokeshi, & Kitamori, 2005). In all these studies, the cells are incubated on the chip and constantly perfused. The perfusate flows into a channel wherein reactants are added. The resulting mixture flows through a reactant channel where the products are detected. The concentration of fatty acids and glycerol secreted by adipocytes is typically monitored off-line by reaction with commercially available enzyme assays; however in a microfluidic device, this reaction can be automatically and continuously performed on-line. This method also has the advantage of lower reagent consumption. Commonly used colorimetric enzyme assays can often be adapted to be measured by fluorescence detection by the addition of the hydrogen peroxide-sensitive dye Amplex UltraRed, to allow for a more sensitive response.

In the present studies, cultured 3T3-L1 adipocytes are loaded onto a reversibly sealed microfluidic cell chamber and alternately perfused with buffer, to measure basal secretion, and isoproterenol, a β-adrenergic agonist that stimulates lipolysis. The perfusate immediately enters micrometer-scale channels where it mixes with the necessary enzyme assay reagents. The fluorescence signal is detected at the outlet using laser-induced fluorescence. The resulting data provides information on adipocyte lipolysis rates under different conditions and can be extended to other studies on adipocyte function.

2. METHODS

2.1. Required reagents

- Free glycerol reagent, glycerol standard solution, Hanks' balanced salt solution (HBSS, Cat. # H6648), dimethyl sulfoxide (DMSO), isoproterenol hydrochloride, and fatty acid-free bovine serum albumin (BSA) were purchased from Sigma-Aldrich (St Louis, MO). Free glycerol reagent was reconstituted using 18.0 MΩ/cm distilled water. All other solutions were prepared with HBSS.

- NEFA-HR(2) kit was obtained from Wako Diagnostics (Richmond, VA). Color reagent A (CR-A) was reconstituted with solvent A, and color reagent B (CR-B) was reconstituted with 25 mL of 18.0 MΩ/cm water.
- Hydrofluoric acid, sulfuric acid, 30% hydrogen peroxide, and ammonium hydroxide were purchased from Fisher Scientific (Pittsburgh, PA).
- Amplex UltraRed reagent and cell culture reagents were obtained from Life Technologies (Carlsbad, CA). Amplex UltraRed was reconstituted with 340 μL DMSO to make a 10 m*M* stock solution.

2.2. Adipocyte culture

1. Glass coverslips (No. 1, Fisher Scientific, Pittsburgh, PA) were cut to the desired size using a diamond-tipped scorer and soaked in ethanol for at least an hour. For experiments performed with the following microfluidic devices, coverslips were cut to either $2.4 \times 40\ mm^2$ or $2 \times 12.5\ mm^2$. The coverslips were transferred to a new petri dish where they were allowed to dry. Three or four sterile coverslips were placed in a 10 cm petri dish before seeding adipocytes.
2. Murine 3T3-L1 preadipocytes were seeded into petri dishes, supplemented with media (DMEM with 10% calf serum), and incubated at 37 °C and 10% CO_2 until confluent. Differentiation was induced 2 days postconfluence with a media consisting of fetal bovine serum, insulin, dexamethasone, and isobutylmethylxanthine. Every 2 days after differentiation, the media was changed using a DMEM solution (high glucose and L-glutamine) containing 10% fetal bovine serum, 1% sodium pyruvate, and 1% penicillin–streptomycin.
3. When a cellular secretion experiment was being performed, sterile tweezers were used to pick up and transfer the coverslip to the microfluidic chip.

2.3. Glass chip fabrication

1. Borofloat glass slides ($2.54 \times 7.62 \times 0.1\ cm^3$, Telic, Valencia, CA) with a 120 nm chrome layer were coated with 530 nm of AZ-1518 photoresist. A photomask (Fineline Imaging, Colorado Springs, CO) with the

appropriate design (created on AutoCAD) was aligned over the glass and was exposed to a collimated UV light for 9 s at 26 mW/cm^2. The exposed photoresist was removed with AZ 726 MIF developer (AZ Electronic Materials, Branchburg, NJ), and the chrome was removed using CEP-200 chrome etchant (HTA Enterprises, San Jose, CA). AZ photoresist is a positive photoresist, meaning the dark patterns on the photomask are the areas that remain covered in photoresist after exposure and developing.

2. Dicing tape was used to cover certain areas of the glass that require less exposure to the etching solution. This was applied to the back side of the slide and in areas where there were shallow features.
3. The glass was etched using a 48:17:55 (v/v/v) HF:HNO_3:H_2O solution, with the etch time-dependent on the channel depth desired (with this solution the etch rate is 1.2 μm/min).
4. After etching, the remaining photoresist was stripped with acetone and the remaining chrome was removed in the CEP-200 etchant. Fluidic access holes were drilled using a diamond-tipped drill bit.
5. For chips with irreversibly bonded glass components, the following procedure was performed.
   a. The glass slides were cleaned in a piranha solution for 20 min. After rinsing, they were further cleaned by a RCA solution at 60 °C for 40 min (refer to Note 3 for mixture specifications).
   b. After rinsing in water, and while still wet, the two slides to be bonded were put in conformal contact and sealed using a 610 °C furnace for 8 h under 400 g weight.
   c. Reservoirs (Upchurch N-124H, IDEX Health & Science, Oak Harbor, WA) were adhered around the access holes using the supplied epoxy ring. A binder clip was used to clamp the reservoir to the chip, and the complete device was placed in the oven, per manufacturer's instructions.

2.4. Cell perfusion and enzyme assay

2.4.1 Glycerol assay chip

For the analysis of glycerol secretion from adipocytes, a dual-chip system was used (Clark, Sousa, Jennings, MacDougald, & Kennedy, 2009). The cell chamber and enzyme assay mixing chip were on separate devices connected by fused silica capillary:

1. The first chip housed the reversibly sealable cell chamber chip with dimensions of 4 cm × 0.5 cm. Two glass slides were etched to make the top and

bottom cell chambers as shown in Fig. 11.1. The bottom cell chamber was etched 450 μm deep and the top cell chamber was etched 100 μm deep. The top cell chamber also contained etched fluidic access channels and a moat (both 5 μm deep) surrounding the rectangular chamber.

2. The second chip contained the mixing channels where the enzyme assay can react with the cell perfusate. The channels, shown in Fig. 11.2, were etched into a glass slide and then bonded to a blank glass slide to create an irreversibly sealed channel network. The channels are of different widths along the fluid network, but all of the channels were etched to 60 μm deep. After the introduction of the reagents and cell perfusate, the channels are 100 μm wide to facilitate liquid mixing (see Note 2). The channels then extend into 600 μm wide channels so the reagents can incubate for longer times before detection.

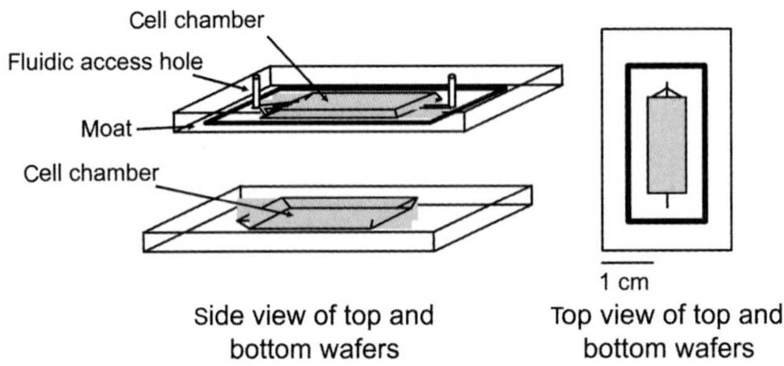

Figure 11.1 A diagram of the perfusion cell chip depicts the two separate wafers employed in this work. The wafers were reversibly sealed with the aid of an in-house-built compression frame. *Adapted with permission from Clark et al. (2009). Copyright 2009 American Chemical Society.*

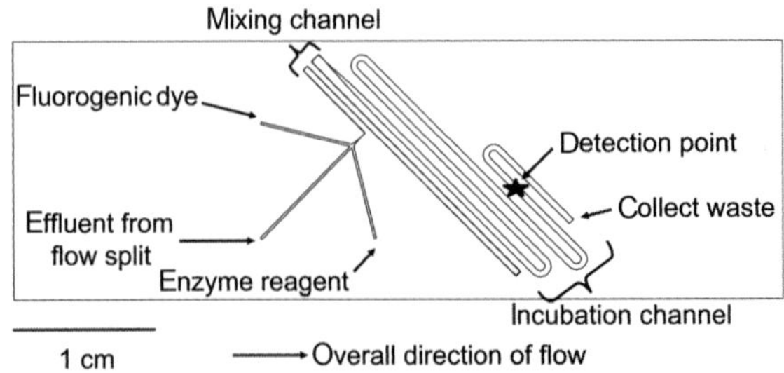

Figure 11.2 The enzyme assay chip was capable of performing online mixing of three solutions and online detection of the enzymatic product. The layout shows the initial mixing channel connected to the incubation channel. *Adapted with permission from Clark et al. (2009). Copyright 2009 American Chemical Society.*

3. The $2.4 \times 40\ mm^2$ coverslip with adhered adipocytes was loaded into the lower cell chamber and additional culture medium was added to fill the chamber. High-vacuum grease was carefully applied to both pieces of glass of the cell chamber chip around the chamber using a scalpel blade. A 125 μm thick sheet of poly(tetrafluoroethylene) (PTFE), with a hole cut out the size of the cell chamber, was placed on the vacuum grease of the lower cell chamber around the cells and pressed to seal. The top cell chamber glass was aligned over the lower chamber and pressed on to seal.
4. An in-house-built compression frame, made from two sheets of acrylic plastic with symmetrical holes drilled through both sheets, was tightened around the chip using screws.
5. A thin-film resistive heater maintained at 37 °C was placed under the compression frame to keep the cells at a physiological temperature.
6. HBSS buffer or 20 μ*M* isoproterenol in HBSS was perfused at 80 μL/min, using pressure-driven flow, through the cell chamber. The perfusate that exited the cell chamber chip was split using a Valco tee (Houston, TX), and 0.31% of the flow was directed to the inlet of the enzyme assay mixing chip via fused silica capillary. The resulting 250 nL/min perfusate flow was mixed in a 1:1:1 ratio with the two reagents: free glycerol reagent and 300 μ*M* Amplex UltraRed solution in DMSO (fluorogenic dye). The reagents were delivered to the chip by 100 μL Hamilton syringes on a syringe pump (CMA 402 pump, CMA Microdialysis, Holliston, MA). The resulting reactions that result in a fluorescent product are shown in Fig. 11.3.

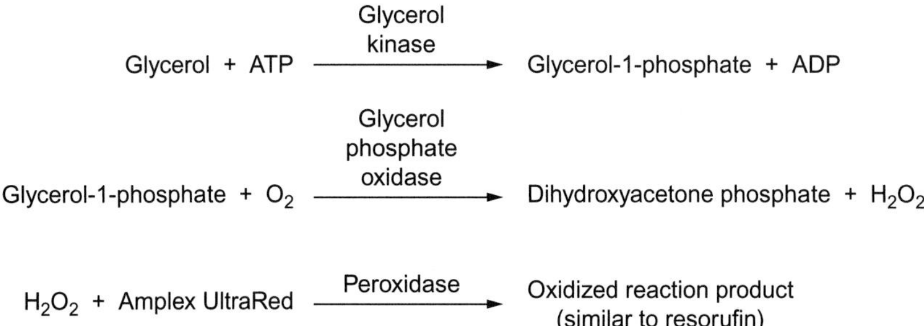

Figure 11.3 Glycerol enzyme assay reaction with the addition of the fluorogenic dye, Amplex UltraRed. The resulting propriety product is fluorescent at 543 nm.

2.4.2 Fatty acid assay chip

To monitor fatty acid secretion from adipocytes, a chip containing both the cell chamber and enzyme assay mixing channels on one device was developed (Clark, Sousa, Chisolm, MacDougald, & Kennedy, 2010). The smaller dimensions of this device allow for a reduced volume of cells and reagents and improved temporal resolution compared to the previously described dual chip:

1. The chip design of the integrated cell chamber and enzyme assay mixing chip comprised of etching three pieces of glass, shown in Fig. 11.4.
   a. The top glass slide contained the upper portion of the cell chamber (100 μm deep), a moat around the chamber (100 μm deep), and fluidic access channels (5 μm deep) leading in and out of the cell chamber. Holes were drilled at the ends of the access channels of the top slide using a 1 mm diamond-tipped drill bit.

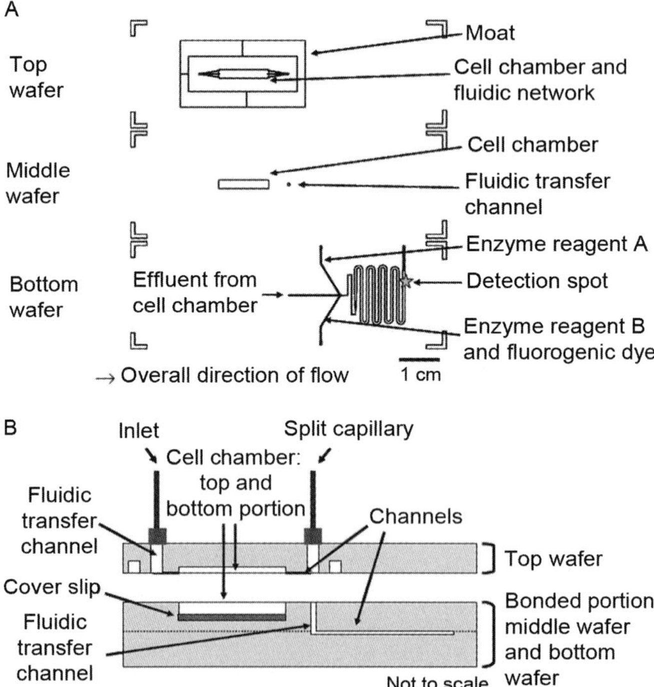

Figure 11.4 Chip design. (A) The multilayer device was comprised of three separately etched glass wafers that integrated a cell perfusion chamber and fluidic channels for online mixing of the fluorescence-based enzyme assay. (B) A side view of the cell chamber depicts the bonded and reversibly sealed portions of the device. *Adapted with permission from Clark et al. (2010). Copyright 2010 Springer.* (See color plate.)

   b. The middle slide contained the lower cell chamber (450 μm deep) and an access hole to allow fluid to enter the lower slide. The access hole was drilled to 360 μm wide.
   c. The bottom glass slide contained the fluidic network for mixing the assay reagents with the cell perfusate (60 μm deep). The middle and lower slides were irreversibly bonded to enclose and prevent leaking from the mixing channels.
2. Holes with a 360 μm diameter were drilled into the side of the bonded chip to allow access to the reagent inlets. Reservoirs were glued to the top of the top slide, and capillaries (50 μm inner diameter/360 μm outer diameter) were inserted into the enzyme reagent inlets and sealed with epoxy.
3. A 2×12.5 mm^2 coverslip with adhered adipocytes was loaded into the lower cell chamber, and the top glass slide was sealed on with vacuum grease as described previously with the dual chip. The entire chip was placed in an in-house-built compression frame. A thin-film resistive heater set to 37 °C was placed over the cell chamber.
4. HBSS buffer or 20 μ*M* isoproterenol in HBSS (both containing 2% BSA) was perfused at 8 μL/min through the cell chamber using a syringe pump. Solution changes were performed by pumping the two solutions through a 6-port valve.
5. The perfusate from the cell chamber was split by exiting either the top of the chip through the 1 mm drill hole (followed by 6.5 cm of 150 μm i.d. capillary) or the 360 μm access hole leading to the mixing channels in the lower chip. The flow entering the lower chip was 3% of the flow through the cell chamber. The resulting 250 nL/min flow was mixed in a 1:1:1 ratio with the reagents entering the other two inlets. One inlet perfused in CR-A and the other inlet perfused in a 2:1 mixture of CR-B:8.3 m*M* Amplex UltraRed in DMSO. The selected flow rates allowed a total mixing time of 5 min before detection. The on-chip reaction is shown in Fig. 11.5.

2.5. Detection

Laser-induced fluorescence was used for detection of the fluorescent resorufin-like product in the chip near the outlet of the enzyme assay mixing channels. A 543 nm HeNe Melles Griot laser was used for excitation in an epifluorescence optics configuration. Light was focused and collected using a 40×, 0.6 numerical aperture objective (Cat. # 440864, Carl Zeiss

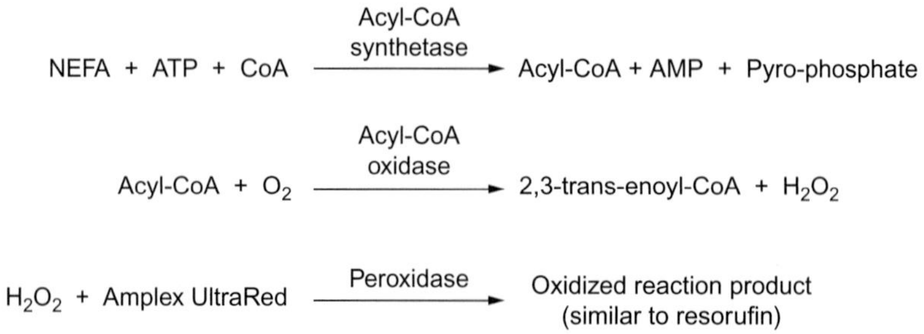

$$\text{NEFA} + \text{ATP} + \text{CoA} \xrightarrow{\text{Acyl-CoA synthetase}} \text{Acyl-CoA} + \text{AMP} + \text{Pyro-phosphate}$$

$$\text{Acyl-CoA} + O_2 \xrightarrow{\text{Acyl-CoA oxidase}} \text{2,3-trans-enoyl-CoA} + H_2O_2$$

$$H_2O_2 + \text{Amplex UltraRed} \xrightarrow{\text{Peroxidase}} \text{Oxidized reaction product (similar to resorufin)}$$

Figure 11.5 NEFA enzyme assay reaction. The addition of the fluorogenic dye, Amplex UltraRed, creates a fluorescent product similar to resorufin that is fluorescent at 543 nm.

Microscopy, Thornwood, NY), along with the appropriate dichroic mirror and filter set. A 1 mm pinhole was used in front of the photon-counting detector and data were collected at 2 Hz. The data output was collected and analyzed on in-house-written LabView programs.

3. DISCUSSION

Online calibration curves were created for both the glycerol and fatty acid assays by perfusing standards through the cell chamber in the absence of adipocytes. The glycerol assay had a limit of detection (LOD) of 4 μM and the fatty acid assay had a LOD of 5 μM. Both assays had linear calibrations past the range that was needed to detect the amount secreted by the adipocytes. The RSD values were maintained under 5%, indicating a stable response, and reproducibility was observed after multiple calibrations. After transfer of the coverslip with a monolayer of differentiated adipocytes onto the chip, the cells were visually inspected to ensure they remained adhered to the glass and were not disrupted.

Representative traces of glycerol secretion from ~50,000 adipocytes are shown in Fig. 11.6. A stable basal response was observed during perfusion of buffer. When isoproterenol was applied, stimulation of lipolysis was observed by an increase in glycerol secretion. An initial spike in concentration was observed threefold higher than basal, but slowly decreased to about 40% over basal concentrations and that was maintained for an hour or more. Control experiments were performed to ensure that isoproterenol did not

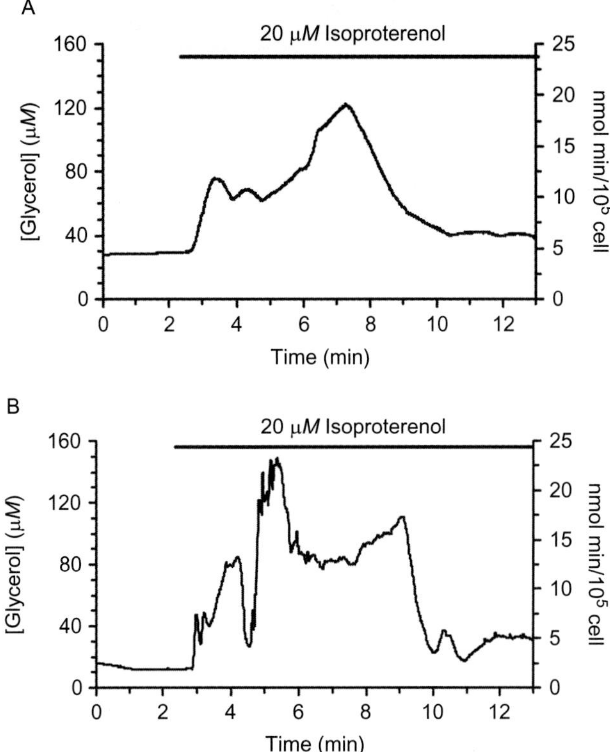

Figure 11.6 A & B: Representative traces of glycerol release from differentiated adipocytes and response upon 20 μ*M* isoproterenol treatment are shown. The bars above traces represent exposure to 20 μ*M* isoproterenol. The traces were shortened to depict only the time surrounding the initial exposure to isoproterenol. *Adapted with permission from Clark et al. (2009). Copyright 2009 American Chemical Society.*

interfere with the enzyme assay and also that the fluorogenic dye was not responsible for the signal observed (data not shown).

The chip used for the fatty acid assay reduced the amount of cells required to about 6200. As a result, the flow rate was decreased, which allowed for less analyte dilution and more of the sample to enter the enzyme mixing portion of the chip instead of being diverted to waste. The plots of the on-chip fatty acid release from adipocytes are shown in Fig. 11.7. Stable basal concentrations were observed, but in contrast to the glycerol assay, when isoproterenol was applied, a stable elevated response was monitored. This elevated concentration was sevenfold higher than basal levels and was maintained over time.

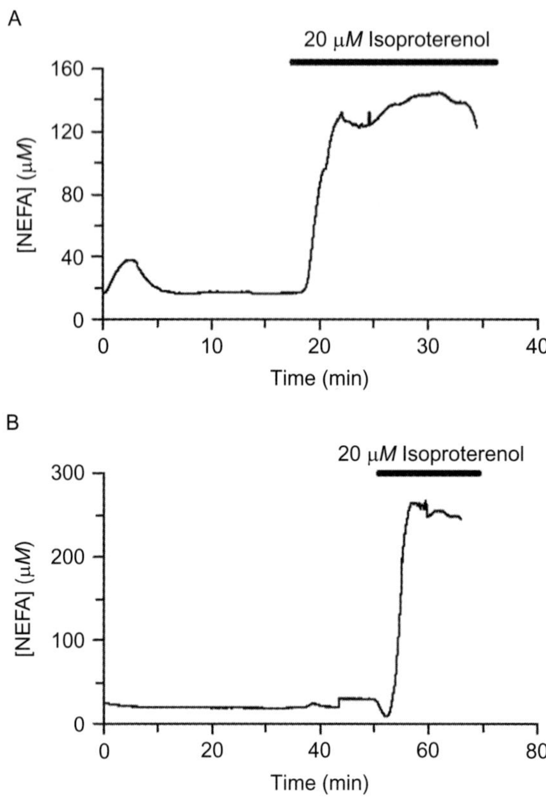

Figure 11.7 A & B: Representative traces of NEFA release from differentiated adipocytes and response from treatment with 20 μ*M* isoproterenol (indicated by the bar) are shown. *Adapted with permission from Clark et al. (2010). Copyright 2010 Springer.*

4. DESIGN AND OPERATION LIMITATIONS

Maintaining flow stability is a major challenge in microfluidic devices. Many factors of the setup can influence the fluid flow of the system: bubbles, connection leaks, pressure fluctuations from breaking connections to change solutions, and stability of the solution infusion pump. The integration of an on-chip reversibly sealed cell chamber adds an extra source of error, as the high flow rates through the chamber can cause leaking of the reversible seal. Proper application of the vacuum grease around the cell chamber is crucial in creating a good seal. A thin layer should be spread, enough to form a seal, but not so much that would cause it to enter the cell chamber or channels when compressed (which could cause clogging). Moreover, the addition of extra medium in the chamber before sealing aided in preventing bubbles from forming in the chamber. As a result of the large volume of the cell chambers used in these experiments, a flow split was required, which is another source of potential flow instability. If any part of the system becomes clogged or contains a bubble, the flow split ratio will change. The removal of the flow split would require a smaller cell chamber with lower flow rates through the chamber.

In the aforementioned experiments, the switch from basal perfusion buffer to a solution containing isoproterenol required either the manual changing of capillary connections or the use of an external 6-port valve. Manually changing connections creates a large disruption in flow because the inlet has to be stopped and moved, followed by an additional time period before stable flow is restored. The use of the 6-port valve reduced the pressure fluctuations but still required long connection capillaries. A more automated, on-chip valve system would improve the system operation.

The design of the chamber and the fluidic connection channels leading in and out of the chamber can affect the temporal resolution of the system. The length of the chamber will influence the paracrine response of cells at the end of the chamber. Additionally, the longer the chamber, the more parabolic the flow becomes at the end of the chamber, increasing temporal resolution. Computer-assisted simulations demonstrated how the number and angle of channels leading to and from the cell chamber can change the flow profile through the chamber, generally with the goal of creating less of a parabolic profile.

5. CONTINUED IMPROVEMENTS

In an effort to improve the chip automation and ease of assembly, a multilayer chip made entirely of PDMS has been developed. The inherent elastomeric properties of PDMS allow for the reversible sealing of a cell chamber without the use of vacuum grease or a sheet of PTFE. Automated switching of perfusion buffers can be performed on chip by integrating valves, similar to those developed by Unger, Chou, Thorsen, Scherer, and Quake (2000). Additionally, chamber dimensions and flow rates have been optimized to improve temporal resolution and allow the entire cell perfusate to enter the mixing channels instead of being directed to waste.

In order to adapt the assays and cell chamber to be compatible with the new material, some changes were made to the previously described procedures. PDMS swells in the presence of most organic solvents, so instead of diluting the Amplex UltraRed stock with DMSO, solutions were made in 35% DMSO. Since fatty acids can absorb into the hydrophobic PDMS, a glass coverslip was bonded onto the upper portion of the cell chamber to minimize the exposed surfaces to fatty acids. In the chips used for fatty acid assays, CR-A was supplemented with Triton X-100 prior to introduction on the chip, and the mixing channel walls were dynamically coated with sodium dodecyl sulfate to prevent fatty acid interactions with the chip.

Through off-line assays, it was determined that improved fatty acid assay sensitivity could be achieved when the fluorogenic dye and CR-B are introduced in separate channels and downstream of the CR-A inlet. Essentially, the fluorogenic dye reacts quickly with CR-B and increases the background signal. The fatty acid reaction with CR-A is relatively slow, and increased sensitivity is observed when there is a delay before the introduction of CR-B.

6. NOTES

1. All chip fabrication was done in a clean room.
2. In microfluidic systems, mixing is only achieved through molecular diffusion (Beebe, Mensing, & Walker, 2002). To have sufficient mixing without the fabrication of complex mixers on-chip, the easiest solution is to have channels thin enough to allow complete diffusion across the channel using the desired flow rate and channel length, hence the two different channel widths seen in the chips.
3. Piranha solution is 3:1 H_2SO_4:H_2O_2 (v/v). This is a very strong oxidizing agent and should be handled with care. RCA solution is 5:1:1 H_2O: NH_4:H_2O_2 (v/v/v).
4. Ideally, the coverslips should be cut with a dicing saw, which will result in more precise measurements than cutting by hand can provide.

REFERENCES

Beebe, D. J., Mensing, G. A., & Walker, G. M. (2002). Physics and applications of microfluidics in biology. *Annual Review of Biomedical Engineering*, *4*, 261–286.

Clark, A., Sousa, K., Chisolm, C., MacDougald, O., & Kennedy, R. (2010). Reversibly sealed multilayer microfluidic device for integrated cell perfusion and on-line chemical analysis of cultured adipocyte secretions. *Analytical and Bioanalytical Chemistry*, *397*, 2939–2947.

Clark, A. M., Sousa, K. M., Jennings, C., MacDougald, O. A., & Kennedy, R. T. (2009). Continuous-flow enzyme assay on a microfluidic chip for monitoring glycerol secretion from cultured adipocytes. *Analytical Chemistry, 81*, 2350–2356.

Dishinger, J. F., Reid, K. R., & Kennedy, R. T. (2009). Quantitative monitoring of insulin secretion from single islets of Langerhans in parallel on a microfluidic chip. *Analytical Chemistry, 81*, 3119–3127.

Duncan, R. E., Ahmadian, M., Jaworski, K., Sarkadi-Nagy, E., & Sul, H. S. (2007). Regulation of lipolysis in adipocytes. *Annual Review of Nutrition, 27*, 79–101.

El-Ali, J., Sorger, P. K., & Jensen, K. F. (2006). Cells on chips. *Nature, 442*, 403–411.

Goto, M., Sato, K., Murakami, A., Tokeshi, M., & Kitamori, T. (2005). Development of a microchip-based bioassay system using cultured cells. *Analytical Chemistry, 77*, 2125–2131.

Huh, D., Matthews, B. D., Mammoto, A., Montoya-Zavala, M., Hsin, H. Y., & Ingber, D. E. (2010). Reconstituting organ-level lung functions on a chip. *Science, 328*, 1662–1668.

Lai, N., Sims, J. K., Jeon, N. L., & Lee, K. (2012). Adipocyte induction of preadipocyte differentiation in a gradient chamber. *Tissue Engineering. Part C, Methods, 18*, 958–967.

Millward, C. A., DeSantis, D., Hsieh, C. W., Heaney, J. D., Pisano, S., Olswang, Y., et al. (2010). Phosphoenolpyruvate carboxykinase (Pck1) helps regulate the triglyceride/fatty acid cycle and development of insulin resistance in mice. *Journal of Lipid Research, 51*, 1452–1463.

Roper, M. G., Shackman, J. G., Dahlgren, G. M., & Kennedy, R. T. (2003). Microfluidic chip for continuous monitoring of hormone secretion from live cells using an electrophoresis-based immunoassay. *Analytical Chemistry, 75*, 4711–4717.

Rosen, E. D., & Spiegelman, B. M. (2006). Adipocytes as regulators of energy balance and glucose homeostasis. *Nature, 444*, 847–853.

Samuel, V. T., & Shulman, G. I. (2012). Mechanisms for insulin resistance: Common threads and missing links. *Cell, 148*, 852–871.

Shackman, J. G., Reid, K. R., Dugan, C. E., & Kennedy, R. T. (2012). Dynamic monitoring of glucagon secretion from living cells on a microfluidic chip. *Analytical and Bioanalytical Chemistry, 402*, 2797–2803.

Tumarkin, E., Tzadu, L., Csaszar, E., Seo, M., Zhang, H., Lee, A., et al. (2011). High-throughput combinatorial cell co-culture using microfluidics. *Integrative Biology, 3*, 653–662.

Unger, M. A., Chou, H. P., Thorsen, T., Scherer, A., & Quake, S. R. (2000). Monolithic microfabricated valves and pumps by multilayer soft lithography. *Science, 288*, 113–116.

van Midwoud, P. M., Janse, A., Merema, M. T., Groothuis, G. M. M., & Verpoorte, E. (2012). Comparison of biocompatibility and adsorption properties of different plastics for advanced microfluidic cell and tissue culture models. *Analytical Chemistry, 84*, 3938–3944.

Wan, J., Ristenpart, W. D., & Stone, H. A. (2008). Dynamics of shear-induced ATP release from red blood cells. *Proceedings of the National Academy of Sciences of the United States of America, 105*, 16432–16437.

Wyne, K. L. (2003). Free fatty acids and type 2 diabetes mellitus. *American Journal of Medicine, 115*, 29–36.

CHAPTER TWELVE

Methods for Performing Lipidomics in White Adipose Tissue

Lee D. Roberts[*,†], James A. West[*,†], Antonio Vidal-Puig[‡], Julian L. Griffin[*,†,1]

[*]MRC Human Nutrition Research, The Elsie Widdowson Laboratory, Cambridge, United Kingdom
[†]Department of Biochemistry and Cambridge Systems Biology Centre, University of Cambridge, Cambridge, United Kingdom
[‡]Metabolic Research Laboratories, Level 4, Institute of Metabolic Science, Addenbrooke's Hospital, Cambridge, United Kingdom
[1]Corresponding author: e-mail address: jules.griffin@mrc-hnr.cam.ac.uk

Contents

Abstract

Lipid metabolism is central to the function of white adipose tissue, with the tissue having a central role in storing triacylglycerides following feeding and releasing free fatty acids and monoacylglycerides during periods of fasting. In addition, lipid species have been suggested to play a role in lipotoxicity and as signaling molecules during adipose tissue inflammation. This chapter details how mass spectrometry (MS) can be used to profile a range of lipid species found in adipose tissue. The initial step required in any MS-based approach is to extract the lipid fraction from the tissue. We detail one commonly used method based on the Folch extraction procedure. The total fatty acid composition of the lipid fraction can readily be defined using gas chromatography–MS, and we provide a method routinely used for rodent and human adipose tissue samples.

Methods in Enzymology, Volume 538
ISSN 0076-6879
http://dx.doi.org/10.1016/B978-0-12-800280-3.00012-8

However, such approaches do not provide insight into what lipid classes the various fatty acids are associated with. To better understand the global lipid profile of the tissue, we provide a general-purpose liquid chromatography–MS-based approach useful for processing phospholipids, free fatty acids, and triacylglycerides. In addition, we provide a method for profiling eicosanoids, a class of important lipid-signaling molecules, which have been implicated in white adipose tissue inflammation in rodent models of obesity, insulin resistance, and type 2 diabetes.

1. INTRODUCTION

One of the primary roles of adipose tissue is to store energy in the form of triacylglycerides. These are hydrolyzed during periods of fasting within the body to release free fatty acids and glycerol; however, for normal cellular functions, adipocytes also require phospholipids for plasma membranes and intracellular organelles. Furthermore, with obesity or in other states where the storage capacity of adipocytes may be exceeded (Medina-Gomez et al., 2007), reactive lipid species such as lysophospholipids and diacylglycerides can be detected within adipose tissue. Such lipid species may be responsible for the development of insulin resistance within adipose tissue as part of lipotoxicity (Virtue & Vidal-Puig, 2010).

Recently, there has been increased interest in both brown adipose tissue and the browning of white adipose tissue as a potential target for treating type 2 diabetes, obesity, dyslipidemia, and atherosclerosis (Virtue et al., 2012; Roberts et al., 2011). In both tissues, fatty acid oxidation can be stimulated by either physiological or pharmacological interventions. To oxidize fatty acids, they must be first transported across the inner mitochondrial membrane across the carnitine shuttle. The oxidation of fatty acids can readily be followed by measuring acylcarnitines using mass spectrometry (MS), and we detail an approach for following this class of compounds.

Furthermore, adipocytes are not the only cell type in adipose tissue. Macrophages play an important role in adipose tissue inflammation and potentially may cause insulin resistance in this tissue. This activation is in part brought about by the production of eicosanoids. We detail a method for measuring these signaling compounds using a targeted liquid chromatography (LC)–MS method.

This chapter deals with a series of lipidomic methods for the profiling of different lipid species important for understanding adipose tissue biology. These tools use both gas chromatography–mass spectrometry (GC–MS) and LC–MS to profile the lipidome in both adipose tissue and adipocytes. While there are a variety of analytic methods used to profile lipids using MS in the literature, the techniques detailed here have been validated for adipose tissue (Atherton et al., 2006; Atherton et al., 2009; Roberts et al., 2011) and should provide a good starting place for the development of new MS approaches.

2. TISSUE EXTRACTION OF ADIPOSE TISSUE AND ADIPOCYTES

2.1. Required materials

The following is from Sigma-Aldrich (Gillingham, Dorset, United Kingdom):

High-performance liquid chromatography (HPLC)-grade (or higher) chloroform

HPLC-grade (or higher) methanol

HPLC-grade water

The following is from Qiagen (Venlo, Netherlands):

Qiagen TissueLyser Beads

The following is from Eppendorf (Hamburg, Germany):

Ultrasonic bath

Benchtop centrifuge

2.2. Method of tissue extraction for general lipid profiling of tissue and cells

1. Frozen white adipose tissue (approximately 100 mg) was pulverized in methanol/chloroform (2:1, 600 μl) using a TissueLyser (Qiagen). For cell culture, cells (10^6) were lysed in situ using 400 μl methanol and scraping prior to addition of 200 μl chloroform.
2. Samples were sonicated for 15 min.
3. Chloroform–water (1:1) was added (200 μl of each).
4. Samples were centrifuged (16,100 × *g*, 20 min) and the organic (upper layer) and aqueous phases (lower layer) were separated. This step should produce two distinct layers with a layer of protein at the interface. If for any reason two distinct layers are not formed, step 3 should be repeated.
5. The organic phase was transferred to glass vials and dried under a stream of nitrogen gas. To maximize lipid recovery, steps 1–5 can be repeated two or three times.

6. Glass vials were stored at −80 °C until analysis. It is important that storage is in glass vials wherever possible as organic solvents can leach plasticizers out of plasticware and severely affect the outcome of MS-based assays.

3. GC–MS OF TOTAL FATTY ACIDS

3.1. Required material

The following material was obtained from Sigma-Aldrich (Gillingham, Dorset, United Kingdom):

Supelco 37-Component FAME Mix
D_{25}-tridecanoic acid in chloroform
HPLC-grade (or better) chloroform
BF_3-methanol (30%)
Analytic-grade hexane
HPLC-grade (or better) water

Note the grade of the solvent is very important as lower-grade solvents can contain contaminants that significantly interfere with the MS.

The following were obtained from Thermo Scientific (Hemel Hempstead, United Kingdom):

GC Ultra coupled to a Trace DSQ 2 GC–MS

Note all GC–MS would be suitable for this analysis. We use a single quadrupole for this work although triple quadrupoles could be used for more robust quantification, particularly in samples with lots of fatty acids where chromatographic peaks may overlap. GC-flame ionization detectors (GC-FID) are also commonly used although they require separation of individual fatty acid methyl esters (FAMEs) by retention time and so may require longer chromatographic runs to provide peak separation for quantification.

30 m × 0.25 mm 70% cyanopropyl polysilphenylene-siloxane 0.25 μm TR-FAME stationary phase column.

3.2. Method for GC–MS for the analysis of total fatty acid content in a lipid extract

1. The organic fraction was dissolved in methanol/chloroform (750 μl, ratio 1:1).
2. D_{25}-tridecanoic acid in chloroform (internal standard, at 200 μ*M*, 50 μl) was added.
3. Acid-catalyzed esterification was used to derivatize the organic phase samples. BF_3-methanol (10%, 0.125 ml) was added to the organic phase and incubated at 90 °C for 90 min. This step cleaves fatty acids

from any lipid head group and carries out a methylation to form a FAME. It is these FAMEs that are detected by the GC–MS.

4. Water (0.15 ml) and hexane (0.3 ml) were added and the samples vortex mixed for 1 min and left to form a bilayer.
5. The aqueous phase was discarded and the organic layer evaporated to dryness under a stream of nitrogen prior to reconstitution in analytic-grade hexane (100 μl) before GC–MS analysis.
6. For our analyses, GC–MS was performed using a Trace GC Ultra coupled to a Trace DSQ II mass spectrometer (see earlier note on GC–MS and GC-FID).
7. The derivatized organic samples were injected splitless (no dilution with carrier gas) (primary adipocytes) or with a split ratio of 20 for white adipose tissue onto a 30 m × 0.25 mm 70% cyanopropyl polysilphenylene-siloxane 0.25 μm TR-FAME stationary phase column (Thermo Scientific).
8. GC–MS chromatograms were processed using Xcalibur (version 2.0; Thermo Scientific) or equivalent vendor's software. Each individual peak was integrated and then normalized. Overlapping peaks were separated using traces of single ions.
9. Peak assignment was based on mass fragmentation patterns matched to the National Institute of Standards and Technology (the United States) library and to previously reported literature.
10. Identification of metabolites from organic phase GC–MS analysis was supported by comparison with a FAME standard mix (Supelco 37-Component FAME Mix; Sigma-Aldrich) and retention time matching.
11. Typical results are shown in Fig. 12.1.

4. LC–MS OF INTACT LIPIDS

4.1. General lipid profiling of intact lipids by LC–MS

This method allows the detection of individual lipids from a range of classes including free fatty acids (negative mode), phospholipids (positive and negative), and glycerolipids (positive).

4.1.1 Required material

The following material was obtained from Sigma-Aldrich (Gillingham, Dorset, United Kingdom):

HPLC-grade (or better) isopropyl alcohol (IPA)

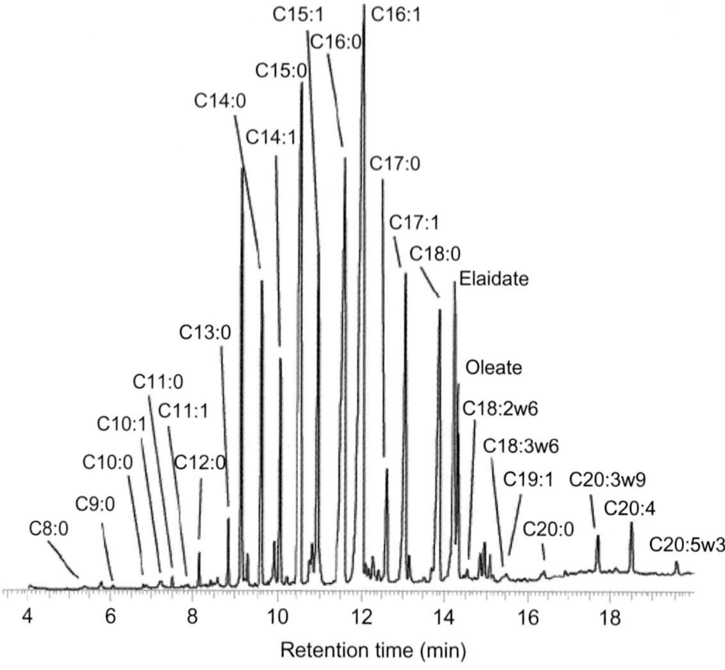

Figure 12.1 Typical GC–MS chromatogram of total fatty acids extracted from 3T3-L1 adipocytes and analyzed according to the methods earlier. The major fatty acid species detected are labeled.

HPLC-grade water
HPLC-grade (or better) methanol
HPLC-grade (or better) acetonitrile
Ammonium formate

The following was obtained from Avanti Lipids (Alabaster, AL, United States):

Lysophosphocholine C17:0
Phosphatidylcholine C34:0 (C17:0, C17:0)

The following was obtained from Waters Ltd. (Milford, MA, United States):

Acquity ultra performance liquid chromatogram (UPLC)
Quadrupole time-of-flight (QToF) Xevo mass spectrometer (for intact lipids)
1.7 μm bridged ethyl hybrid C8 column (2.1 × 100 mm)
1.7 μm bridged ethyl hybrid amide hydrophobic interaction liquid chromatography (HILIC) column (100 × 2.1 mm)
Leucine enkephalin

A note on equipment for LC–MS: We have used an ultra performance LC unit for our analyses. While regular HPLC units can be used, the UPLC provides superior separation and also concentrates a given lipid species to a narrower chromatographic peak, hence making the analysis more sensitive.

We have used a high-resolution mass spectrometer as this allows the identification of lipid species from the exact mass. Typically, mass accuracies of less than 5 parts per million (ppm) can be obtained on the newer QToF instruments (equivalent to 3 decimal places in terms of the mass accuracy). This allows the ready identification of lipids in terms of their class, number of carbon atoms present, and number of double bonds. The QToF also allows the detection of a relatively large number of points across each chromatographic peak, allowing better discrimination of individual lipid species. In addition, Orbitrap LC–MS and Fourier transform MS have been used widely in the literature.

4.1.2 General lipid profiling of intact lipids by LC–MS

1. IPA/methanol/water (1 ml, of volume ratio 2:1:1) was used to reconstitute the organic fraction (1/4 of the original lipid extract).
2. A lysophosphocholine C17:0 internal standard (in IPA/methanol/water ratio 2:1:1) was spiked into each sample to give a final concentration of 20 μ*M*.
3. Analysis of intact lipids was performed using a Waters QToF Xevo (Waters Corporation, Manchester, United Kingdom) in combination with an Acquity UPLC.
4. 2 μl of each extract was injected onto a 1.7 μm bridged ethyl hybrid C8 column (2.1 × 100 mm; Waters Corporation, Manchester, United Kingdom) held at 55 °C.
5. The binary solvent system (flow rate 0.400 ml/min) employed a gradient of solvent A (HPLC-grade acetonitrile/water 60:40, 10 m*M* ammonium formate) and solvent B (LC–MS-grade acetonitrile/isopropanol 10:90, 10 m*M* ammonium formate). The gradient started at 30% B, reached 99% B over 18 min, and then returned to the starting conditions for the next 2 min.
6. The data were collected in positive and negative modes with a mass range of 100–1200 m/z, a scan duration of 0.2 s, and an interscan delay of 0.014 s. The source temperature was set at 120 °C and nitrogen was employed as the desolvation gas (600 l/h) at 280 °C. Sampling cone and capillary voltages were 30 V and 3 kV. The collision energy was 4.6 V. A lock-mass solution of 2 ng/μl (50:50 acetonitrile/water)

leucine enkephalin (*m/z* 556.2771) was infused into the instrument at 3 μl/min.

7. LC–MS spectra and chromatograms were analyzed using the MarkerLynx Application MassLynx (version 4.1; Waters corporation). Each peak was detected based on a mass window of 0.05 Da and retention time window of 6 s. The peaks were normalized to the internal standard and deisotoped. Other vendor software can be used as well as open-source software such as XCMS (http://metlin.scripps.edu/xcms/) and MZmine 2 (http://mzmine.sourceforge.net/).
8. Typical results are shown in Fig. 12.2.

The reverse-phase chromatography is just one option available to researchers to provide separation of lipid species prior to analysis by MS. An alternative approach we have employed in the laboratory is to make use of HILIC. While this approach does not separate out triacylglycerides as well as the reverse-phase chromatography detailed earlier, it does separate out better lipid classes. Some have advocated the use of multiple chromatographic methods to maximize the detection of intact lipids. The chromatographic method is as follows:

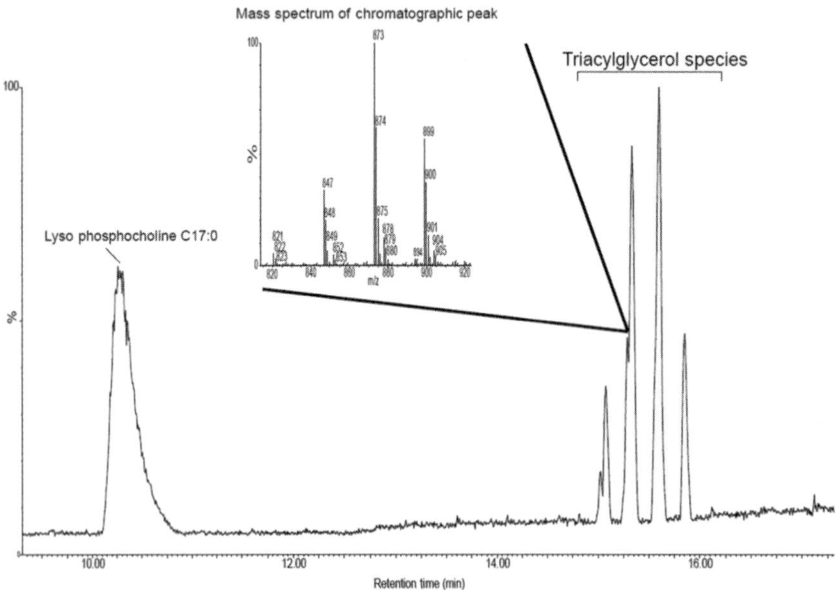

Figure 12.2 Typical LC–MS chromatogram of intact lipids extracted from white adipose tissue of rats and analyzed according to the methods earlier for reverse-phase chromatography (1.7 μm bridged ethyl hybrid C8 column). Triacylglycerol species are highlighted with a panel displaying the mass spectrum for a selected chromatographic peak.

1. The entire lipid extract was dissolved in 500 μl 1:1 methanol/chloroform containing 100 μ*M* (PC 34:0) phosphatidylcholine (Avanti Lipids Inc., Alabaster, AL, Unites States).
2. This solution was then further diluted 1 in 5 in acetonitrile and transferred into a 300 μl vial for analysis.
3. A BEH amide HILIC column (100 × 2.1 mm, 1.7 μm) was used with the following gradient: 5% 10 m*M* ammonium acetate adjusted to pH 9.0 using ammonia in acetonitrile was held for 2 min and increased to 20% over 7 min with further reequilibration for 3 min.
4. All MS parameters were as described earlier.
5. Typical results are shown in Fig. 12.3.

4.2. Profiling of acylcarnitines

4.2.1 Required material

The following material was purchased from Sigma-Aldrich (Gillingham, Dorset, United Kingdom):

3 *M* HCl in butanol

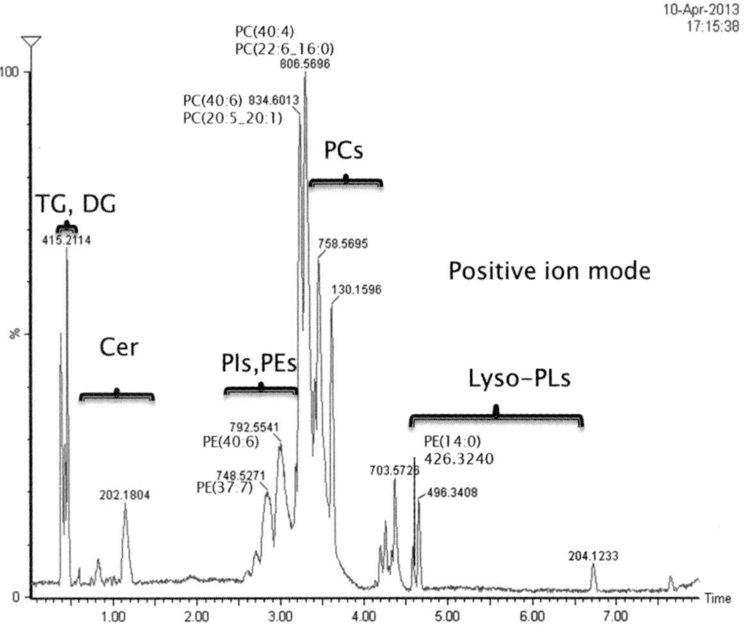

Figure 12.3 Total ion chromatogram of the intact lipids detected in cardiac tissue using amide HILIC chromatography. Notice the different order of lipid classes separated compared with the reverse-phase method in Fig. 12.2. (See color plate.)

The following material was purchased from Cambridge Isotope Laboratories (Andover, MA, Unites States):

Mixed standard of eight deuterated carnitines

The following material was purchased from Phenomenex (Warrington, United Kingdom):

Phenomenex Synergi Polar-RP column

The following material was purchased from Waters (Micromass) Ltd. (Atlas Park, Manchester, United Kingdom):

A Quattro Premier XE quadrupole mass spectrometer

The following material was purchased from AB Sciex Ltd. (Warrington, United Kingdom):

A QTRAP 5500 triple quadrupole mass spectrometer

The results described here were obtained on a Waters Quattro Premier triple quadrupole or an AB Sciex QTRAP 5500. However, it should be noted that with only minor modifications, the method could be performed on any triple quadrupole mass spectrometer. Indeed, these methods should also work on any instrument capable of tandem mass spectrometry (MS/MS), although limits of detection and reproducibility of quantification will vary between MS types (i.e., QToF and ion trap instruments will be in general less sensitive compared with comparable triple quadrupoles in this targeted assay).

In addition, we used a Waters Acquity UPLC for the chromatography. This instrument is capable of ultrahigh-pressure liquid chromatography. This allows a relatively short chromatographic run time and results in relatively narrow chromatographic peaks, the latter increasingly sensitivity by concentrating species in these chromatographic peaks. However, regular HPLC should still be useable and in our experience sensitivity is not usually limiting with this assay.

It is also possible to not use chromatography prior to analysis and run a direct infusion method. This has the advantage of providing a very fast assay and is often used for screening of inborn errors of metabolism. However, we have found such methods to be affected by a process referred to as ion suppression. In this phenomenon, polar species (e.g., phospholipids) use up a proportion of the energy available for ionization at the source of the mass spectrometer and reduce the efficiency of ionization of the analytes (in this case, the butylated acylcarnitines). For this reason, we recommend some chromatography to separate the individual species prior to MS.

4.2.2 Optimization of the method for the particular mass spectrometer

This needs to be only performed once when the assay is set up.

1. Compounds were optimized for tandem MS analysis by preparing individual standard solutions at 10 μ*M* for the described Quattro Premier and 100 n*M* for the QTRAP 5500. While we cannot provide concentrations of individual machines, this should provide a good starting point for most mass spectrometers.
2. Standards were dried down under a stream of nitrogen gas and then derivatized by heating with 3 *M* HCl in butanol for 15 min at 65 °C.
3. The standards were dried again under a stream of nitrogen gas and then reconstituted in 0.1% formic acid to the earlier concentrations.
4. Samples were directly infused into the mass spectrometer using a syringe pump.
5. Optimum MS parameters and mass transitions were obtained by using the automatic optimization protocols of Analyst® (Version 1.6, AB Sciex) and MassLynx™ (Version 1.4, Waters). Alternatively, other vendor's software will be capable of this analysis.
6. For situations where no standards were available, mass transitions and MS parameters were inferred from the parameters of known analogs. See Table 12.1 for typical values obtained using the Waters Quattro Premier LC–MS system.

4.2.3 Analysis of acylcarnitines in tissue extracts

1. 200 μl of the mixed standard of eight deuterated carnitines was diluted into 25 ml of acetonitrile to form the standard solution used throughout the analysis.
2. 200 μl of the standard solution was added to one-half of the organic and aqueous fractions obtained from the tissue extraction described earlier. Thus, in each sample, the carnitine complement should be the equivalent of that in 20–50 mg of wet weight adipose tissue. The organic fraction will contain acylcarnitines with more than two carbons on the acyl chain. The aqueous fraction contains a large proportion of the free carnitine and acetylcarnitine from the tissue.
3. This solution was dried down under nitrogen and derivatized with 3 *M* HCl in butanol for 15 min at 65 °C.
4. The extract was reconstituted in 4:1 acetonitrile/0.1% formic acid in water followed by sonication to dissolve all species present.

Table 12.1 Compound-specific mass spectrometry parameters for the analysis of acylcarnitines

Compound	Ion mode	Parent mass (*m/z*)	Daughter mass (*m/z*)	Declustering potential (V)	Collision energy (eV)
C10 carnitine butyl ester	Positive	372.3	85.0	35	25
C10:1 carnitine butyl ester	Positive	370.3	85.0	35	25
C10:2 carnitine butyl ester	Positive	368.3	85.0	35	25
C12 carnitine butyl ester	Positive	400.3	85.0	35	25
C12:1 carnitine butyl ester	Positive	398.3	85.0	35	25
C14 carnitine butyl ester	Positive	428.4	85.0	35	25
C14:1 carnitine butyl ester	Positive	426.4	85.0	35	25
C14:2 carnitine butyl ester	Positive	424.3	85.0	35	25
C14-OH carnitine butyl ester	Positive	444.4	85.0	35	25
C16 carnitine butyl ester	Positive	456.4	85.0	35	25
C16:1 carnitine butyl ester	Positive	454.4	85.0	35	25
C16:1-OH carnitine butyl ester	Positive	470.4	85.0	35	25
C16:2 carnitine butyl ester	Positive	452.4	85.0	35	25
C16-OH carnitine butyl ester	Positive	472.4	85.0	35	25
C18 carnitine butyl ester	Positive	484.4	85.0	35	25
C18:1 carnitine butyl ester	Positive	482.4	85.0	35	25

Table 12.1 Compound-specific mass spectrometry parameters for the analysis of acylcarnitines—cont'd

Compound	Ion mode	Parent mass (*m/z*)	Daughter mass (*m/z*)	Declustering potential (V)	Collision energy (eV)
C18:1-OH carnitine butyl ester	Positive	498.4	85.0	35	25
C18:2 carnitine butyl ester	Positive	480.4	85.0	35	25
C18:2-OH carnitine butyl ester	Positive	496.4	85.0	35	25
C18-OH carnitine butyl ester	Positive	500.4	85.0	35	25
C2 carnitine butyl ester	Positive	260.2	85.0	35	25
C20 carnitine butyl ester	Positive	512.4	85.0	35	25
C20:1 carnitine butyl ester	Positive	510.4	85.0	35	25
C20:2 carnitine butyl ester	Positive	508.4	85.0	35	25
C3 carnitine butyl ester	Positive	274.2	85.0	35	25
C4 carnitine butyl ester	Positive	288.2	85.0	35	25
C4 dicarboxyl carnitine dibutyl ester	Positive	374.3	85.0	35	25
C5 carnitine butyl ester	Positive	302.3	85.0	35	25
C5 dicarboxyl carnitine dibutyl ester	Positive	388.3	85.0	35	25
C5:1 carnitine butyl ester	Positive	300.2	85.0	35	25
C5-OH carnitine butyl ester	Positive	318.2	85.0	35	25

Continued

Table 12.1 Compound-specific mass spectrometry parameters for the analysis of acylcarnitines—cont'd

Compound	Ion mode	Parent mass (*m/z*)	Daughter mass (*m/z*)	Declustering potential (V)	Collision energy (eV)
C6 carnitine butyl ester	Positive	316.3	85.0	35	25
C6 dicarboxyl carnitine dibutyl ester	Positive	402.3	85.0	35	25
C8 carnitine butyl ester	Positive	344.3	85.0	35	25
C8 dicarboxyl carnitine dibutyl ester	Positive	430.4	85.0	35	25
C8:1 carnitine butyl ester	Positive	342.3	85.0	35	25
C8-OH carnitine butyl ester	Positive	361.3	85.0	35	25
d3 C16 carnitine butyl ester	Positive	459.4	85.0	35	25
d3 C2 carnitine butyl ester	Positive	263.2	85.0	35	25
d3 C3 carnitine butyl ester	Positive	277.2	85.0	35	25
d3 C4 carnitine butyl ester	Positive	291.2	85.0	35	25
d3 C8 carnitine butyl ester	Positive	347.3	85.0	35	25
d9 C14 carnitine butyl ester	Positive	437.4	85.0	35	25
d9 C5 carnitine butyl ester	Positive	311.3	85.0	35	25
d9 carnitine butyl ester	Positive	227.2	85.0	35	25

The table shows ionization mode, mass transitions (parent and daughter masses), and declustering potentials and the collision energies required for each analyte.

5. Samples were analyzed using a Waters Quattro Premier XE or an AB Sciex QTRAP 5500 both coupled to Acquity UPLC systems (see note previous for more information on LC–MS systems).
6. The strong mobile phase used for analysis was acetonitrile with 0.1% formic acid (B) and the weak mobile phase was 0.1% formic acid in water (A). The analytical UPLC gradient used a Synergi Polar-RP phenyl ether column (100×2.1 mm, 2.5 μm) from Phenomenex with 30% B in 0.1% formic at 0 min followed by a linear gradient to 100% B for 3 min and held at 100% B for the next 5 min with a further 2 min reequilibration. The total run time was 10 min and the flow rate was 0.5 ml/min with an injection volume of 2 μL.
7. For the Quattro Premier, the MS parameters were source temperature 150 °C, desolvation temperature 350 °C, capillary voltage 3.5 kV, and 500 l/h of desolvation gas; all other parameters were compound-specific and are detailed in Table 12.1.
8. Data were processed using QuanLynx within Masslynx (version 1.4; Waters Corp.) and Quantitation Wizard within Analyst (version 1.6; AB Sciex Ltd.).
9. Typical results are shown in Fig. 12.4.

4.3. Profiling of oxylipins and eicosanoids

4.3.1 Required materials

The following materials were purchased from Waters Ltd. (Atlas Park, Manchester, United Kingdom):

Waters Oasis HLB cartridges
Acquity UPLC

The following materials were purchased from Sigma-Aldrich (Gillingham, Dorset, United Kingdom):

Ethyl acetate
HPLC-grade methanol
Analytic-grade acetic acid
Glycerol

The following material was purchased from Macherey-Nagel Inc. (Bethlehem, PA, United States):

Chromabond vacuum manifold for 24 samples

The following were purchased from Cayman Chemical (Ann Arbor, MI, United States):

12*S*-hydroxyeicosatetraenoic-5,6,8,9,11,12,14,15-d8 acid (12(S)-HETE-d8)
Prostaglandin E_2-d4 (PGE_2-d4)

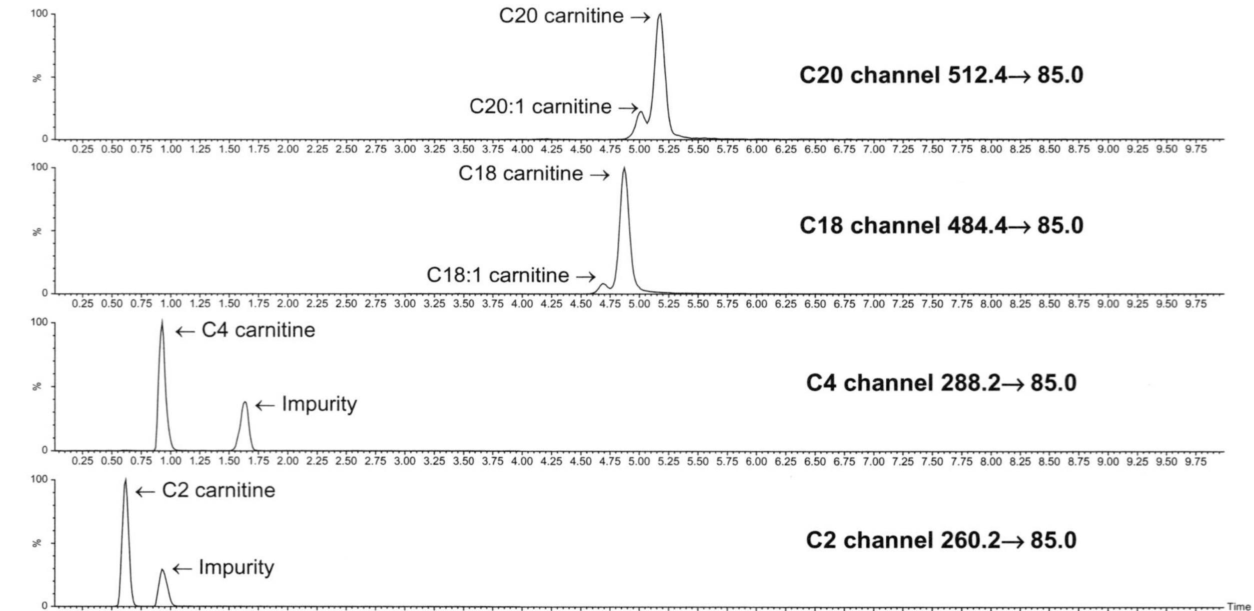

Figure 12.4 Four extracted ion chromatograms of a tissue extract measured using a Phenomenex Synergi Polar-RP column with the acylcarnitine gradient described in the methods section earlier. This figure demonstrates the need for specificity when conducting acylcarnitine analysis.

The following material was purchased from Varian Inc. (Palo Alto, CA, United States):

Pursuit Plus C18 2.0 × 150 mm, 5 μ*M* column

The following material was purchased from AB Sciex Ltd. (Warrington, United Kingdom):

A QTRAP 4000 triple quadrupole mass spectrometer

4.3.2 Solid-phase extraction of eicosanoids from adipose tissue

1. Adipose tissue (100 mg) was homogenized using a TissueLyser (Qiagen Ltd., Manchester, United Kingdom; 10 min at 30 Hz) in 1.5 ml 15% methanol with 0.1% acetic acid.
2. The samples were centrifuged (17000 × *g*, 2 min).
3. Waters Oasis HLB cartridges were loaded into the vacuum manifold and washed with ethyl acetate (2 mL), methanol (2 × 2 ml), and 95:5 v/v water/methanol with 0.1% acetic acid (2 ml). 10 μl of 300 n*M* 12(S)-HETE-d8 internal standard was spiked into the tissue extract. The samples were then loaded onto the cartridges. Cartridges were washed with 1 ml 95:5 v/v water/methanol with 0.1% acetic acid. Solid-phase extraction (SPE) cartridges were dried for 20 min using a vacuum manifold. SPE cartridges were then eluted with 0.5 ml of methanol followed by 1 ml of ethyl acetate into 2 ml tubes containing 6 μl of 30% glycerol in methanol as a trap solution.
4. Volatile solvents were removed using a SpeedVac until only the trap solution glycerol remained. Residues were reconstituted in 40 μl methanol containing 70 n*M* PGE_2-d4 internal standard, mixed, and transferred to autosampler vials with low-volume inserts.

4.3.3 LC–MS/MS analysis of eicosanoids

1. Analysis was performed using an Acquity UPLC pump (Waters Ltd.; Milford, MA, United States). The autosampler was maintained at 4°C.
2. LC separation was performed on a Pursuit Plus C18 2.0 × 150 mm, 5 μ*M* column (Varian Inc. Palo Alto, CA). Mobile phase A was water with 0.1% acetic acid. Mobile phase B was acetonitrile/methanol (84:16) with 0.1% acetic acid.

Table 12.2 The chromatography elution gradient for the LC–MS oxylipin and eicosanoid method

Time (min)	Flow rate (μl/min)	B%
0.00	400	15
0.75	400	15
1.50	400	30
3.50	400	47
5.00	400	54
6.00	400	55
10.50	400	60
15.00	400	70
16.00	400	80
17.00	400	100
19.00	400	100
19.30	400	15
21.00	400	15

The table shows the incremental time in the 21 min run alongside the flow rate and the percentage of liquid phase solvent B (acetonitrile/methanol 84:16 with 0.1% acetic acid).

3. 10 μl of sample was injected onto the column. Gradient elution was performed using a flow rate of 400 μl/min over a period of 21 min. The chromatography gradient is outlined in Table 12.2.
4. MS was performed using a 4000 QTRAP triple quadrupole mass spectrometer (AB Sciex) coupled to an electrospray source (Turbo V). The instrument was operated in negative MRM mode. The curtain gas was set at 20 psi. Source temperature was 550°C, source gas 1 was 50 psi, and source gas 2 was 30 psi. The CAD gas was high, the ion spray voltage was −4.5 kV, the declustering potential was −60, the entrance potential was −10, and the interface heater was on. The optimized mass transitions and mass spectrometric parameters are outlined in Table 12.3.
5. Analyst software (Version 1.6; AB Sciex) was used for peak integration and peaks were manually reviewed for quality of integration. Internal standard peak areas were monitored for quality control.

Table 12.3 The optimized mass transitions and mass spectrometric parameters for lipid metabolites in the LC–MS eicosanoid screening method

Compound	Q1 (*m/z*)	Q3 (*m/z*)	Collision energy
6-keto-PGF1α	369.3	163.2	36
TXB2	369.2	169.1	25
9,12-13-TriHOME	329.2	211.1	28
9,10-13-TriHOME	329.2	171.1	32
PGF2α	353.2	309.3	28
PGE2-d4	355.3	275.3	27
PGE2	351.2	271.3	28
PGD2	351.2	271.3	26
11,12,15 THET	353.2	167.1	32
Lipoxin A4	351.2	115.2	20
PGB2/PGJ2	333.2	235.3	28
THF diols	353.2	167.1	32
LTB4	335.2	195.1	23
12,13-DHOME	313.2	183.2	32
9, 10-DHOME	313.2	201.2	30
14,15-DHET	337.2	207.1	24
11,12-DHET	337.2	167.1	28
8,9-DHET	337.2	127.1	30
15-deoxy PGJ2	315.2	271.3	20
19-HETE	319.2	275.1	24
20-HETE	319.2	275.2	23
5,6-DHET	337.2	145.1	26
13-HODE	295.2	195.0	25
9-HODE	295.2	171.0	25
15-HETE	319.2	301.4	18
13-oxo ODE	293.2	113.0	29
11-HETE	319.2	167.2	23

Continued

Table 12.3 The optimized mass transitions and mass spectrometric parameters for lipid metabolites in the LC–MS eicosanoid screening method—cont'd

Compound	Q1 (*m/z*)	Q3 (*m/z*)	Collision energy
15-oxo-EET	317.2	113.1	24
9-oxo ODE	293.2	185.1	28
12-HETE	319.2	179.2	20
8-HETE	319.2	301.2	17
9-HETE	319.2	123.1	20
5-HETE	319.2	115.1	21
12(13)-EpOME	295.2	195.2	20
14(15)-EET	319.2	219.3	18
9(10)-EpOME	295.2	171.1	22
11(12)-EET	319.2	167.0	20
5-oxo-EET	317.2	273.2	20
8(9)-EET	319.2	123.0	20
5(6)-EET	319.2	191.0	20
Arachidonic acid	303.3	259.1	20
LTD4	495.3	176.9	23
8-isoPGF2α	353.1	193.2	28
Docosahexaenoic acid	327.2	283.1	20
12-OxoETE	317.2	272.9	20
F2 isoprostanes	353	309	40
15(S)-HpETE	335.3	139	40
12(S)-HpETE	335.16	59.16	40
5(S)-HpETE	335	59.1	40
12(S)-HETE-d8	327.21	184	22

The table shows mass transitions (parent (Q1) and daughter masses (Q3)) and the collision energies required for each analyte.

4.3.4 Optimization of the method parameters

The following materials were purchased from Cayman Chemical (Ann Arbor, MI, United States):

Eicosanoid and oxylipin standards

The following materials were purchased from Santa Cruz Biotechnology (Dallas, TX, United States):

Eicosanoid and oxylipin standards

These steps should be performed when developing the method:

1. Compounds were optimized for tandem MS analysis by preparing individual standard solutions at 100 n*M* for the QTRAP 4000. While we cannot provide concentrations of individual machines, this should provide a good starting point for most mass spectrometers.
2. Samples were directly infused into the mass spectrometer using a syringe pump.
3. Optimum MS parameters and mass transitions were obtained by using the automatic optimization protocols of Analyst® (Version 1.6, AB Sciex).

5. CONCLUSIONS

We have outlined four methods for analyzing the lipidome of adipose tissue—one using GC–MS and three using LC–MS. These methods provide two general profiling methods and two targeted analyses for specific lipid classes. They should serve as a useful starting point for lipidomics before more specific assays are developed.

REFERENCES

Atherton, H. J., Bailey, N. J., Zhang, W., Taylor, J., Major, H., Shockcor, J., et al. (2006). A combined 1H-NMR spectroscopy- and mass spectrometry-based metabolomic study of the PPAR-alpha null mutant mouse defines profound systemic changes in metabolism linked to the metabolic syndrome. *Physiological Genomics*, *27*(2), 178–186.

Atherton, H. J., Gulston, M. K., Bailey, N. J., Cheng, K. K., Zhang, W., Clarke, K., et al. (2009). Metabolomics of the interaction between PPAR-alpha and age in the PPAR-alpha-null mouse. *Molecular Systems Biology*, *5*, 259.

Medina-Gomez, G., Gray, S. L., Yetukuri, L., Shimomura, K., Virtue, S., Campbell, M., et al. (2007). PPAR gamma 2 prevents lipotoxicity by controlling adipose tissue expandability and peripheral lipid metabolism. *PLoS Genetics*, *3*(4), e64.

Roberts, L. D., Murray, A. J., Menassa, D., Ashmore, T., Nicholls, A. W., & Griffin, J. L. (2011). The contrasting roles of PPARδ and PPARγ in regulating the metabolic switch between oxidation and storage of fats in white adipose tissue. *Genome Biology*, *12*(8), R75.

Virtue, S., Feldmann, H., Christian, M., Tan, C. Y., Masoodi, M., Dale, M., et al. (2012). A new role for lipocalin prostaglandin d synthase in the regulation of brown adipose tissue substrate utilization. *Diabetes*, *61*(12), 3139–3147.

Virtue, S., & Vidal-Puig, A. (2010). Adipose tissue expandability, lipotoxicity and the metabolic syndrome—An allostatic perspective. *Biochimica et Biophysica Acta*, *1801*(3), 338–349.

CHAPTER THIRTEEN

Measuring Respiratory Activity of Adipocytes and Adipose Tissues in Real Time

Anne Bugge, Lea Dib, Sheila Collins[1]

Metabolic Signaling and Disease Program, Diabetes and Obesity Research Center, Sanford-Burnham Medical Research Institute, Orlando, Florida, USA

[1]Corresponding author: e-mail address: scollins@sanfordburnham.org

Contents

Abstract

The realization that obesity and its associated diseases have become one of modern society's major challenges to the health of the world's population has fueled much effort to understand white adipocyte biology and elucidate pathways to increase energy expenditure. One strategy has been to increase the oxidative capacity and activity of the adipocytes themselves. This has the advantage that free fatty acids (FAs) would not be released into the circulation in copious amounts, which can have detrimental effects. This is particularly true for obese individuals, who often already display severe dyslipidemia, putting them at increased risk for cardiovascular diseases. It was recently discovered that adult humans, in addition to infants, possess active brown adipocytes, characterized by expression of the mitochondrial electron gradient dissipater uncoupling protein 1 (UCP1). This has generated renewed interest in finding ways to

Methods in Enzymology, Volume 538
ISSN 0076-6879
http://dx.doi.org/10.1016/B978-0-12-800280-3.00013-X

"convert" or "adapt" white adipocytes into a more brown adipocyte-like state by increasing mitochondrial content and expression of UCP1 and activating UCP1 via lipolysis-mediated free FAs. Another approach to consider is elevating the activity of the not insignificant amount of mitochondria found in white adipocytes. The invention of the XF Flux Analyzer by Seahorse Bioscience has revolutionized this line of research as it allows for real-time measurements of respiration in multiple samples simultaneously. In this chapter, we describe our approaches and experience with employing this technology to study the metabolism of mouse and human primary and immortalized cells and mouse white adipose tissue.

1. INTRODUCTION

The evolution of adipose tissue permitted organisms to store almost unlimited amounts of calories at high density in the form of triglycerides. This unique feature of adipose tissue allowed our ancestors to stock up on nutrients in times of plenty, thus increasing their chance of surviving at times of famine. However, the constant accessibility to high-caloric food in modern society has turned adipose tissue into a foe, as obesity now occurs at epidemic rates and is associated with several serious diseases such as type II diabetes, cardiovascular disorders, and some cancers (Zafon & Simo, 2011). The revived appreciation for the existence of brown adipose tissue (BAT) in adult humans (Cypess et al., 2009; Nedergaard, Bengtsson, & Cannon, 2007; Saito et al., 2009; van Marken Lichtenbelt et al., 2009; Virtanen et al., 2009; Zingaretti et al., 2009) has sparked renewed interest in the idea that mitochondrial uncoupling could lead to increased fatty acid (FA) oxidation and thus weight reduction. It is not yet clear whether there are sufficient numbers of "brown adipocytes" to have a significant impact on body weight and energy expenditure, but some human data show correlations between amounts of detectable brown fat and body weight/insulin sensitivity and energy expenditure (Orava et al., 2013; Ouellet et al., 2012). Another study in human subjects concluded that the contribution of BAT to daily energy expenditure was very low, ~20 kcal/day (Muzik et al., 2013). What will be interesting to see is whether weight loss over time via brown adipocyte activity will be significant, just as a low daily calorie intake beyond energy requirements eventually results in significant weight gain over time. Nevertheless, most of the adipose tissue in adult humans consists of white adipocytes, and these cells are now appreciated to (i) have a greater complement of mitochondria than previously thought (Wilson-Fritch et al., 2004) and (ii) be able to oxidize FAs *in situ*

(Cho et al., 2009; Maassen, Romijn, & Heine, 2008; Stenson et al., 2009; Yehuda-Shnaidman, Buehrer, Pi, Kumar, & Collins, 2010). In addition, it was observed in earlier studies that rodent white adipocytes can exhibit mitochondrial uncoupling after catecholamine stimulation (Davis & Martin, 1982; Hepp, Challoner, & Williams, 1968), and this has further increased interest in the potential for energy expenditure in white adipocytes. Here, we discuss experimental approaches that we have used in cultured human and mouse white adipocytes, as well as freshly isolated adipose tissues, to measure the oxygen consumption rate (OCR) as an indicator of mitochondrial respiration and extracellular acidification rate (ECAR) as an indicator of glycolysis. These studies evolved out of efforts to determine whether catecholaminergic stimulation could increase "brown adipocyte" markers and metabolic activity in human subcutaneous or visceral adipocytes and included measuring changes in the expression of "brown fat thermogenic genes" such as uncoupling protein 1 and PGC-1α, as well as oxygen consumption and metabolism. As one of the early users of the Seahorse Bioscience (North Billerica, MA, United States) technology (http://www.seahorsebio.com), here we explain how we developed these methods and interpret the results.

2. BASIS OF THE METHODS AND THE XF24

For our studies, we used the XF24 Flux Analyzer from Seahorse Bioscience. Depending on the sample and the experimenter's objectives, there are various microplate sizes to choose from. In our studies, we used XF24 V7 (#100777-004) and XF Islet Capture Microplates (#101122-100). In the latter, there is an additional depression of 3.17 mm diameter width and 0.25 mm height in the center of each well, in which tissue can be contained. Seahorse also offers a third plate-type XF24 V28 (#100882-004), which has increased dynamic OCR range. This plate type allows studies of cells with high respiratory activity by decreasing the risk of them reaching a hypoxic state, but the trade-off is that the sensitivity of ECAR measurements is reduced. Because of the rapid technical evolution in this area based on experimental needs, interested investigators should consult the Seahorse Bioscience website (http://www.seahorsebio.com) and contact application specialists to accommodate your needs. The Seahorse technology is unique in allowing measurement of OCR and ECAR in multiple samples simultaneously. In the XF24-well microplates, each well has a wider opening at the top and sloping walls that result in a 96-well-size bottom surface area of ~0.32 cm^2. Each measurement takes approximately 2–5 min and is made in a very small volume (7 μl) of media above the cells or tissue explants. The small measurement volume is created when the FluxPak that houses the O_2- and pH-sensitive fluorophores is lowered into the wells. In the 2–5 min measurement time, each well is measured every 22 s for its oxygen and pH levels. OCR and ECAR are calculated from the rate of change in oxygen and pH, respectively, within the microenvironment in the well. Once completed, the small volume of media subject to the measurement is reequilibrated with the larger volume of media present in the well, and another 2–5 min measurement is performed. This allows for repeated OCR/ECAR measurements of the same cell population over time, of which an example is shown in Fig. 13.1. A temperature control system maintains the XF24 internal environment at 37 °C, and drugs or nutrients can be added (up to four agents/well) at user-defined intervals. Therefore, a sensitive measurement of the small medium volume around the cells/tissue can be used to determine rates of cellular metabolism with precision and in a noninvasive, label-free manner (Wu et al., 2007). Data analysis is integrated into Excel or within the Seahorse software, and there is no need to clean the instrument thanks to the disposable cartridges.

3. CELL CULTURE AND TISSUE ISOLATION

3.1. Human primary adipocytes

The method here is based on studies described by Yehuda-Shnaidman et al. (2010). Culture human white preadipocytes (we use cells from Zen-Bio Inc.; Halvorsen et al., 2001) in a 10 cm diameter plate. When confluent, trypsinize the cells into 3 ml of media (provided by Zen-Bio Inc.). Count the cells and seed 13,000 cells in 100 μl media per well of an XF24 V7 plate. Four wells/plate are left empty to correct for positional temperature variations. In our earliest experiments, we covered the bottom growing surface of 24-well XF24 V7 cell culture plates with sterile 0.2% gelatin solution to increase adherence of the differentiated adipocytes (Yehuda-Shnaidman et al., 2010). However, more recently, we have not found this step to be necessary. Users should experiment with and without the gelatin to determine which approach they prefer. The gelatin solution is made in a sterile biosafety cabinet. Preheat gelatin 2% solution (G1393, Sigma-Aldrich) at 37 °C for 30 min, and dilute it to 0.2% in sterile PBS. Add 50 μl of the diluted gelatin to each well of the XF24 plate and leave for 30 min. Then, remove excess liquid and leave the plate open in the hood for 1 h. The plate is then ready for seeding the cells. When the cells are attached

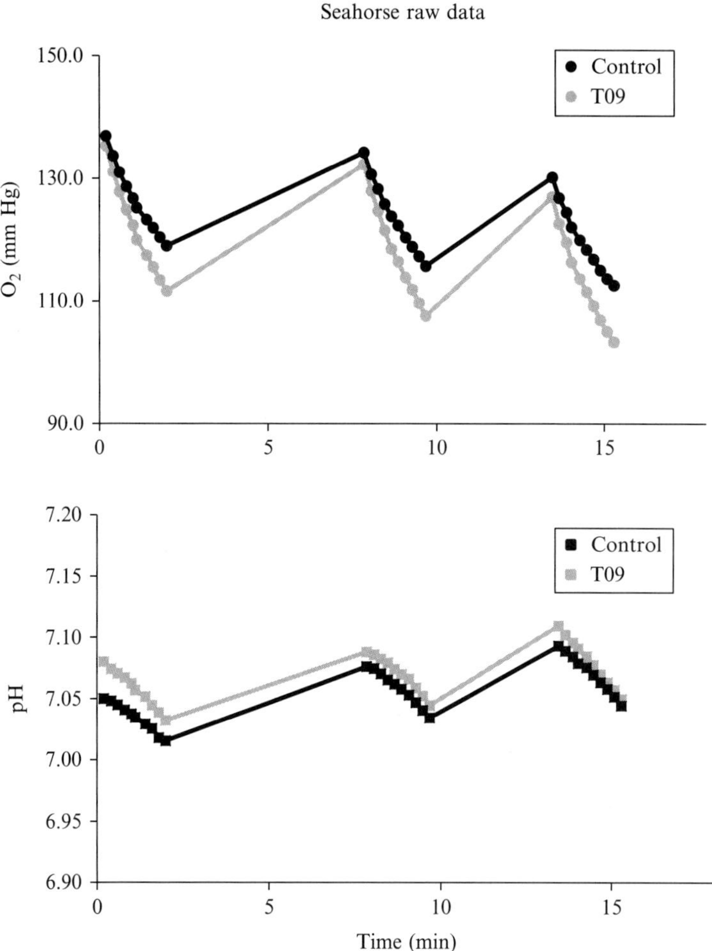

Figure 13.1 XF24 representative raw data of repeated oxygen consumption and pH measurements from which OCR or ECAR is calculated, respectively, in mouse primary inguinal adipocytes. The measurements shown here are from two different wells of which one was treated for 3 days with the synthetic LXR ligand T0901317 (1 μM) prior to being analyzed in the XF24. Note that by using an appropriate cell number and measurement time, the decrease in the oxygen level during each measurement stays well above the 10 mm Hg O_2 hypoxia level. If the oxygen level in the chamber drops to hypoxic levels during a measurement period, the cells could be damaged or it could result in a lower initial oxygen level when starting the next consecutive measurement.

to the plate (about an hour after the seeding), add an additional 150 μl media. The reason for this two-step seeding protocol is to get an even distribution of the cells in one monolayer. This is highly important for the XF24 measurement since the oxygen and pH sensors are located over the center of each well. The surface tension of the media against the tiny tissue culture well will result in the cells being more likely to settle down along the rim instead of the center if plated in a larger volume. After the cells reach confluence, expose them to differentiation medium (Zen-Bio Inc.) containing rosiglitazone, 3-isobutyl-1-methylxanthine (IBMX), dexamethasone, and insulin for 3 days. Cells are then fed with adipocyte medium (Zen-Bio Inc.), which is replaced every other day. Ten to 14 days after the induction of differentiation, cells are fully differentiated to adipocytes and ready to be analyzed in the XF24. Note: due to the shape of the wells in the XF24 plate, the media can rapidly evaporate during longer-term culture. Consider adding higher medium volumes (500–1000 μl).

3.2. Mouse primary adipocytes

The isolation method is based in general on earlier methods (Nechad et al., 1983; Rehnmark et al., 1989; Rodbell, 1964). Isolate and pool the adipose depot of interest from two to three mice that are preferably below the age of 6 weeks. Immediately proceed by mincing them to a very fine consistency followed by digestion for 45–70 min at 37 °C and shaking (115 rpm) in 3 ml Krebs-Ringer bicarbonate buffer (Sigma K4002) supplemented with 2% BSA and 1600U type I collagenase (Worthington). Vortex vigorously for a few seconds every 5–10 min in order to lyse the mature adipocytes. Add 9 ml Krebs-Ringer bicarbonate buffer with 2% BSA and centrifuge for 5 min at 350 × *g*. Then, remove the supernatant and gently resuspend the cell pellet in 4 ml high-glucose DMEM (GIBCO #11995) supplemented with 10% fetal bovine serum (FBS) and penicillin/streptomycin and pass the suspension through a 40 μm cell strainer into a 6 cm diameter cell culture dish. The following day, replenish with fresh media and do so every second day until the cells reach 90% confluence; this should take approximately 4–5 days. At this point, there should be approximately 400,000 cells, which are sufficient for distribution into 20 wells of an XF24 plate, leaving four wells/plate empty to correct for positional temperature variations. Depending on the quality of your primary cell prep, you can attempt to expand the cells further, but we find that with each passage and extended time in culture, the percentage differentiation decreases.

For mouse primary adipocytes, we have obtained the best results by differentiating them directly in the Seahorse XF24 assay plates, but in our hands, we find that the cells do not proliferate well in the XF24 plates so they should be plated at the final density that allows for optimal differentiation and acceptable OCR/ECAR levels. Detach the cells using 200 μl of trypsin and resuspend in 1 ml media. Count the cells and plate 20,000 cells/well in 100 μl media. When the cells are attached to the plate, add an additional 150 μl media. The next day, differentiation is induced by 48 h exposure to DMEM supplemented with 10% FBS, penicillin/streptomycin, 0.5 μg/ml insulin, 0.5 μ*M* dexamethasone, 250 μ*M* IBMX, 60 μ*M* indomethacin, and 2 μ*M* rosiglitazone, followed by 48 h in media supplemented only with 0.5 μg/ml insulin. The differentiation process is now complete but in general, the cells are only healthy for a few days following completion of differentiation.

3.3. Mouse adipocyte cell lines (3T3-L1, 3T3-F442A)

Grow 3T3-L1/3T3-F442A cells in 10 cm diameter plates in high-glucose DMEM with penicillin/streptomycin and 10% FBS. When 90% confluent, trypsinize the cells and count. Seed around 15,000–20,000 cells/well in 100 μl media directly into the Seahorse XF24 assay plate. Again, four wells/plate are left empty to correct for positional temperature variations. When the cells are attached to the plate, add an additional 150 μl medium. Usually, they are ready for differentiation 2 days later when they will be confluent. Remove the media and replace with the adipocyte differentiation media—DMEM supplemented with 10% FBS, 0.5 m*M* IBMX, 1 μ*M* dexamethasone, 1 μg/ml insulin, and 2 μ*M* rosiglitazone. Three days later, the medium is replaced with DMEM supplemented with 10% FBS and 1 μg/ml insulin for an additional 2 days. At this time, the cells are ready for Seahorse analysis or they can be maintained with medium change every 2 days and/or subjected to further treatment according to the investigator's experimental protocol.

3.4. Fresh adipose tissue pieces

3.4.1 Isolate the tissue

Working quickly but carefully in a sterile environment, collect the mouse adipose tissue of interest and rinse it briefly in the unbuffered Seahorse assay DMEM (sodium carbonate and serum-free, #D5030, Sigma-Aldrich), supplemented in-house with 1 m*M* sodium pyruvate, 2 m*M* GlutaMAX-1™,

25 mM glucose, 1.85 g/l NaCl, and 15 mg/l phenol red (pH 7.4) (this medium is also sold by the company) heated to 37 °C in a water bath or in an incubator *without* CO_2. Clean the tissue of any blood vessels and nonadipose material and cut into small (~2–5 mg) pieces and immediately deposit them separately in 500 μl assay DMEM. Gently wash each tissue piece twice with 500 μl assay DMEM. Place one piece of tissue in the center of each well of an XF24 Islet Capture Microplate and cover with the customized screens that allow free perfusion while minimizing tissue movement during measurements. Seahorse has invented an islet capture screen insert tool (#101135-100) that significantly aids in completing this step. Upon positioning of the screen, immediately add at least 500 μl of the assay DMEM to the well.

3.4.2 Optimize your experiment

In a preliminary experiment, compare between different sizes (e.g., 2–5 mg) of tissue for their OCR levels to make sure that the measurements are in the dynamic range (Fig. 13.2). Determine the tissue size that gives acceptable variations between replicates, but expect that the variability will be higher for tissue pieces than for cell experiments. Analyze 5–10 replicate wells for each experimental condition and leave four wells empty to correct for positional temperature variations. Note: take into account that one piece of 5 mg tissue has lower surface area (and thus lower OCR level) than several pieces that sum up to 5 mg. For the same reason, it is recommended to choose

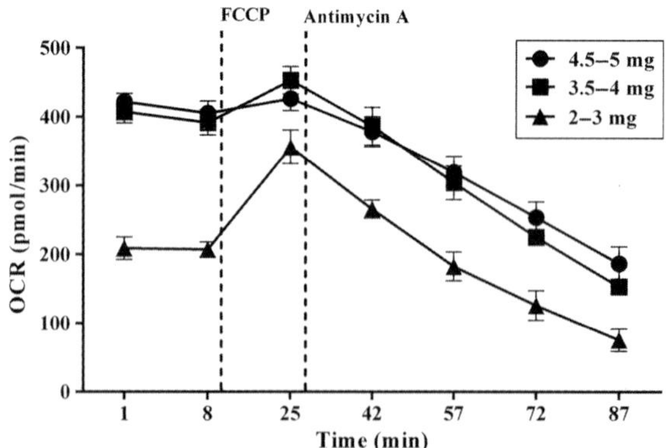

Figure 13.2 OCR of different sizes of inguinal adipose tissue pieces from C57BL/6J mice. FCCP and antimycin A were injected at the specified time points. Pieces weighing 2–3 mg display OCR levels within the dynamic range of the XF24 Flux Analyzer, where the full effects of *in vitro* or *in vivo* treatments can be evaluated with confidence. $n=3$–4, standard deviation is noted by vertical bars.

pieces of similar shape. The OCR of the tissue pieces should be analyzed on the XF24 immediately as they are only viable for a few hours after euthanasia of the mouse. The duration of viability will depend on the origin and size of the tissue pieces, so an important part of setting up a protocol for adipose tissue XF analysis is to determine the extent of time where you can safely study the OCR before it starts declining. If different mouse genotypes or treatments are to be compared, it is of paramount importance that the animals are euthanized and the tissue isolated and cut simultaneously and subsequently analyzed side by side on the same plate.

4. PREPARING THE XF24 ANALYZER SETUP

At least 12 h before the XF analysis, hydrate an XF24 FluxPak (#101174-100) by adding 1 ml of XF Calibrant Solution (#100840-000, Seahorse Bio) to each well while taking care not to damage or scratch the bottom of the electrodes. Store the FluxPak closed overnight at 37 °C, *without* CO_2.

4.1. Prepare the samples for analysis

4.1.1 For cells

Wash the cells with 1 ml unbuffered Seahorse assay DMEM heated to 37 °C. Then, add 675 μl of assay DMEM to each well. Fully differentiated adipocytes lyse and detach very easily from the plates so avoid removing all the media and aspirate it gently, perhaps using a pipette. Similarly, care should be taken when adding media. Incubate the cells at 37 °C, *without* CO_2, for 30–60 min to allow medium temperature and pH to reach equilibrium before the first rate measurement. Just prior to initiating the experiment, inspect the cells under a microscope and note the wells where a significant number of cells have detached. Also pay attention to whether the cells display an adverse reaction to the Seahorse assay medium. Some cells do not tolerate the absence of serum or other supplements very well, in which case it is necessary to experiment with adding these back at the lowest possible concentration to avoid interference with the assay.

4.1.2 For tissues

Aspirate the media and replace it with 675 μl 37 °C Seahorse assay DMEM and proceed as quickly as possible with loading the plate into the XF analyzer and initiating the assay. If there is a short wait time, the tissues are kept at 37 °C, *without* CO_2.

4.2. Prepare the drugs to deliver

If the experiment involves testing the response of cells/tissue to different drugs, then load the ports of the hydrated FluxPak with the desired compounds. Each XF24 FluxPak contains four injection ports per probe. Depending on the specific experimental design, each port can be preloaded with Seahorse assay DMEM containing a 10 × solution of the compound(s) of interest. Remember to take into account that the medium volume in your wells increases with each injection. As an example, consider a classical experiment assessing mitochondrial function by studying effects of oligomycin, FCCP, and rotenone/antimycin A on OCR (Fig. 13.3). Oligomycin inhibits the ATP synthase so any oxygen consumption remaining in the presence of this compound is due to uncoupled and nonmitochondrial respiration. FCCP disperses the proton gradient by permeabilizing the inner mitochondrial membrane, thus revealing the maximum respiratory capacity of the cells. Finally, the level of nonmitochondrial respiration is assessed by the addition of rotenone, which abrogates the

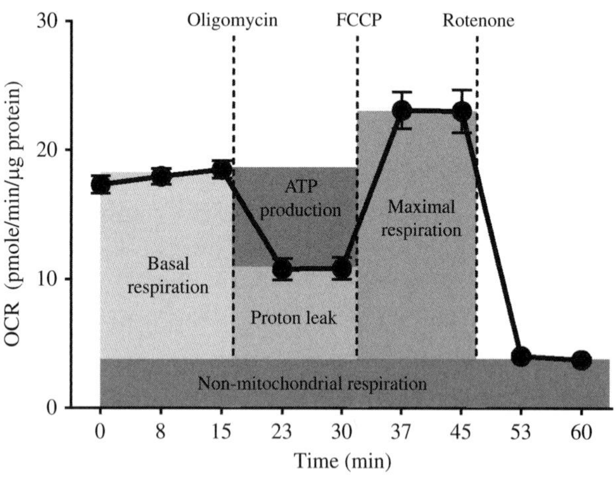

Figure 13.3 Interpretation of an XF24 cell mitochondrial stress experiment investigating mitochondrial function in mouse primary inguinal adipocytes, represented as by Seahorse Bioscience (http://seahorsebio.com/products/consumables/kits/cell-mito-stress.php). Maximal respiratory capacity and the contribution of ATP producing and uncoupled respiration (proton leak) to the basal respiration are determined by observing how the OCR changes in response to drugs that modulate mitochondrial activity. Oligomycin inhibits the ATP synthase, FCCP disperses the proton gradient by permeabilizing the inner mitochondrial membrane, and rotenone abrogates the mitochondrial electron transport chain through the inhibition of complex I ($n=10$, SEM is denoted by vertical bars). (See color plate.)

mitochondrial electron transport chain through the inhibition of complex I. More recently, the complex III inhibitor antimycin A has been included in this part of the assay to ensure complete inhibition of the electron transport chain in cells that display high levels of $FADH_2$-mediated oxidative phosphorylation. For this type of experiment on human or mouse adipocytes, the FluxPak would be loaded with 75 μl assay DMEM containing 10 μg/ml oligomycin in port A for a final concentration of 1 μg/ml in the assay, 83 μl assay DMEM containing 6 μ*M* FCCP in port B for a final concentration of 0.6 μ*M* in the assay, and 93 μl assay DMEM containing 20 μ*M* rotenone and antimycin A in port C for a final concentration of 2 μ*M* of both compounds in the assay. Importantly, always adjust the pH of the injected drug solutions to 7.4. We have successfully manipulated the mitochondrial respiration in inguinal white adipose tissue pieces with FCCP and antimycin A, using a final concentration of 20 μ*M* for each compound (Fig. 13.2).

4.3. Set up the XF protocol

Set up the XF protocol to have the optimal measuring time of the OCR and ECAR levels, according to the cell/tissue respiratory capacity. Cells and tissues with a high basal OCR are at risk of consuming most of the oxygen that is available in the very small volume of media generated when the probes are lowered to perform the measurements, especially if treatment with FCCP is part of the experiment. Different cell types display different vulnerability to hypoxia, and OCR cannot be safely interpreted under circumstances where the cell is attempting to adapt to such conditions. The XF analyzer software discards measurements of O_2 levels below 10 mm Hg, which Seahorse considers a hypoxic state for most cell types (personal correspondence from Seahorse Bioscience). We prefer to stay above O_2 levels of 25 mm Hg either by reducing the cell number/well or, if this is not an option, by reducing measurement time to 2 min.

4.3.1 For cells

For experiments with cells, set up the protocol to perform 2–5 successive OCR/ECAR measurements at 2–5 min intervals.

4.3.2 For tissues

For tissue experiments, it may take much longer for the compounds to penetrate into a sufficient number of cell layers for the effects to show, and the limited survival time of the tissue pieces may force you to have to reduce the number of OCR measurements obtained or drugs studied.

4.4. Start the XF run

Insert the FluxPak plate into the XF24 analyzer and start running the XF experiment. After equilibrating the temperature of the cells or tissues, basal OCR and ECAR levels are determined, followed by measurements of these rates in response to each drug injection. Each condition/treatment should be measured in at least five replicates of independent wells for cells and preferentially more for tissues.

4.5. Postexperiment assessment and normalizations

Once the run is complete, observe the cells under a microscope to evaluate if significant numbers have detached during the assay, as this will have an impact on your results. By observing the oxygen consumption curves (mm Hg; Fig. 13.1), you can evaluate if measurements have been affected by the cells/tissue being exposed to hypoxia and whether there was ample time for the media to reequilibrate to the initial O_2 between successive measurements. The pH curve should also display reequilibration between the measurements (a good example is shown in Fig. 13.1).

4.5.1 For cells

For cells, the OCR and ECAR values are normalized to either cell number (following trypsinization) or protein concentration. The latter can be done by carefully aspirating the media and lysing the cells in 100 μl 50 m*M* NaOH, followed by the application of the Bradford method for the determination of protein concentration.

4.5.2 For tissues

In tissue experiments, the average OCR/ECAR of the replicates is normalized to the total amount of tissue in those wells. This method is valid only when there are no major differences in adipose tissue composition or adipocyte volume among different treatment groups or genotypes, which can be assessed by performing histology. In the event that adipocyte volume differs significantly, this can be taken into account by determining the protein concentration per ~10 mg of tissue and factoring the difference into the normalization.

5. IMPORTANT REMARKS

1. When using a new cell line or tissue, always start with a preliminary XF experiment to find the best cell seeding number or tissue size and the best drug concentrations to be injected. The preliminary protocol should include (A) seeding different cell numbers with at least five replicates, in order to find the optimal cell density in the wells that balances achieving good differentiation and maintaining the OCR/ECAR levels within the dynamic range, and (B) injecting different concentrations of each drug to determine the highest cell/tissue response with minimal toxicity. For example, injection of the optimal FCCP concentration will reveal the maximum respiratory capacity of the cell, while higher or lower concentrations will elicit a lesser response because of acute cytotoxicity or insufficient levels of the drug, respectively.
2. Large variation between replicates and high standard variation occur when the OCR level is too high, using too many or too few cells; the percentage differentiation varies greatly between the wells; or the cells are not distributed evenly in the wells. Cells that are overconfluent can detach from the plates during the measurements, while cells that are not distributed uniformly will not be "seen" by the XF sensors evenly.
3. The effects of gene "knockdown" by RNAi on respiratory capacity of differentiated adipocytes can be explored by transfecting the adipocytes as previously described (Jiang et al., 2003; Yehuda-Shnaidman et al., 2010).
4. Studies of isolated mitochondria from various tissues allow more detailed understanding of bioenergetics. We have not embarked on such methods, preferring to stay closer to the physiological status of the intact cells and tissues. A protocol by Klingenspor and colleagues can be found on the Seahorse Bioscience website.

ACKNOWLEDGMENT

We would like to thank Drs. David Ferrick and Min Wu of Seahorse Bioscience for their helpful discussions and for providing to us plates and resources that were in the development phase of their product line. We also acknowledge Drs. Einav Yehuda-Shnaidman and Jingbo Pi for their early work developing strategies to measure OCR in freshly isolated tissues.

REFERENCES

Cho, S. Y., Park, P. J., Shin, E. S., Lee, J. H., Chang, H. K., & Lee, T. R. (2009). Proteomic analysis of mitochondrial proteins of basal and lipolytically (isoproterenol and TNF-alpha)-stimulated adipocytes. *Journal of Cellular Biochemistry*, *106*(2), 257–266.

Cypess, A. M., Lehman, S., Williams, G., Tal, I., Rodman, D., Goldfine, A. B., et al. (2009). Identification and importance of brown adipose tissue in adult humans. *The New England Journal of Medicine, 360*(15), 1509–1517.

Davis, R. J., & Martin, B. R. (1982). The effect of beta-adrenergic agonists on the membrane potential of fat-cell mitochondria in situ. *The Biochemical Journal, 206*(3), 611–618.

Halvorsen, Y. D., Bond, A., Sen, A., Franklin, D. M., Lea-Currie, Y. R., Sujkowski, D., et al. (2001). Thiazolidinediones and glucocorticoids synergistically induce differentiation of human adipose tissue stromal cells: Biochemical, cellular, and molecular analysis. *Metabolism, 50*(4), 407–413.

Hepp, D., Challoner, D. R., & Williams, R. H. (1968). Respiration in isolated fat cells and the effects of epinephrine. *The Journal of Biological Chemistry, 243*(9), 2321–2327.

Jiang, Z. Y., Zhou, Q. L., Coleman, K. A., Chouinard, M., Boese, Q., & Czech, M. P. (2003). Insulin signaling through Akt/protein kinase B analyzed by small interfering RNA-mediated gene silencing. *Proceedings of the National Academy of Sciences of the United States of America, 100*(13), 7569–7574.

Maassen, J. A., Romijn, J. A., & Heine, R. J. (2008). Fatty acid-induced mitochondrial uncoupling in adipocytes as a key protective factor against insulin resistance and beta cell dysfunction: Do adipocytes consume sufficient amounts of oxygen to oxidise fatty acids? *Diabetologia, 51*(5), 907–908.

Muzik, O., Mangner, T. J., Leonard, W. R., Kumar, A., Janisse, J., & Granneman, J. G. (2013). 15O PET measurement of blood flow and oxygen consumption in cold-activated human brown fat. *Journal of Nuclear Medicine, 54*(4), 523–531.

Nechad, M., Kuusela, P., Carneheim, C., Bjorntorp, P., Nedergaard, J., & Cannon, B. (1983). Development of brown fat cells in monolayer culture. I. Morphological and biochemical distinction from white fat cells in culture AND II. Ultrastructural characterization of precursors, differentiating adipocytes and their mitochondria. *Experimental Cell Research, 149*(1), 105–118, 119–127.

Nedergaard, J., Bengtsson, T., & Cannon, B. (2007). Unexpected evidence for active brown adipose tissue in adult humans. *American Journal of Physiology Endocrinology and Metabolism, 293*, E444–E452.

Orava, J., Nuutila, P., Noponen, T., Parkkola, R., Viljanen, T., Enerback, S., et al. (2013). Blunted metabolic responses to cold and insulin stimulation in brown adipose tissue of obese humans. *Obesity, 21*, 2279–2287.

Ouellet, V., Labbe, S. M., Blondin, D. P., Phoenix, S., Guerin, B., Haman, F., et al. (2012). Brown adipose tissue oxidative metabolism contributes to energy expenditure during acute cold exposure in humans. *The Journal of Clinical Investigation, 122*(2), 545–552.

Rehnmark, S., Kopecky, J., Jacobsson, A., Nechad, M., Herron, D., Nelson, B. D., et al. (1989). Brown adipocytes differentiated in vitro can express the gene for the uncoupling protein thermogenin: Effects of hypothyroidism and norepinephrine. *Experimental Cell Research, 182*(1), 75–83.

Rodbell, M. (1964). Metabolism of isolated fat cells. Effects of hormones on glucose metabolism and lipolysis. *The Journal of Biological Chemistry, 239*, 375–380.

Saito, M., Okamatsu-Ogura, Y., Matsushita, M., Watanabe, K., Yoneshiro, T., Nio-Kobayashi, J., et al. (2009). High incidence of metabolically active brown adipose tissue in healthy adult humans: Effects of cold exposure and adiposity. *Diabetes, 58*(7), 1526–1531.

Stenson, B. M., Ryden, M., Steffensen, K. R., Wahlen, K., Pettersson, A. T., Jocken, J. W., et al. (2009). Activation of liver X receptor (LXR) regulates substrate oxidation in white adipocytes. *Endocrinology, 150*(9), 4103–4113.

van Marken Lichtenbelt, W. D., Vanhommerig, J. W., Smulders, N. M., Drossaerts, J. M., Kemerink, G. J., Bouvy, N. D., et al. (2009). Cold-activated brown adipose tissue in healthy men. *The New England Journal of Medicine, 360*(15), 1500–1508.

Virtanen, K. A., Lidell, M. E., Orava, J., Heglind, M., Westergren, R., Niemi, T., et al. (2009). Functional brown adipose tissue in healthy adults. *The New England Journal of Medicine, 360*(15), 1518–1525.

Wilson-Fritch, L., Nicoloro, S., Chouinard, M., Lazar, M. A., Chui, P. C., Leszyk, J., et al. (2004). Mitochondrial remodeling in adipose tissue associated with obesity and treatment with rosiglitazone. *The Journal of Clinical Investigation, 114*(9), 1281–1289.

Wu, M., Neilson, A., Swift, A. L., Moran, R., Tamagnine, J., Parslow, D., et al. (2007). Multiparameter metabolic analysis reveals a close link between attenuated mitochondrial bioenergetic function and enhanced glycolysis dependency in human tumor cells. *American Journal of Physiology Cell Physiology. 292*(1), C125–C136.

Yehuda-Shnaidman, E., Buehrer, B., Pi, J., Kumar, N., & Collins, S. (2010). Acute stimulation of white adipocyte respiration by PKA-induced lipolysis. *Diabetes, 59*(October), 2474–2483.

Zafon, C., & Simo, R. (2011). The current obesity epidemic: Unravelling the evolutionary legacy of adipose tissue. *The Open Obesity Journal, 3*, 98–106.

Zingaretti, M. C., Crosta, F., Vitali, A., Guerrieri, M., Frontini, A., Cannon, B., et al. (2009). The presence of UCP1 demonstrates that metabolically active adipose tissue in the neck of adult humans truly represents brown adipose tissue. *The FASEB Journal, 23*(9), 3113–3120.

CHAPTER FOURTEEN

Detecting Protein Carbonylation in Adipose Tissue and in Cultured Adipocytes

Qinghui Xu, Wendy S. Hahn, David A. Bernlohr[1]
Department of Biochemistry, Molecular Biology and Biophysics, The University of Minnesota, Minneapolis, Minnesota, USA
[1]Corresponding author: e-mail address: bernl001@umn.edu

Contents

Abstract

Reactive oxygen species-mediated attack of the acyl chains of polyunsaturated fatty acids and triglycerides leads to the formation of lipid hydroperoxides. Lipid hydroperoxides are subject to nonenzymatic Fenton chemistry producing a variety of reactive aldehydes that covalently modify proteins in a reaction referred to as protein carbonylation. Given the significant content of triglycerides in fat tissue, adipose proteins are among the most heavily carbonylated. The laboratory has utilized two methodologies for the detection of protein carbonylation in tissue- and cell-based samples. The first utilizes biotin coupled to a hydrazide moiety and takes advantage of the numerous biotin detection systems. The second method utilizes an anti 4-hydroxy-*trans*-2,3-nonenal (4-HNE)-directed antibody that can detect both 4-HNE and the corresponding 4-oxo derivative when the samples are reduced. Using such methods, we have evaluated the profile of carbonylated proteins in epididymal white adipose tissue and 3T3-L1 adipocytes using both methods. In addition, we have investigated the effects of two antidiabetic drugs, pioglitazone and metformin, on protein carbonylation in 3T3-L1

Methods in Enzymology, Volume 538
ISSN 0076-6879
http://dx.doi.org/10.1016/B978-0-12-800280-3.00014-1

adipocytes. Overall, the biotin hydrazide method is rapid, inexpensive, and easy to use, but its usefulness is limited because it detects a wide variety of carbonylated derivatives, which makes assignments of individual proteins difficult. Compared to the biotin hydrazide method, the anti-HNE antibody method detects specific proteins more readily but identifies only a subset of carbonylated proteins. As such, the combination of both methods allows for a comprehensive evaluation of protein carbonylation plus provides a means towards identification of specific carbonylation targets.

1. INTRODUCTION

Reactive oxygen species (ROS; superoxide anion, hydrogen peroxide, and hydroxyl radical) oxidize a variety of cellular components resulting in modification of intracellular proteins, DNA, RNA, carbohydrates, and lipids. Of the ROS, a hydroxyl radical can initiate peroxidation of polyunsaturated fatty acyl chains of numerous lipids, and given the high content of triglyceride in adipocytes, fat cells are particularly susceptible to oxidative damage. With respect to lipid peroxidation, the oxidized acyl chains undergo nonenzymatic Hock cleavage (Uchida, 2003) to produce a variety of lipid-derived aldehydes. Recent work from Long et al. (2013) has shown that the major aldehydes produced by adipose tissue of obese C57Bl/6J or Lep$^{ob/ob}$ mice are 4-hydroxy-*trans*-2,3-nonenal (4-HNE) and 4-oxo-*trans*-2,3-nonenal (4-ONE). 4-HNE and 4-ONE are capable of covalently modifying the side chains of histidine, cysteine, and lysine residues in a process generically termed protein carbonylation (Sayre, Lin, Yuan, Zhu, & Tang, 2006). High-fat feeding of C57Bl/6 mice led to a two- to threefold increase in total adipose protein carbonylation compared to the lean mice (Grimsrud, Picklo, Griffin, & Bernlohr, 2007). Conjugation of the highly reactive aldehydes 4-HNE or 4-ONE with glutathione is catalyzed by glutathione S-transferase A4 (GSTA4) and constitutes a major route of detoxification (Grimsrud, Xie, Griffin, & Bernlohr, 2008), and silencing of GSTA4 was shown to increase protein carbonylation in 3T3-L1 adipocytes (Curtis et al., 2010). More recently, Curtis et al. (2012) had used a proteomic approach to identify carbonylated proteins and identified the mitochondrial phosphate carrier and two components of complex I of the respiratory chain (NDUFA2 and NDUFA3) as critical carbonylation targets.

Given the importance of redox biology to cellular signaling, oxidative stress (and therefore protein carbonylation) is regulated by a variety of metabolic processes. Treatment of 3T3-L1 adipocytes with tumor necrosis factor (TNF-α) increases ROS production, lipid peroxidation, and protein carbonylation and collectively leads to a mitochondrial dysfunction phenotype (Chen et al., 2010). In contrast, pioglitazone, a member of the thiazolidinedione class of insulin-sensitizing drugs widely used to ameliorate insulin sensitivity in patients with type 2 diabetes (Nawrocki & Scherer, 2005), works in part via its action on the nuclear peroxisome proliferator-activated receptor-gamma and reduces oxidative stress in humans and animal models (Dobrian, Schriver, Khraibi, & Prewitt, 2004; Ishida et al., 2004).

Metformin is widely used for the treatment of type 2 diabetes (Kirpichnikov, McFarlane, & Sowers, 2002) as it ameliorates hyperglycemia without stimulating insulin secretion, promoting weight gain, or causing hypoglycemia (Stumvoll, Nurjhan, Perriello, Dailey, & Gerich, 1995). Metformin has been shown to attenuate the intracellular levels of ROS induced by palmitic acid in human aortic endothelial cells through the activation of the AMPK–FOXO3 pathway (Hou et al., 2010). In addition, it has been shown that metformin can inhibit cell respiration via specifically targeting respiratory chain complex I (El-Mir et al., 2000), thereby potentially activating the AMP-activated protein kinase (AMPK).

Herein, we describe two methods for the analysis of protein carbonylation and demonstrate how such complementary techniques may be used for the analysis of protein carbonylation in response to factors that increase (TNF-α) or reduce (pioglitazone and metformin) oxidative stress.

2. METHODS TO ANALYZE PROTEIN CARBONYLATION

2.1. Sample preparation

2.1.1 Preparation of extracts from epididymal white adipose tissue

One gram of epididymal white adipose tissue (EWAT) from lean or obese mice is incubated on ice in 1 ml pH 5.5 biotin hydrazide buffer (100 m*M* sodium acetate; 20 m*M* NaCl; 0.1 m*M* EDTA, supplemented with protease inhibitors; and 0.25 m*M* butylated hydroxytoluene (BHT)) and homogenized for 30 s using an electronic homogenizer (PowerGen 125). Samples are vortexed briefly and centrifuged at ~1200 rpm for 10 min at 4 °C in a microcentrifuge to float the lipid cake. The lower aqueous phase and any pellet are transferred to a new centrifuge tube and supplemented with SDS to a final concentration of 2%. The samples are then heated at 65 °C for 5 min and subjected to ultracentrifugation at 100,000 × *g* for 1 h at 4 °C to remove insoluble residue. The concentration of detergent-solubilized protein is determined using the bicinchoninic acid assay (Sigma-Aldrich, St Louis, MO).

2.1.2 Preparation of extracts from 3T3-L1 adipocytes

3T3-L1 cells are cultured and differentiated into adipocytes as described previously (Student, Hsu, & Lane, 1980). Monolayers are washed with phosphate-buffered saline (PBS) and scraped into homogenization buffer (50 m*M* Tris–HCl pH 7.4; 50 m*M* NaCl; 50 m*M* NaF; 1 m*M* NaP_2O_4; 1 m*M* EDTA; 1 m*M* EGTA, supplemented with protease inhibitors; and 0.25 m*M* BHT). Samples analyzed for protein carbonylation using the anti 4-hydroxy-*trans*-2,3-nonenal (anti-HNE) method are sonicated for 10 s at 4 °C and centrifuged for 10 min at 700 × *g* to separate the lipid layer from the aqueous proteins. The lower phase and any pellet are transferred to a new centrifuge tube, supplemented with detergent (2% SDS), and centrifuged at 100,000 × *g* for 1 h at 4 °C to remove insoluble residue. Samples destined for carbonylation analysis via the biotin hydrazide method are dialyzed into biotin hydrazide buffer and subsequently treated identically.

2.2. Detection of carbonylation

2.2.1 Analysis of protein carbonylation via using anti-HNE antibody

Protein is separated by SDS–polyacrylamide gel electrophoresis (PAGE) (Mini-PROTEAN®TGX™ 4–20% gradient precast gel) and transferred to Immobilon-FL membranes (Millipore, Darmstadt, Germany) according to the manufacturer's instructions. The membrane is reduced for 1 h at room temperature with 50 m*M* sodium borohydride ($NaBH_4$) in PBS followed by washing three times with PBS to remove the excess reductant. The membrane is then blocked using LI-COR Odyssey Imaging Systems blocking buffer (LI-COR Biosciences, Lincoln, NE) for 1 h at room temperature prior to overnight incubation at 4 °C with the anti-HNE Michael adduct polyclonal primary antibody (Millipore, Catalog No. 393207, Billerica, MA) in PBST (PBS containing 0.2% Tween-20) containing 3% BSA. The membrane is washed four times with PBST and then incubated with LI-COR goat anti-rabbit IR800 secondary antibody for 1 h at room temperature while protected from light, washed again, and visualized using an Odyssey Infrared Imager (LI-COR).

To determine the effect of $NaBH_4$ reduction on carbonylation detection, two identical protein extracts from 3T3-L1 adipocytes were resolved by SDS–PAGE and transferred to PVDF. One sample was reduced in $NaBH_4$ for 1 h before blocking, while the other one was blocked directly. Surprisingly, only minor differences in the carbonylation profile were revealed, suggesting that most of the bands are derived from 4-HNE-modified proteins (results not shown).

2.2.2 Analysis of protein carbonylation via the biotin hydrazide method

Grimsrud et al. (2007) had previously reported optimized conditions for detecting carbonylation with respect to the amount of protein and hydrazide reagent needed, the time of reaction, and the temperature dependence. Utilizing those optimized conditions, the soluble protein (30 μg/sample) is diluted with an equal volume in pH 5.5 biotin hydrazide homogenization buffer and incubated with a final concentration of freshly prepared 0.5 m*M* EZ-link biotin hydrazide (Pierce, Catalog No. 21339, Rockford, United States) in homogenization buffer for 2 h at room temperature. After coupling, the samples are separated by SDS–PAGE and transferred to Immobilon-FL membranes. The membrane is blocked with LI-COR Odyssey Imaging Systems (LI-COR Biosciences, Lincoln, NE) blocking buffer for 1 h at room temperature and rinsed with PBST. The membrane is incubated with IR800-conjugated Streptavidin (Pierce, Rockford, United States) in PBST (1:15,000 dilution) for 1 h at room temperature while protected from light. Figure 14.1 presents a schematic of the isolation and derivatization methods.

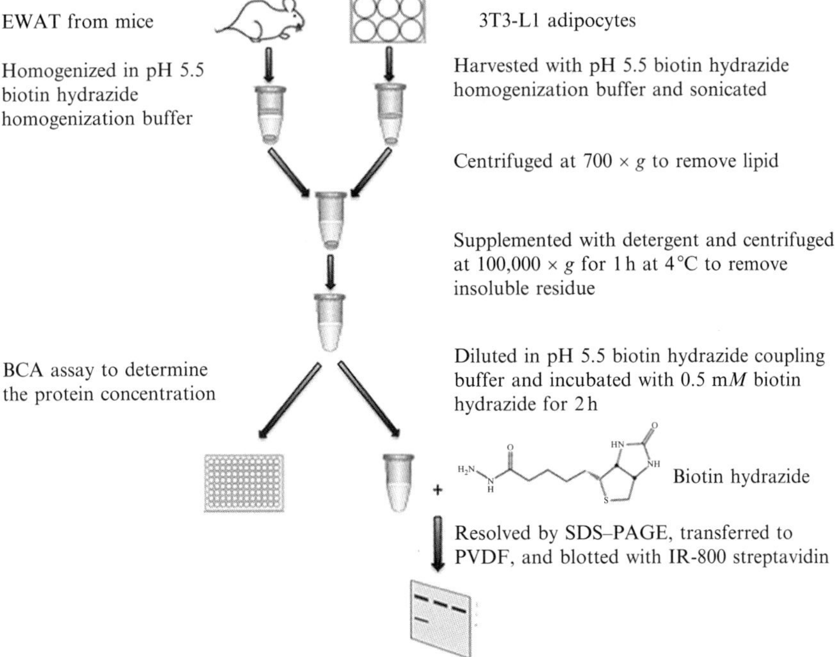

Figure 14.1 Schematic diagram for assessing protein carbonylation in adipose tissue and 3T3-L1 adipocytes using the biotin hydrazide method.

3. RESULTS

3.1. Comparison of the anti-HNE antibody method to the biotin hydrazide method for evaluating protein carbonylation in EWAT of C57Bl/6J mice

To compare the profile of carbonylated proteins detected using the anti-HNE method to those detected using the biotin hydrazide method, soluble proteins from EWAT of lean mice were analyzed in parallel. Figure 14.2 shows the resulting images from the two different methods. Similar but non-identical patterns were observed when comparing the two blots with additional bands detected using anti-HNE antibody relative to biotin hydrazide method. It should be noted that the biotin hydrazide method requires coupling under mildly acidic conditions (pH 5.5), while the immunoblotting method is carried out at more neutral pH. Moreover, if the soluble proteins prepared at neutral pH are dialyzed into biotin hydrazide coupling buffer, many proteins precipitate suggesting that one reason that the biotin hydrazide profile shown in Fig. 14.2 is less complex may be due to the different pHs used in the analyses.

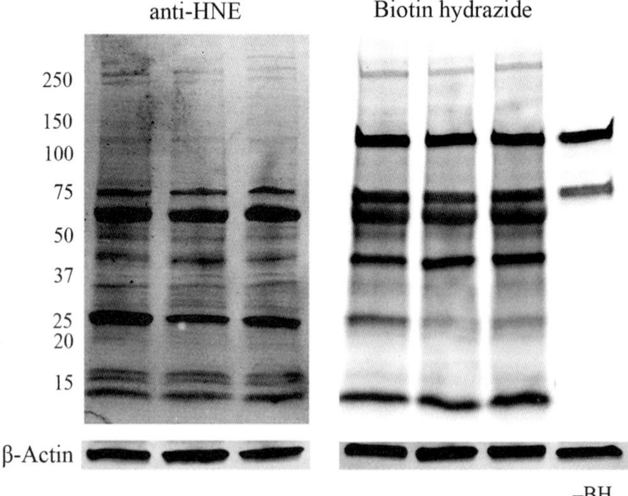

Figure 14.2 Comparison of the anti-HNE antibody method with the biotin hydrazide method for assessing protein carbonylation. Soluble proteins (30 μg) from EWAT of lean mice were either immunoblotted with anti-HNE antibody or coupled with biotin hydrazide followed by secondary detection using IR800 streptavidin. Images were collected utilizing the identical LI-COR detection settings according to the manufacturer's instructions. Molecular mass (in kDa) of protein standards is indicated on the left.

3.2. Using anti-HNE antibody to assess protein carbonylation in GSTA4-silenced or TNF-α-treated 3T3-L1 adipocytes

To further characterize protein carbonylation and to evaluate the anti-HNE method as a means to detect regulated changes in carbonylation, proteins from GSTA4-silenced and scrambled (Scr) cell lines were evaluated (Curtis et al., 2010). GSTA4 is the major antioxidant enzyme responsible for glutathionylation of reactive aldehydes, and previous studies have demonstrated significant changes in mitochondrial function in GSTA4-silenced 3T3-L1 adipocytes and GSTA4-null mice (Curtis et al., 2010). GSTA4-silenced and Scr 3T3-L1 cells were cultured and differentiated into adipocytes as described and protein samples harvested on day 8 at pH 7.4 in western homogenization buffer. Samples were separated using SDS–PAGE and transferred to membranes, and carbonylation profiles detected using the anti-HNE antibody. As shown in Fig. 14.3A, several proteins exhibited a prominent increase in carbonylation in samples from the GSTA4 knockdown cells (Fig. 14.3A), and quantitation of the total intensity from the GSTA4-silenced and control cells (Fig. 14.3B) revealed a significant increase in total carbonylation. Similar results were obtained by Grimsrud et al. (2008) who used the biotin hydrazide method to profile carbonylation. Curtis et al. (2012) extended this analysis to identify carbonylated proteins in the mitochondrion and use biotin hydrazide and avidin affinity chromatography to capture the carbonylated proteins and identify them using iTRAQ-based mass spectrometry methods.

To extend the analysis of protein carbonylation using the anti-HNE method to the analysis of inflammation, 3T3-L1 adipocytes were treated with 1 n*M* TNF-α and harvested 24 h later. Treatment of 3T3-L1 adipocytes with TNF-α also elevated carbonylation of several specific proteins (Fig. 14.4A) and increased total protein carbonylation significantly (Fig. 14.4B). Increased carbonylation of a subset of proteins (60–75 kDa) was observed in both GSTA4-silenced and TNF-α-treated 3T3-L1 adipocytes. These results suggest that different oxidative stress-inducing conditions can lead to the carbonylation of the same proteins supporting the chemical nature of the modification.

3.3. Effects of pioglitazone and metformin on protein carbonylation in 3T3-L1 adipocytes

In contrast to protein carbonylation increasing in response to pro-oxidative conditions, the biotin hydrazide method has been applied to the analysis of

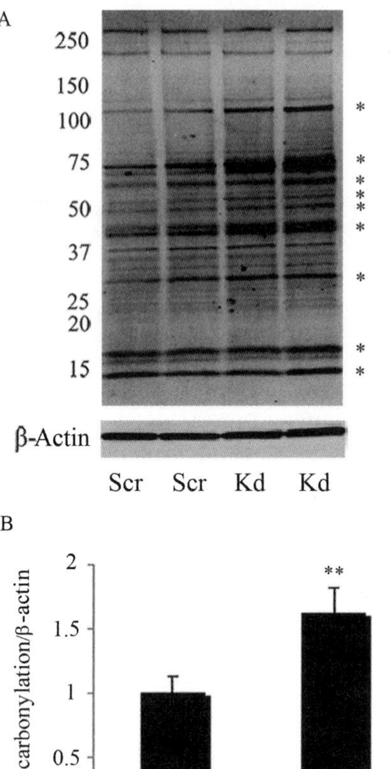

Figure 14.3 Protein carbonylation in GSTA4 knockdown adipocytes detected using the anti-HNE antibody. (A) Protein carbonylation in scrambled (Scr) control cells and GSTA4 knockdown adipocytes (Kd). (B) Total carbonylation as determined from the blot in (A) normalized to β-actin. Protein bands exhibiting increased carbonylation are indicated with an asterisk (*). Molecular mass (kDa) of protein standards is indicated on the left. Values are expressed as mean ± SE ($n = 6$ per group; *, $p < 0.05$ relative to control group; **, $p < 0.01$ relative to control group).

antioxidative stimuli. Pioglitazone, the antidiabetic drug, has been shown to reduce oxidative stress as measured by Ishida et al. (2004) and Dobrian et al. (2004). Similarly, metformin reduces oxidative conditions by Hou et al. (2010). To determine if pioglitazone and/or metformin can affect protein carbonylation, 3T3-L1 adipocytes were treated with 100 μ*M* pioglitazone (Kanda et al., 2008) or 4 m*M* metformin (Anedda, Rial, & González-Barroso, 2008) on day 6 and harvested with pH 5.5 biotin hydrazide homogenization buffer on day 8. Carbonylation was detected via biotin hydrazide

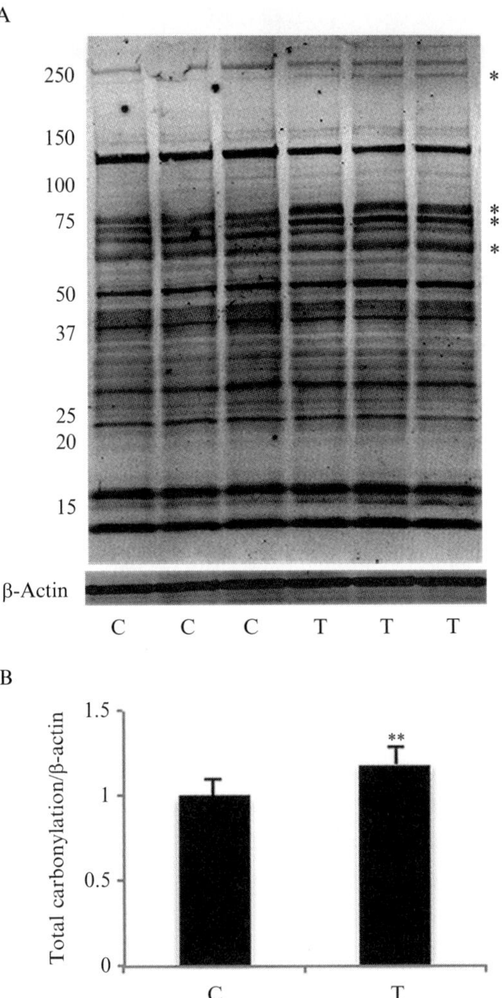

Figure 14.4 Protein carbonylation in TNF-α-treated 3T3-L1 adipocytes as detected using the anti-HNE antibody method. (A) Protein carbonylation in control adipocytes (C) and adipocytes after 24 h treatment of 1 n*M* TNF-α (T). (B) Total carbonylation was determined from the blot in (A) normalized to β-actin. Protein bands exhibiting increased carbonylation are indicated with an asterisk (*). Molecular mass (kDa) of protein standards is indicated on the left. Values are expressed as mean ± SE ($n = 6$ per group; *, $p < 0.05$ relative to control group; **, $p < 0.01$ relative to control group).

modification as described and quantified by the analysis of the signal intensity. Decreased carbonylation of specific proteins was noted in both pioglitazone- and metformin-treated adipocytes compared to control cells (Fig. 14.5). In the pioglitazone-treated cells, two prominent bands that were

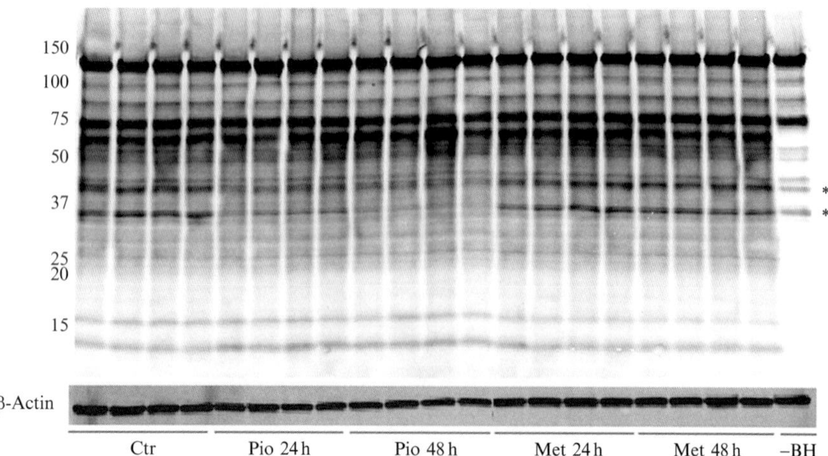

Figure 14.5 Effect of pioglitazone and metformin on protein carbonylation in 3 T3-L1 adipocytes assessed using the biotin hydrazide method. Protein carbonylation in control adipocytes and adipocytes after 24 h and 48 h treatment of 100 μ*M* pioglitazone or 4 m*M* metformin as indicated. Protein bands with decreased carbonylation after treatment are indicated with an asterisk (*). Molecular mass (kDa) of protein standards is indicated on the left.

near 42 and 34 kDa were significantly reduced in their carbonylation level. Treatment of 3T3-L1 adipocytes with metformin showed a modest total protein carbonylation relative to control cells (Fig. 14.5) with the same 42 and 34 kDa polypeptides being most markedly affected. Interestingly, carbonylation upregulated by molecular (GSTA4 silencing) or pharmacologic (TNF-α) methods identified a subset of proteins (60–75 kDa) that were different than those downregulated by pioglitazone or metformin (42 and 34 kDa; compare Figs. 14.4 and 14.5). This may be due to true molecular differences or methodological considerations (pH 5.5 vs. pH 7.4).

4. DISCUSSION

The biotin hydrazide method or other similar hydrazide-based detection systems (e.g., 2,4-dinitrophenol hydrazide) are widely used to assess protein carbonylation (Yoo & Regnier, 2004). Biotin hydrazide detects all sources of carbonylated proteins, including not only carbonyls due to 4-HNE and 4-ONE but also direct oxidation of side chains of lysine, arginine, proline, and threonine residues (Suzuki, Carini, & Butterfield, 2010). In contrast, the anti-HNE antibody only recognizes 4-HNE + 4-ONE adducts since 4-ONE is converted to 4-HNE through $NaBH_4$ reduction

(Abdel-Magid, Maryanoff, & Carson, 1990). Therefore, it is not surprising that the two methods detect different band patterns. Moreover, since the two methods are operationally defined by different pHs, the two methods would be expected to reveal similar, but not identical proteins. Compared to the biotin hydrazide method, the anti-HNE antibody method produces an image with more distinct bands where hydrazide-based methods frequently produce gel images with a smeared appearance (Grimsrud et al., 2007). This is particularly apparent when analyzing membrane proteins where the anti-HNE method produces images indistinguishable from the analysis of soluble proteins. In terms of cost and time of analysis, the biotin hydrazide method is less expensive and can be scaled easily to classroom teaching applications, whereas the anti-HNE method is more laborious. It should be noted that while pH 5.5 biotin hydrazide homogenization buffer is required when protein carbonylation is assessed by the biotin hydrazide method, either pH 5.5 biotin hydrazide homogenization buffer or pH 7.4 western homogenization buffer can be used to harvest adipocytes using the anti-HNE antibody (data not shown).

Consistent with the previous work in which the biotin hydrazide method was used to detect protein carbonylation (Curtis et al., 2010), increased protein carbonylation was identified in GSTA4 knockdown adipocytes and TNF-α-treated 3T3-L1 adipocytes using anti-HNE antibody. The reduced abundance of GSTA4, which catalyzes the conjugation of the highly reactive aldehydes (with greatest specificity for 4-HNE) to glutathione, results in increased levels of reactive aldehydes in adipocytes (Engle et al., 2004). It has been shown that TNF-α treatment leads to decreased mitochondrial membrane potential and reduced production of intracellular ATP, as well as accumulation of significant amounts of ROS, which further contributes to increased protein carbonylation (Chen et al., 2010).

The mechanisms on how pioglitazone and metformin decrease protein carbonylation are complex, and multiple explanations have been suggested. However, it is clear that factors that decrease ROS and ameliorate insulin resistance decrease protein carbonylation. Pioglitazone markedly down-regulated total protein carbonylation, consistent with reduced oxidative stress after treatment (Dobrian et al., 2004; Ishida et al., 2004). Pioglitazone has also been shown to reduce TNF-α expression (Shimizu et al., 2006), which might contribute to the upregulation of GSTA4 and the further decrease in carbonylation via glutathionylation. Previous studies have shown that metformin treatment increases UCP2 expression in 3T3-L1 adipocytes (Anedda et al., 2008) as part of the antioxidant defense response to minimize

ROS levels, thus potentially providing a mechanism for reduced protein carbonylation. Metformin has also been shown to upregulate expression of the antioxidant thioredoxin through the activation of the AMPK–FOXO3 pathway resulting in decreased ROS level (Hou et al., 2010). Metformin-induced reduction in ROS level may also result from induction of Mn-superoxide dismutase and promotion of mitochondrial biogenesis through the activation of AMPK–PGC-1α pathway (Kukidome et al., 2006).

Overall, the two methods provide complementary information and may be used for the analysis of protein carbonylation in response to any number of cellular or genetic factors. While the methods described herein are focused on adipose tissue, similar procedures have been used successfully for the analysis of protein carbonylation in a variety of other cell/tissue samples. The methods are relatively rapid and inexpensive and provide the investigator with experimental opportunities to explore the role of protein carbonylation in cellular control.

REFERENCES

Abdel-Magid, A. F., Maryanoff, C. A., & Carson, K. G. (1990). Reductive amination of aldehydes and ketones by using sodium triacetoxyborohydride. *Tetrahedron Letters*, *31*, 5595–5598.

Anedda, A., Rial, E., & González-Barroso, M. M. (2008). Metformin induces oxidative stress in white adipocytes and raises uncoupling protein 2 levels. *Journal of Endocrinology*, *199*, 33–40.

Chen, X. H., Zhao, Y. P., Xue, M., Ji, C. B., Gao, C. L., Zhu, J. G., et al. (2010). TNF-α induces mitochondrial dysfunction in 3 T3-L1 adipocytes. *Molecular and Cellular Endocrinology*, *328*, 63–69.

Curtis, J. M., Grimsrud, P. A., Wright, W. S., Xu, X., Foncea, R. E., Graham, D. W., et al. (2010). Down regulation of adipose glutathione S-transferase A4 leads to increased protein carbonylation, oxidative stress, and mitochondrial dysfunction. *Diabetes*, *59*, 1132–1142.

Curtis, J. M., Hahn, W. S., Long, E. K., Burrill, J. S., Arriaga, E. A., & Bernlohr, D. A. (2012). Protein carbonylation and metabolic control systems. *Trends in Endocrinology & Metabolism*, *23*, 399–406.

Dobrian, A. D., Schriver, S. D., Khraibi, A. A., & Prewitt, R. L. (2004). Pioglitazone prevents hypertension and reduces oxidative stress in diet-induced obesity. *Hypertension*, *43*, 48–56.

El-Mir, M. Y., Nogueira, V., Fontaine, E., Avéret, N., Rigoulet, M., & Leverve, X. (2000). Dimethylbiguanide inhibits cell respiration via an indirect effect targeted on the respiratory chain complex I. *Journal of Biological Chemistry*, *275*, 223–228.

Engle, M. R., Singh, S. P., Czernik, P. J., Gaddy, D., Montague, D. C., Ceci, J. D., et al. (2004). Physiological role of mGSTA4-4, a glutathione S-transferase metabolizing 4-hydroxynonenal: Generation and analysis of mGsta4 null mouse. *Toxicology and Applied Pharmacology*, *194*, 296–308.

Grimsrud, P. A., Picklo, M. J., Griffin, T. J., & Bernlohr, D. A. (2007). Carbonylation of adipose proteins in obesity and insulin resistance identification of adipocyte fatty acid-binding protein as a cellular target of 4-hydroxynonenal. *Molecular & Cellular Proteomics*, *6*, 624–637.

Grimsrud, P. A., Xie, H., Griffin, T. J., & Bernlohr, D. A. (2008). Oxidative stress and covalent modification of protein with bioactive aldehydes. *Journal of Biological Chemistry*, *283*, 21837–21841.

Hou, X., Song, J., Li, X. N., Zhang, L., Wang, X., Chen, L., et al. (2010). Metformin reduces intracellular reactive oxygen species levels by upregulating expression of the antioxidant thioredoxin via the AMPK-FOXO3 pathway. *Biochemical and Biophysical Research Communications*, *396*, 199–205.

Ishida, H., Takizawa, M., Ozawa, S., Nakamichi, Y., Yamaguchi, S., Katsuta, H., et al. (2004). Pioglitazone improves insulin secretory capacity and prevents the loss of beta-cell mass in obese diabetic db/db mice: Possible protection of beta cells from oxidative stress. *Metabolism, Clinical and Experimental*, *53*, 488.

Kanda, Y., Matsuda, M., Tawaramoto, K., Kawasaki, F., Hashiramoto, M., Matsuki, M., et al. (2008). Effects of sulfonylurea drugs on adiponectin production from 3 T3-L1 adipocytes: Implication of different mechanism from pioglitazone. *Diabetes Research and Clinical Practice*, *81*, 13–18.

Kirpichnikov, D., McFarlane, S. I., & Sowers, J. R. (2002). Metformin: An update. *Annals of Internal Medicine*, *137*, 25–33.

Kukidome, D., Nishikawa, T., Sonoda, K., Imoto, K., Fujisawa, K., Yano, M., et al. (2006). Activation of AMP-activated protein kinase reduces hyperglycemia-induced mitochondrial reactive oxygen species production and promotes mitochondrial biogenesis in human umbilical vein endothelial cells. *Diabetes*, *55*, 120–127.

Long, E. K., Olson, D. M., & Bernlohr, D. A. (2013). High fat diet Induces changes in adipose tissue trans 4-oxo-2-nonenal and trans 4-hydroxy 2-nonenal levels in a depot-specific manner. *Free Radical Biology and Medicine*, *63*, 390–398.

Nawrocki, A. R., & Scherer, P. E. (2005). Keynote review: The adipocyte as a drug discovery target. *Drug Discovery Today*, *10*, 1219–1230.

Sayre, L. M., Lin, D., Yuan, Q., Zhu, X., & Tang, X. (2006). Protein adducts generated from products of lipid oxidation: Focus on HNE and ONE. *Drug Metabolism Reviews*, *38*, 651–675.

Shimizu, H., Oh-I, S., Tsuchiya, T., Ohtani, K. I., Okada, S., & Mori, M. (2006). Pioglitazone increases circulating adiponectin levels and subsequently reduces TNF-α levels in type 2 diabetic patients: A randomized study. *Diabetic Medicine*, *23*, 253–257.

Student, A. K., Hsu, R. Y., & Lane, M. D. (1980). Induction of fatty acid synthetase synthesis in differentiating 3 T3-L1 preadipocytes. *Journal of Biological Chemistry*, *255*, 4745–4750.

Stumvoll, M., Nurjhan, N., Perriello, G., Dailey, G., & Gerich, J. E. (1995). Metabolic effects of metformin in non-insulin-dependent diabetes mellitus. *New England Journal of Medicine*, *333*, 550–554.

Suzuki, Y. J., Carini, M., & Butterfield, D. A. (2010). Protein carbonylation. *Antioxidants & Redox Signaling*, *12*, 323–325.

Uchida, K. (2003). 4-Hydroxy-2-nonenal: A product and mediator of oxidative stress. *Progress in Lipid Research*, *42*, 318–343.

Yoo, B. S., & Regnier, F. E. (2004). Proteomic analysis of carbonylated proteins in two-dimensional gel electrophoresis using avidin-fluorescein affinity staining. *Electrophoresis*, *25*, 1334–1341.

CHAPTER FIFTEEN

Use of Fluorescence Microscopy to Probe Intracellular Lipolysis

Emilio P. Mottillo[*], **George M. Paul**[*], **Hsiao-Ping H. Moore**[†,1], **James G. Granneman**[*,‡,1]

[*]Center for Integrative Metabolic and Endocrine Research, Wayne State University School of Medicine, Detroit, Michigan, USA
[†]College of Arts and Sciences, Lawrence Technological University, Southfield, Michigan, USA
[‡]John D. Dingell Veterans Affairs Medical Center, Detroit, Michigan, USA
[1]Corresponding authors: e-mail address: hmoore@ltu.edu; jgranne@med.wayne.edu

Contents

Abstract

Intracellular lipolysis is an important cellular process in key metabolic tissues, and while much is known about the enzymatic basis of lipolysis, our understanding of how these processes are organized and regulated within cells is incomplete. Lipolysis takes place on the surface of intracellular lipid droplets, which are now recognized as bona fide organelles, and a large number of proteins have been found to change their associations with lipid droplets in response to lipolytic stimulation. Intracellular lipolysis has critical spatial and temporal domains that can be investigated using high-resolution imaging of fixed and live cells. Here, we describe techniques for high-resolution imaging of native lipid droplet proteins, of dynamic trafficking and interaction of these proteins in model systems, and of intracellular fatty acid production using fluorescent reporters in live adipocytes.

Methods in Enzymology, Volume 538
ISSN 0076-6879
http://dx.doi.org/10.1016/B978-0-12-800280-3.00015-3

1. INTRODUCTION

Intracellular lipolysis is an important cellular process in key metabolic tissues, including adipose tissue, liver, and muscle. A considerable amount is known about the enzymatic basis for lipogenesis and lipolysis; however, our understanding of how these processes are organized and regulated within cells is incomplete. Until recently, cellular triglyceride was considered to be stored in droplets lacking biological structure or organization. Growing evidence, however, suggests that lipid droplets are specialized, heterogeneous organelles that perform distinct roles in lipid biosynthesis, in transport and mobilization (Brasaemle & Wolins, 2012; Murphy, Martin, & Parton, 2009; Wolins, Brasaemle, & Bickel, 2006), and more recently in cell signaling (Mottillo, Bloch, Leff, & Granneman, 2012; Zechner et al., 2012). A large number of proteins have been found to change their associations with lipid droplets in response to lipolytic stimulation (Brasaemle, Dolios, Shapiro, & Wang, 2004). How these proteins are coordinated to control lipid storage and utilization remains an important area of investigation.

The size and protein composition of lipid droplets vary among these key metabolic tissues (Bell et al., 2008; Granneman, Moore, Mottillo, & Zhu, 2009; Wolins et al., 2006) and likely reflect their specialized functions. A substantial body of evidence indicates that members of the perilipin family of proteins play a central role in coordinating the trafficking of lipolytic effectors at the surface of lipid droplets, where fatty acid mobilization takes place (Moore, Silver, Mottillo, Bernlohr, & Granneman, 2005). The expression pattern of certain perilipin family members varies among metabolic tissues, which likely reflects functional specialization (Granneman, Moore, Krishnamoorthy, & Rathod, 2009; Granneman, Moore, Mottillo, Zhu, & Zhou, 2011; Subramanian et al., 2004; Wang et al., 2011).

Intracellular lipolysis has critical spatial and temporal domains. For example, lipolysis takes place at the surface of lipid droplets, yet several critical effector proteins, like hormone-sensitive lipase (HSL) and adipose triglyceride lipase (ATGL), are largely cytosolic (Sztalryd et al., 2003). Furthermore, unperturbed lipid droplets can vary in diameter from 200 to 300 nm in cardiac muscle to >50 μm in white adipocytes (Bosma et al., 2013; Granneman et al., 2007; Moore et al., 2005). In adipocytes, hormones and neurotransmitters stimulate lipolysis within seconds of application, and this activation involves substantial changes in trafficking and protein–protein interactions at the lipid droplet surface. The molecular details of these critical processes have not been fully developed; however, given the key role of lipolysis in metabolism, a fuller understanding of the mechanisms involved could lead to new therapies for cardiovascular and metabolic diseases. Advancing this goal requires the ability to perform high-resolution imaging of native lipid droplets and monitoring the dynamic trafficking and specific protein–protein interactions on physiologically relevant timescales. In the succeeding text, we describe the techniques for high-resolution imaging of native lipid droplet proteins, dynamic trafficking of these proteins in model systems, and imaging intracellular fatty acid production using fluorescent reporters in live cells.

2. METHODS TO IMAGE AND COLOCALIZE ENDOGENOUS LIPOLYTIC EFFECTORS IN ADIPOCYTES

2.1. Materials

2.1.1 Fat cell isolation from adipose tissue, fat cells floated to Matrigel, and microdissected muscle fibers

- Collagenase type I (Worthington Biochemical, Cat. # CLS-1)
- Bovine serum albumin (BSA) Fraction V, fatty acid-free (Roche, Cat. # 03117057001)
- (−)-N^6-(2-phenylisopropyl)adenosine (PIA; Sigma-Aldrich, Cat. # P4532)
- Matrigel (BD Biosciences, Cat. # 354234)

Isolation of mature adipocytes from mouse adipose tissue is carried out according to the method of Rodbell (1964). Briefly, perigonadal fat pads from two 3-month-old C57/Bl6 mice are removed and finely minced with scissors. The tissue is then digested with 1.5 mg/ml collagenase in isolation buffer (DMEM/10 m*M* HEPES, pH 7.4/1% BSA/100 n*M* PIA) for 30 min at 37 °C with vigorous shaking. The addition of PIA suppresses lipolysis and improves adipocyte yield (Viswanadha & Londos, 2006). Adipocytes are separated from stromal cells by gentle centrifugation, and the floating adipocytes are then washed three times with the isolation buffer. For imaging purposes, it is best to embed the cells in Matrigel. Isolated cells are resuspended in 1 ml ice-cold Matrigel and plated on 25 mm glass coverslips in a six-well culture plate and allowed to solidify at 37 °C for 10 min. The cells are then incubated overnight in a 1:1 mixture of isolation buffer and growth medium (DMEM/10%FCS/1% penicillin and streptomycin) at 37 °C under 5% CO_2 atmosphere. Alternatively, a single 50 μl drop of adipocytes, isolated as described earlier, is applied to the center of a Matrigel-coated coverslip. The coverslip is then inverted to create a hanging drop that allows floating fat cells to contact the surface of the Matrigel. Cells can then be fixed to the coverslips with 4% paraformaldehyde and processed for immunofluorescence, as detailed in the succeeding text.

2.1.2 Cell culture of differentiated adipocytes

Three-dimensional cultures of 3T3-L1 adipocytes (American Type Culture Collection (ATCC), Cat. # CL-173) are prepared as follows. 3T3-L1 preadipocytes are differentiated in 10 cm plates for 5 days as described (Zebisch, Voigt, Wabitsch, & Brandsch, 2012). Adipocytes are trypsinized and resuspended in 1 ml of ice-cold Matrigel. Forty microliters of the cell suspension are placed on 12 mm glass coverslips in a 24-well culture plate and allowed to solidify at 37 °C for 10 min. Growth medium is added and the cells are maintained in this medium for two more weeks. At the end of this period, over 75% of the cells will develop 2–3 major lipid droplets with diameters >20 μm.

2.1.3 Lipolytic activation

Typical experiments examine the effects of agents that activate lipolysis via stimulation of the protein kinase A (PKA) signaling pathway. A variety of agents are effective in this regard, including the nonselective beta adrenergic agonist isoproterenol (100 n*M*; Sigma-Aldrich, Cat. # I2760) or the beta3-selective agonist CL 316,243 (100 n*M*; Sigma-Aldrich, Cat. # C5976). Strong activation of lipolysis can be achieved by activating adenylyl cyclase and inhibiting phosphodiesterases with a mixture of forskolin (10 μ*M*; Sigma-Aldrich, Cat. # F6886) and 3-isobutyl-1-methylxanthine (IBMX, 1 m*M*; Sigma-Aldrich, Cat. # I5879), respectively. In a typical experiment, 3T3-L1 adipocytes or isolated mature adipocytes are rinsed with PBS and preincubated with DMEM/20 m*M* HEPES, pH 7.4/3 n*M* PIA for 1 h at 37 °C. A 10 × concentrated stock of lipolytic-stimulating reagents is added to one set of wells to achieve appropriate final concentrations of activators. The other set of wells (unstimulated control) receives additional PIA at a final concentration of 200 n*M*, which serves to reduce PKA activation and inhibit lipolysis. Cells are incubated at 37 °C in a water bath for 5–30 min and then fixed with 4% paraformaldehyde for 20 min and processed for indirect immunofluorescence microscopy.

2.2. Methods

2.2.1 Fluorescence and immunofluorescence labeling

2.2.1.1 Materials

Antibodies to lipolytic effector proteins can be purchased from (or produced by) a variety of commercial sources: anti-phospho-HSL (Cell Signaling

Technology, Cat. # 4139), guinea pig anti-PLIN2 (Fitzgerald Industries International), goat anti-PLIN1 (Everest Biotech, EB07728), rabbit anti-ATGL (Cell Signaling Technology, Cat. # 2439), and rabbit anti-phospho-(Ser/Thr) PKA substrate (Cell Signaling Technology, Cat. # 9621). Our lab has found it cost-effective to contract the production of affinity-purified polyclonal antibodies from commercial sources (Proteintech, Chicago, IL). HCS LipidTOX Deep Red (Cat. # H34477), Bodipy FL C_{16} (Cat. # D3821), and MitoTracker Red CMXRos (Cat. # M-7512) are purchased from Invitrogen.

2.2.1.2 Fluorescence staining procedures

To preserve lipid droplet structures, cells are typically fixed in 4% paraformaldehyde prepared in PBS for 15 min at room temperature before immunocytochemistry. It has been reported that other common fixation procedures, such as cold methanol or acetone, may induce lipid droplet fusion and thus should be avoided (DiDonato & Brasaemle, 2003). For colocalization experiments, it is best to use primary antisera raised in different species. This, however, is not always possible. To examine colocalization of two proteins using antibodies raised in the same species, we use the procedure of Negoescu et al. (1994). The key to this approach is the application of saturating levels of a monovalent secondary antibody when detecting the first antigen. Fixed cells are permeabilized with PBS containing 5% normal goat serum and 0.03% saponin (permeabilization buffer) for 1 h. Cells are then incubated for 1 h with primary antibodies against the first antigen. Cells are washed four times with PBS over a 40 min period and then incubated for 1–3 h with 3 μg/ml Cy3-conjugated goat anti-rabbit Fab fragment (Jackson Immunoresearch, Cat. # 111-167). Coverslips are washed as before and then incubated for 1 h with primary antibodies against the second antigen. The coverslips are again washed and incubated for 45 min with 2 μg/ml Alexa Fluor 488-conjugated goat anti-rabbit $(Fab)_2$ antibodies (Invitrogen, Cat. # A11070). Washed slides are postfixed with 1% paraformaldehyde for 15 min. Affinity-purified primary antibodies are typically applied at a concentration of 1–3 μg/ml. All antibodies are diluted in permeabilization buffer, and incubations are at room temperature. Depending on the antigens, it may be necessary to adjust the antibody concentrations and time of incubation to achieve optimal results. For each pair of antibodies used, control experiments must be performed in which one of the primary antibodies is omitted in order to assure that the observed signals are not due to cross-reactivity. Omission of a primary antibody should completely eliminate fluorescent signals in its corresponding channel.

It is often desirable to counterstain the coverslips with organelle markers. For lipid droplets, we use either 0.25 μg/ml BODIPY FL C_{16} or a 1:200 fresh dilution of HCS LipidTOX Deep Red neutral lipid stain. These stains may be included in the last secondary antibody incubation during the immunofluorescence procedure. Although Nile Red can also be used to label lipid droplets, its use as a counterstain is limited due to the broad spectra that overlap with other fluorophores. If Nile Red is to be used, it must be applied after other fluorescent images have been acquired. Live cells may also be counterstained with 100–500 n*M* of freshly diluted MitoTracker CMXRos.

2.2.2 Microscopy and image analysis

2.2.2.1 Microscopes and lenses

A variety of high-quality confocal microscopes are commercially available for static imaging of lipolytic effectors. The exact choice depends on specific applications. Spinning disc confocal microscopes offer high-speed imaging that minimizes photobleaching and phototoxicity and are suitable for imaging protein translocation and changes in protein–protein interactions in live cells (see Section 3 in the succeeding text). A laser-scanning confocal microscope is useful for high-quality static imaging of fixed samples and dynamic imaging in live cells where precise laser control is needed, for example, in fluorescence recovery after photobleaching (FRAP) and fluorescence activation studies. In the majority of our studies, images are acquired with the Olympus IX81 microscope equipped with automated filter controls and a spinning disc confocal unit. The use of efficient high-numerical-aperture (NA) lenses is recommended.

A number of special considerations should be taken into account when imaging lipid droplet proteins. Due to the thickness of the droplets, it is important to use confocal imaging mode with multiple z sections. Lipid droplet proteins reside on the droplet surface, so optical sections that capture the surface provide the best information regarding the distribution of proteins. For freshly isolated adipocytes, excellent images can be obtained by stacking a series of 20–30 individual confocal slices captured at 0.4 μm intervals using a 60 × 1.2 NA water objective. The approach avoids imaging the fluorescence at the droplet equator, where fluorescence intensity is greatest but spatial resolution is poor. It may not be feasible to perform surface imaging of droplets smaller than 1–2 μm in diameter. Under these conditions, the most intense fluorescence signal is obtained at the droplet equator, and the resulting signals are nearly perfectly round rims of fluorescence. An example of images captured with the earlier considerations is shown in Fig. 15.1.

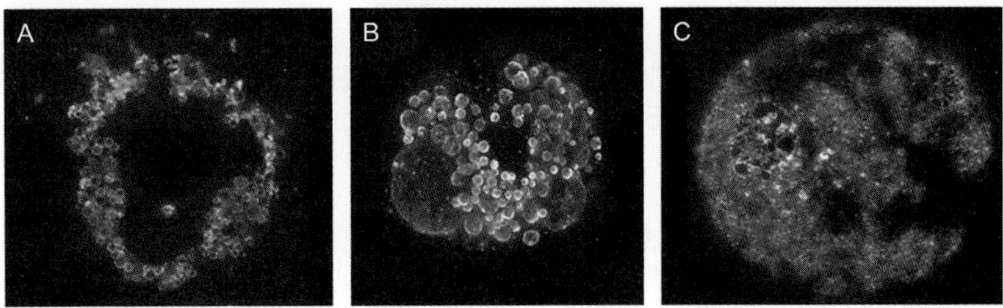

Figure 15.1 Confocal images showing distribution of perilipin 1 in (A) 2D culture of 3T3-L1 adipocytes grown on glass coverslips, (B) 3D culture of 3T3-L1 adipocytes in Matrigel, and (C) mature adipocytes isolated from perigonadal fat pads of mice. Shown are stacked images of 10–30 confocal z-slices. See text for details.

Figure 15.1A illustrates a 2D culture of 3T3-L1 adipocytes grown on glass coverslip and stained with anti-PLIN1. Most of the droplets are small and the fluorescent staining appears as perfectly round rims. When grown in 3D Matrigel (Fig. 15.1B), 3T3-L1 adipocytes develop two to three large central droplets surrounded by numerous smaller droplets. PLIN1 is preferentially localized to the smaller droplets. In isolated adipocytes (Fig. 15.1C), PLIN1 is not uniformly distributed on the surface of the central droplet and instead shows a patchy distribution.

2.2.2.2 Image analysis

Quantitation of fluorescence intensities and colocalization of fluorescence is determined by line scan analysis of merged confocal images using IPlabs software (Scanalytics, BD Biosciences), with the analyst blind to the labeling conditions. For colocalization studies, it is important to avoid displacement artifacts resulting from inadequate color correction of the objective or poorly aligned filters. This can often be recognized by a loss of pixel registration and the uniform displacement of color between fluorescence signals during color merging. Many of the available software packages can be used to correct displacement artifacts. Multicolor fluorescent latex microspheres added at a concentration of several particles per view field to the specimen provide a convenient reference point to ensure proper pixel registration. Pearson's linear regression analyses are performed for each scan using the GraphPad Prism 5 software (GraphPad Software), and coefficients of determination (r^2) are averaged for each cell, with n = number of cells examined. Lipid droplet size is determined by measuring the cross-sectional length of respective immunofluorescence signals.

3. IMAGING PROTEIN TRAFFICKING AND PROTEIN–PROTEIN INTERACTIONS IN LIVE MODEL CELLS

3.1. Materials

3.1.1 Fluorescent fusion proteins

The panel of fluorescent fusion proteins that we have developed for imaging lipolytic trafficking utilizes fusions with enhanced yellow fluorescent protein (EYFP), enhanced cyan fluorescent protein (ECFP), and mCherry. The fluorescence emissions of each of these are readily resolved using standard filter sets, and the ECFP and EYFP can be used as fluorescence resonance energy transfer (FRET) donor and acceptors for monitoring close molecular interactions (Day & Davidson, 2012). Because tagging with fluorescent proteins can influence targeting, it is important to confirm correct targeting by comparing subcellular distribution to that of endogenous protein (e.g., in 3T3-L1 adipocytes). This also allows for assessing magnitude of overexpression. Fluorescent fusion proteins used in our lab are based on the Clontech ECFP/EYFP C1 and N1 vectors that allow placing the fluorescent protein on the N- or C-terminus of the targeted protein, respectively. We have found that the following fusions work well in a variety of cell types: PLIN1–EYFP, PLIN1–ECFP, and PLIN1–mCherry; ECFP–HSL, EYFP–HSL, and mCherry–HSL; ECFP–ABHD5, EYFP–ABHD5, mCherry–ABHD5, and ABHD5–mCherry; ECFP–ATGL, EYFP–ATGL, mCherry–ATGL, and ATGL-EYFP.

3.1.2 Bimolecular fluorescence complementation constructs

Bimolecular fluorescence complementation is a protein complementation methodology that assesses protein–protein interactions in live cells (Hu, Chinenov, & Kerppola, 2002; Magliery et al., 2005). The basic methodology involves fusing N-terminal (Yn = amino acid 1–158) and C-terminal (Yc = amino acids 155–239) fragments of EYFP (or Venus) to proteins of interest. Close interaction of test proteins allows spontaneous folding of the reporter fragments into an active protein. We found that the following protein pairs produce complemented fluorescence: Yn-ABHD5 + PLIN1-Yc; Yn-ABHD5 + PLIN5-Yc; Yn-ABHD5 + ATGL-Yc; Yn-ABHD5 + S47A ATGL-Yc; Yn-HSL + PLIN1-Yc; PLIN1-Yn + PLIN1-Yc.

3.2. Methods

3.2.1 Considerations

As mentioned earlier, major considerations concern faithful targeting of fusion proteins and analysis in an appropriate cellular background. Fluorescent reporters can be used as tracers to monitor protein trafficking when expressed at relatively low levels. Experiments that involve assessment of protein–protein interactions require the use of cells lacking endogenous expression, since endogenous proteins will compete and reduce signals. For this purpose, one can use 3T3-L1 preadipocytes. The imaging of lipid droplets in transfected preadipocytes is greatly facilitated by supplementing the media with 400 μ*M* oleic acid that has been complexed to BSA. When using transfected cells, it is important to keep expression levels as low as possible to avoid mistargeting. For example, PLIN1 accumulates on the nuclear envelope and ER when moderately overexpressed or when the triglyceride supply is limiting. Lastly, fluorescence imaging of lipid droplet protein trafficking can be complicated by lipase activity of transfected proteins. For example, it is very difficult for cells coexpressing ABHD5 and ATGL to accumulate lipid droplets. Under these conditions, it is useful to use lipase dead (e.g., ATGL S47A) mutants.

3.2.2 Transfection

Two days before imaging, cells are trypsinized and plated into six-well cluster plates containing 25 mm glass coverslips (# 1.5). Cells are allowed to attach overnight and are transfected the following day. Efficient transfection can be achieved using any of several commercially available transfection reagents. For 3T3-L1 cells, the use of Lipofectamine LTX with Plus reagent (Invitrogen, Cat. # 15338-100; following the manufacturer's instructions) yields 10–40% transfection efficiency. For preadipocytes, 400 μ*M* oleic acid/0.5% BSA is added immediately after cells are returned to full media following transfection.

3.2.3 Imaging

Cells are imaged 1–2 days following transfection. A single coverslip is removed from the cluster plate using fine forceps and placed in a recording chamber, the design of which may vary depending on coverslip configuration. For 25 mm coverslips, we use an Attofluor cell chamber (Invitrogen, Cat. # A-7816), which allows imaging in a static bath. Krebs–Henseleit buffer containing 25 m*M* HEPES and 1% BSA (Sigma-Aldrich, Cat. # K4002) is gently added immediately after securing the coverslip into the chamber. Cells are typically imaged at room temperature in a normal atmosphere for up to 150 min.

3.2.3.1 Colocalization of fluorescent proteins

Routine imaging is performed using an Olympus IX81 equipped with a spinning disc confocal unit and 40× (0.9 NA) or 60× (1.2 NA) apochromatic objectives. Static colocalization of fluorescent proteins can be performed as detailed earlier. For typical trafficking experiments using lipolysis activators, images are captured at a rate of 1–2 frames per minute in confocal mode. Excitation and camera gain are set so that each exposure is less than 1 s and is usually on the order of 100–200 ms. It is important to match the capture frequency to the phenomenon under study: too slow will miss important events, whereas too fast may unnecessarily bleach the sample. A typical experimental design is to capture three or more control images, then add lipolysis activators or vehicle, and continue imaging for 10–30 min. Quantification of protein trafficking and colocalization is performed as follows. Lipid droplets are usually found as clusters in the perinuclear region. First, images are merged from the channels of interest, for example, PLIN1–EYFP and ECFP–HSL. The largest clusters of lipid droplets (detected in the PLIN1–EYFP channel) are outlined, and fluorescence intensity values of individual pixels in the ECFP and EYFP channels are indexed and imported into GraphPad Prism 5. Linear regression analyses are performed on indexed intensity values of the same droplet region for each captured frame. The extent of colocalization is evaluated pixel by pixel using Pearson's linear regression. Since PLIN1–EYFP remains bound to the LD surface, the magnitude of translocation to droplets by soluble proteins (e.g., HSL or ATGL) can be assessed by an increase in the slope of the linear regression, that is, the amount of soluble protein bound per unit PLIN1–EYFP. The actual slope values under control conditions are somewhat arbitrary, since fluorescence intensities vary according to capture parameters and amounts of reporters present in any given cell. Therefore, to normalize across cells and experiments, the effect of stimulation is calculated as a ratio of stimulated slope to the control slope.

3.2.3.2 Fluorescence resonance energy transfer

The experimental setup for FRET is identical to the translocation analysis just described. FRET is performed using the three filter method

(Gordon, Berry, Liang, Levine, & Herman, 1998) and net FRET calculated using the FRET extension of the IPlabs software. Calculation of FRET signals involves determining FRET constants, which quantifies the fluorescence bleed-through between donor (ECFP) and acceptor (EYFP) channels. FRET constants are determined for each experimental day using 3T3-L1 cells transfected with single-fluorescent constructs. For a given experimental day, all acquisition parameters (exposure, camera gain, neutral density filters, etc.) are equivalent for channels and all cell records. To quantify the net FRET signals, the subcellular domain containing PLIN–EYFP lipid droplets is defined for each frame using the autosegmentation tool of IPlabs, and the calculated net FRET values of those pixels are summed for each cell in every frame. Values for each experiment are normalized to the maximal FRET values obtained for each cell. The adequacy of the FRET constants in eliminating bleed-through can be verified with independent singly transfected cells. In addition, FRET signals should only be observed at droplets containing ECFP, and photobleaching of the EYFP acceptor should eliminate the calculated FRET signals.

FRET is fundamentally a method for measuring the proximity of fluorescent molecules, and it is possible to observe FRET signals owing to molecular crowding at the droplet surface. It is therefore important to keep expression levels well below the maximum that can be targeted to droplets and to demonstrate that FRET occurs across a range of expression and is not affected by coexpression of untagged, control proteins that are also targeted to droplets.

3.2.3.3 Fluorescence recovery after photobleaching

Lipolytic proteins have different affinities for the lipid droplet surface, and the amount of time that a given protein resides on the lipid droplet surface can vary according to physiological conditions. FRAP is a technique for determining the temporal association of proteins in specific subcellular compartments, like the lipid droplet surface. The basic experimental design is to use laser excitation to bleach a specific patch of lipid droplets and then to record the recovery of fluorescence over time as a measure of dynamic exchange on the droplet surface. FRAP experiments are performed using a Leica TCS SP5 laser-scanning microscope with a 63 × (1.4 NA) oil objective. FRAP is performed using the Leica FRAP wizard module. Briefly, a region of interest (ROI) on a patch of lipid droplets is delineated and bleached by 1–2 scans (2.6 s/scan) with full laser power from the 405, 458, and 496 nm lasers. Fluorescence recovery for proteins with ECFP tags is then monitored for up to 15 min by 5% 405 nm laser excitation and 454–493 nm emission. For each FRAP experiment, fluorescence intensities are monitored in the bleached ROI and an unbleached ROI on droplets in the same focal plane. Monitoring beached and unbleached areas allows for one to correct for bleaching that may occur during the recovery/image acquisition phase of the experiment. Average fluorescence intensity of the bleached ROI is compared to prebleach level (with 100% bleach and 0% recovery = first postbleach scan), with correction for bleaching during acquisition. Rate and extent of recovery are estimated by fitting recovery values to a single exponential association function by nonlinear regression.

3.2.3.4 Bimolecular fluorescence complementation

3T3-L1 preadipocytes are grown on 18 mm coverslips and transfected with 500 ng each of the complementary Yn and Yc fusion constructs, along with 100 ng of an ECFP tracer to identify transfected cells. For qualitative assessment of protein–protein interactions, 3T3-L1 preadipocytes are cultured for 24 h following transfection at 32 °C in growth media supplemented with oleic acid and then fixed in PBS containing 1% paraformaldehyde. Reduced temperature facilitates fluorophore maturation (Hu et al., 2002). Transfected cells are identified by ECFP fluorescence and cells scored as to the presence or absence of EYFP fluorescence. The rate of false positives arising from cellular autofluorescence (i.e., background) is usually <2%.

4. IMAGING INTRACELLULAR FATTY ACID PRODUCTION

4.1. Materials

4.1.1 Intracellular fatty acid bipartite sensor constructs

Intracellular production of fatty acids can be monitored using a bimolecular system (Mottillo et al., 2012) (Fig. 15.2A) consisting of the ligand-binding domain (LBD) of mouse PPAR-α fused to the C-terminus of full-length PLIN1 and the nuclear receptor binding domain of steroid receptor coactivator 1 (SRC-1) fused to the C-terminus of EYFP (Clontech). In this system, the binding of fatty acids to the PPAR-α LBD recruits EYFP to lipid droplets via ligand-dependent interactions of the LxxLL motifs in SRC-1 (Fig. 15.2A). The "sensor" LBD component can be additionally targeted to other intracellular compartments such as ER (McFie, Banman, Kary, & Stone, 2011) and mitochondria (Means et al., 2011).

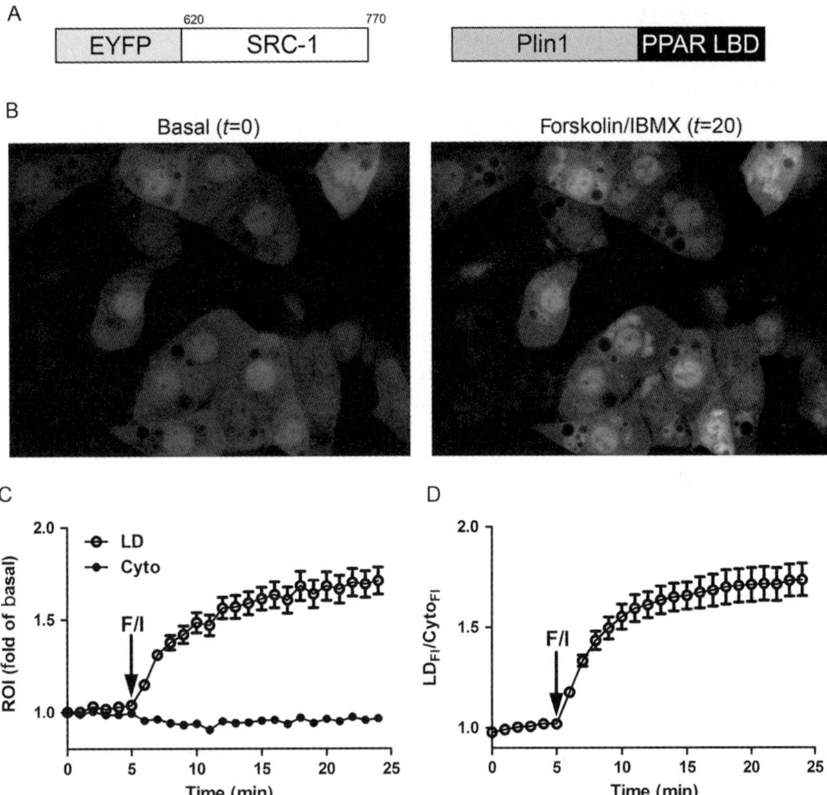

Figure 15.2 Fluorescence sensor for imaging intracellular fatty acid production. (A) A schematic of the constructs used for the fluorescence fatty acid sensor (amino acids for SRC1 are shown). PLIN1, perilipin-1; LBD, ligand-binding domain. (B) EYFP fluorescence images of the lipid droplet fatty acid sensor stably expressed in BA cells. Basal image capture at $t=0$ and at $t=20$ min, 15 min after addition of forskolin/IMBX (F/I). (C) Quantification of the region of interest (ROI) from lipid droplet (LD) or cytoplasm (Cyto) and expressed as a fold of basal. (D) Data were further expressed as a ratio of the fluorescent intensity from the LD (LD_{FI}) over the fluorescence intensity from the cytoplasm ($Cyto_{FI}$). (See color plate.)

4.2. Methods

4.2.1 Cell culture and transfections

Immortalized brown adipocytes are grown and differentiated as previously described (Uldry et al., 2006). Differentiated brown adipocytes (4–5 days postinduction) are transfected with Lipofectamine LTX with Plus reagent. Twenty-five millimeter coverslips are placed in wells of a six-well (35 mm) cluster plate. A transfection mixture is prepared containing (for each well)

0.75 μg of PPAR–PLIN1, 0.75 μg of EYFP-SRC1, 1.5 μl of Plus reagent, and 4 μl of LTX (Invitrogen) in 1 ml of growth media. Differentiated brown adipocyte cultures are trypsinized and resuspended at 12 million cells/ml of full media. The cells (200 μl) are applied to individual wells containing transfection mixture and incubated overnight. The media is changed the following day and cultures are imaged 1–2 days posttransfection.

4.2.2 Imaging and analysis

Images are acquired as described earlier for EYFP fluorescence using a 40× (0.9 NA) or 60× (1.2 NA) water immersion lens. The EYFP reporter is imaged every 30–60 s over a 30 min time course. Typically, basal fluorescence is recorded for the first 5 min, followed by stimulation of lipolysis with forskolin/IBMX over 20 min, after which synthetic PPAR-α ligand (Wy 14,643, 100 μ*M*; Enzo Life Sciences, Cat. # BML-GR200-0050) is added for 5 min to determine maximal response. Examples of images from brown adipocytes that express the lipid droplet reporter are shown before ($t=0$ min; Fig. 15.2B) and after stimulation with forskolin/IBMX ($t=20$ min; Fig. 15.2B). The ROI (i.e., lipid droplets or cytoplasm) from two to three frames for each treatment (i.e., basal, forskolin/IBMX, or ligand) is quantified using an appropriate program such as ImageJ. Data are initially normalized by subtracting basal fluorescence or by expressing as a fold of basal fluorescence (Fig. 15.2C). Depending on the experiment, data can be also normalized to the maximal effect of synthetic PPAR ligand to probe effects such as inhibition of lipolysis (Mottillo et al., 2012). Additionally, data can be expressed as the fluorescence intensity of the LD (LD_{FI}) relative to the intensity of the cytoplasm ($Cyto_{FI}$), which controls for nonspecific changes in fluorescence intensity due to minor changes in excitation or focal plane (Fig. 15.2D).

ACKNOWLEDGMENTS

This work was supported by NIH Grants RO1DK76229 and R21DK90598, and by a grant from the Department of Veterans Affairs.

REFERENCES

Bell, M., Wang, H., Chen, H., McLenithan, J. C., Gong, D. W., Yang, R. Z., et al. (2008). Consequences of lipid droplet coat protein downregulation in liver cells: Abnormal lipid droplet metabolism and induction of insulin resistance. *Diabetes*, *57*(8), 2037–2045.

Bosma, M., Sparks, L. M., Hooiveld, G. J., Jorgensen, J. A., Houten, S. M., Schrauwen, P., et al. (2013). Overexpression of PLIN5 in skeletal muscle promotes oxidative gene expression and intramyocellular lipid content without compromising insulin sensitivity. *Biochimica et Biophysica Acta*, *1831*(4), 844–852.

Brasaemle, D. L., Dolios, G., Shapiro, L., & Wang, R. (2004). Proteomic analysis of proteins associated with lipid droplets of basal and lipolytically stimulated 3T3-L1 adipocytes. *The Journal of Biological Chemistry, 279*(45), 46835–46842.

Brasaemle, D. L., & Wolins, N. E. (2012). Packaging of fat: An evolving model of lipid droplet assembly and expansion. *The Journal of Biological Chemistry, 287*(4), 2273–2279.

Day, R. N., & Davidson, M. W. (2012). Fluorescent proteins for FRET microscopy: Monitoring protein interactions in living cells. *Bioessays, 34*(5), 341–350.

DiDonato, D., & Brasaemle, D. L. (2003). Fixation methods for the study of lipid droplets by immunofluorescence microscopy. *The Journal of Histochemistry and Cytochemistry, 51*(6), 773–780.

Gordon, G. W., Berry, G., Liang, X. H., Levine, B., & Herman, B. (1998). Quantitative fluorescence resonance energy transfer measurements using fluorescence microscopy. *Biophysical Journal, 74*(5), 2702–2713.

Granneman, J. G., Moore, H. P., Granneman, R. L., Greenberg, A. S., Obin, M. S., & Zhu, Z. (2007). Analysis of lipolytic protein trafficking and interactions in adipocytes. *The Journal of Biological Chemistry, 282*(8), 5726–5735.

Granneman, J. G., Moore, H. P., Krishnamoorthy, R., & Rathod, M. (2009). Perilipin controls lipolysis by regulating the interactions of AB-hydrolase containing 5 (Abhd5) and adipose triglyceride lipase (Atgl). *The Journal of Biological Chemistry, 284*(50), 34538–34544.

Granneman, J. G., Moore, H. P., Mottillo, E. P., & Zhu, Z. (2009). Functional interactions between Mldp (LSDP5) and Abhd5 in the control of intracellular lipid accumulation. *The Journal of Biological Chemistry, 284*(5), 3049–3057.

Granneman, J. G., Moore, H. P., Mottillo, E. P., Zhu, Z., & Zhou, L. (2011). Interactions of perilipin-5 (Plin5) with adipose triglyceride lipase. *The Journal of Biological Chemistry, 286*(7), 5126–5135.

Hu, C. D., Chinenov, Y., & Kerppola, T. K. (2002). Visualization of interactions among bZIP and Rel family proteins in living cells using bimolecular fluorescence complementation. *Molecular Cell, 9*(4), 789–798.

Magliery, T. J., Wilson, C. G., Pan, W., Mishler, D., Ghosh, I., Hamilton, A. D., et al. (2005). Detecting protein-protein interactions with a green fluorescent protein fragment reassembly trap: Scope and mechanism. *Journal of the American Chemical Society, 127*(1), 146–157.

McFie, P. J., Banman, S. L., Kary, S., & Stone, S. J. (2011). Murine diacylglycerol acyltransferase-2 (DGAT2) can catalyze triacylglycerol synthesis and promote lipid droplet formation independent of its localization to the endoplasmic reticulum. *The Journal of Biological Chemistry, 286*(32), 28235–28246.

Means, C. K., Lygren, B., Langeberg, L. K., Jain, A., Dixon, R. E., Vega, A. L., et al. (2011). An entirely specific type I A-kinase anchoring protein that can sequester two molecules of protein kinase A at mitochondria. *Proceedings of the National Academy of Sciences of the United States of America, 108*(48), E1227–E1235.

Moore, H. P., Silver, R. B., Mottillo, E. P., Bernlohr, D. A., & Granneman, J. G. (2005). Perilipin targets a novel pool of lipid droplets for lipolytic attack by hormone-sensitive lipase. *The Journal of Biological Chemistry, 280*(52), 43109–43120.

Mottillo, E. P., Bloch, A. E., Leff, T., & Granneman, J. G. (2012). Lipolytic products activate peroxisome proliferator-activated receptor (PPAR) α and δ in brown adipocytes to match fatty acid oxidation with supply. *Journal of Biological Chemistry, 287*(30), 25038–25048.

Murphy, S., Martin, S., & Parton, R. G. (2009). Lipid droplet-organelle interactions; sharing the fats. *Biochimica et Biophysica Acta, 1791*(6), 441–447.

Negoescu, A., Labat-Moleur, F., Lorimier, P., Lamarcq, L., Guillermet, C., Chambaz, E., et al. (1994). F(ab) secondary antibodies: A general method for double immunolabeling with primary antisera from the same species. Efficiency control by chemiluminescence. *The Journal of Histochemistry and Cytochemistry*, *42*(3), 433–437.

Rodbell, M. (1964). Metabolism of isolated fat cells. I. Effects of hormones on glucose metabolism and lipolysis. *The Journal of Biological Chemistry*, *239*, 375–380.

Subramanian, V., Rothenberg, A., Gomez, C., Cohen, A. W., Garcia, A., Bhattacharyya, S., et al. (2004). Perilipin A mediates the reversible binding of CGI-58 to lipid droplets in 3T3-L1 adipocytes. *The Journal of Biological Chemistry*, *279*(40), 42062–42071.

Sztalryd, C., Xu, G., Dorward, H., Tansey, J. T., Contreras, J. A., Kimmel, A. R., et al. (2003). Perilipin A is essential for the translocation of hormone-sensitive lipase during lipolytic activation. *The Journal of Cell Biology*, *161*(6), 1093–1103.

Uldry, M., Yang, W., St-Pierre, J., Lin, J., Seale, P., & Spiegelman, B. M. (2006). Complementary action of the PGC-1 coactivators in mitochondrial biogenesis and brown fat differentiation. *Cell Metabolism*, *3*(5), 333–341.

Viswanadha, S., & Londos, C. (2006). Optimized conditions for measuring lipolysis in murine primary adipocytes. *Journal of Lipid Research*, *47*(8), 1859–1864.

Wang, H., Bell, M., Sreenevasan, U., Hu, H., Liu, J., Dalen, K., et al. (2011). Unique regulation of adipose triglyceride lipase (ATGL) by perilipin 5, a lipid droplet-associated protein. *The Journal of Biological Chemistry*, *286*(18), 15707–15715.

Wolins, N. E., Brasaemle, D. L., & Bickel, P. E. (2006). A proposed model of fat packaging by exchangeable lipid droplet proteins. *FEBS Letters*, *580*(23), 5484–5491.

Zebisch, K., Voigt, V., Wabitsch, M., & Brandsch, M. (2012). Protocol for effective differentiation of 3T3-L1 cells to adipocytes. *Analytical Biochemistry*, *425*(1), 88–90.

Zechner, R., Zimmermann, R., Eichmann, T. O., Kohlwein, S. D., Haemmerle, G., Lass, A., et al. (2012). FAT SIGNALS—Lipases and lipolysis in lipid metabolism and signaling. *Cell Metabolism*, *15*(3), 279–291.

CHAPTER SIXTEEN

Evaluation of Protein Phosphorylation During Adipogenesis

Xi Li*, Rong Zeng†, Qi-Qun Tang*,[1]

*Key Laboratory of Molecular Medicine, The Ministry of Education, Department of Biochemistry and Molecular Biology, Fudan University Shanghai Medical College, Shanghai, PR China

†Key Laboratory of Systems Biology, Shanghai Institutes for Biological, Sciences, Chinese Academy of Sciences, Shanghai, PR China

[1]Corresponding author: e-mail address: qqtang@shmu.edu.cn

Contents

Methods in Enzymology, Volume 538
ISSN 0076-6879
http://dx.doi.org/10.1016/B978-0-12-800280-3.00016-5

Abstract

Adipocyte differentiation is a complex process that involves the sequential expression of various adipocyte-specific genes controlled by signaling pathways and transcription factors for which phosphorylation plays a crucial regulatory role. CCAAT/enhancer-binding proteins and peroxisome proliferator-activated receptors are the most important transcriptional regulators in adipogenesis, and the functions of these proteins are regulated by various phosphorylation events. Because cultured 3T3-L1 preadipocytes are commonly used as a model for adipocyte differentiation, we used these cells for a proteomic analysis to identify kinases, phosphatases, and phosphosites that participate in adipogenesis. In addition to the phosphoproteomic analysis, we provide a detailed description of Western blotting, an *in vitro* phosphorylation assay, enzyme-linked immunosorbent assay, and phosphorylation site mutagenesis to fully characterize the phosphorylation of proteins and verify their roles in adipogenesis.

1. INTRODUCTION

The reversible phosphorylation of proteins regulates nearly every aspect of cell biology. Phosphorylation and dephosphorylation, catalyzed by protein kinases and protein phosphatases, respectively, can modify the function of a protein in many ways. For example, phosphorylation status of a protein can increase or decrease its biological activity, stabilize it or promote its destruction, facilitate or inhibit movement between subcellular compartments, or initiate or disrupt protein–protein interactions (Cohen, 2002). In eukaryotes, it is mainly serine, threonine, and, to a lesser extent, tyrosine residues that are targeted for phosphorylation (Mann et al., 2002). Regulation of signaling transduction often relies heavily not only on the identity of phosphosites but also on the timing of phosphorylation events. Outputs of signaling, including the transcription of target genes, often rely on the sequential phosphorylation of various signaling molecules.

Adipocyte differentiation is a complex process that requires sequential expression of multiple adipocyte-specific genes driven by signaling events which phosphorylation plays a crucial role. For example, increased phosphorylation of AMP-activated protein kinase (AMPK) on Thr^{172} and insulin-induced phosphorylation AKT on Ser^{473} both impair adipocyte differentiation (Moreno-Navarrete, Ortega, Ricart, & Fernandez-Real, 2009). In white adipose tissue, the formation of new adipocytes from precursor cells contributes to adipose tissue expansion. Numerous studies demonstrate that the extracellular signal-regulated kinase (ERK) pathway regulates adipogenesis at each stage in the process, from multipotent stem cells to differentiated adipocytes (Bost, Aouadi, Caron, & Binetruy, 2005). ERK directly regulates adipocyte differentiation effectors, such as the transcription factors CCAAT/enhancer-binding protein β(C/EBP-β) (Tang et al., 2005) and peroxisome proliferator-activated receptor-γ (PPARγ) (Rosen, Walkey, Puigserver, & Spiegelman, 2000). C/EBPs and PPARs are the most important transcriptional regulators that participate in adipogenesis, and their functions are regulated by phosphorylation.

1.1. PPARγ

PPARs are a group of related ligand-dependent nuclear receptors that function as transcriptional regulators of adipogenic differentiation and lipid metabolism. Similar to other nuclear receptors, the activities of PPARs are modulated by phosphorylation, and the ability of many different kinases and phosphatases to target PPARs makes these transcription factors a point of cross talk between different signaling pathways. Phosphoregulation of PPAR activity is complex and controversial. The major phosphorylation site of PPARs, which is targeted by both the ERK and the c-Jun N-terminal kinase (JNK)–mitogen-activated protein kinase (MAPK) pathways (Camp, Tafuri, & Leff, 1999), was mapped at Ser^{82} in mouse PPARγ1, which corresponds to Ser^{112} in mouse PPARγ2 (Shao, Rangwala, Bailey, et al., 1998). Mutation of the main MAPK site of phosphorylation in the mutant protein $PPAR\gamma2^{S112D}$ results in decreased ligand affinity (Shao et al., 1998). Several growth factors that inhibit adipogenic differentiation promote MAPK-mediated phosphorylation of the dominant adipogenic transcription factor PPARγ at Ser^{112} and reduce its transcriptional activity (Hu, Kim, Sarraf, & Spiegelman, 1996).

Downregulation of the MAPK p38 increases adipogenesis in embryonic stem cells (Aouadi et al., 2006). The activation of the MEK/ERK signaling pathway by high serum concentration promoted PPAR expression and phosphorylation and subsequently enhanced adipogenic differentiation of mouse stem cells (Wu et al., 2010). Activation of AMPK dramatically inhibited the expression of PPARγ and reduced the presence of adipocytes after 10 days of differentiation. Inhibition of AMPK enhanced the expression of PPARγ. AMPK activity is inversely related to adipogenesis in 3T3-L1 cells (Tong et al., 2008). Obesity induced in mice by high-fat feeding activates the protein kinase Cdk5 (cyclin-dependent kinase 5) in adipose tissues, resulting in phosphorylation of PPARγ at Ser^{273}. This modification of PPARγ does not alter its adipogenic capacity, but instead leads to dysregulation of many genes whose expression is altered in obesity, including a reduction in the expression of the insulin-sensitizing adipokine, adiponectin (Choi et al., 2010).

Recent reports demonstrate that adipocyte-specific *nuclear receptor corepressor* (*NCoR*) knockout (AKO) mice were obese but with enhanced insulin sensitivity in the liver, muscle, and adipose tissue. In adipose tissue from AKO mice, PPARγ-responsive genes were upregulated and CDK5-mediated phosphorylation of PPARγ on Ser^{273} was reduced, thus creating a constitutively active PPARγ state. This identifies NCoR as an adaptor protein that enhances the ability of CDK5 to associate with and phosphorylate PPARγ. The dominant function of adipocyte NCoR is to transrepress PPARγ and promote PPARγ Ser^{273} phosphorylation such that the absence of NCoR leads to adipogenesis, reduced inflammation, and enhanced systemic insulin sensitivity, phenocopying the thiazolidinedione-treated state (Li et al., 2011).

1.2. C/EBPs

C/EBPs are required during adipogenesis for insulin-stimulated glucose uptake. Upon induction of differentiation, C/EBPβ is expressed immediately, triggering a transcriptional cascade by transcriptionally inducing the expression of C/EBPα and PPARγ. C/EBPα and PPARγ are pleiotropic transcriptional activators of genes that give rise to the adipocyte phenotype.

Phosphorylation of C/EBPα on Ser^{21} has been implicated in the regulation of C/EBP granulopoiesis (Ross et al., 2004) and hepatic gene expression. When Ser^{21} was mutated to Ala (S21A) to mimic dephosphorylation, the S21A-C/EBPα mutant protein supports normal lipid accumulation and expression of many adipogenic markers in two cell models lacking endogenous C/EBPα expression. However, S21A-C/EBPα had reduced ability to activate the *Glut4* promoter, thereby causing diminished GLUT4 and adiponectin expression, and reduced insulin-stimulated glucose uptake. Mice in which endogenous C/EBPα were replaced with S21A- C/EBPα displayed reduced GLUT4 and adiponectin protein abundance in epididymal adipose tissue and increased blood glucose compared to wild-type littermates. These results suggest that phosphorylation of C/EBPα on Ser^{21} may regulate adipocyte gene expression and whole-body glucose homeostasis (Cha et al., 2008).

Studies have shown that C/EBPβ is sequentially phosphorylated during differentiation of 3T3-L1 adipocytes (Tang et al., 2005). C/EBPβ is rapidly

(≤4 h) expressed and phosphorylated on Thr188 by MAPK after induction of differentiation. MAPK/ERK itself is rapidly (≤1 h) phosphorylated/activated after induction of differentiation. When the cells reenter cell cycle, the phosphorylation/activity of MAPK (phospho-ERK) falls off abruptly after 10–12 h. As cells enter S-phase, cdk2/cyclin A maintains C/EBP-β phosphorylation on Thr^{188} (Li, Kim, Gronborg, Lane, & Tang, 2007). Later, C/EBPβ is phosphorylated by GSK3β on Ser^{184} or Thr^{179} at the onset of S-phase concurrent with the translocation of GSK3β into the nucleus. Phosphorylation of Thr^{188} appears to prime C/EBPβ for subsequent phosphorylation on Ser^{184} or Thr^{179}. Studies indicate that this dual phosphorylation induces a conformational change in C/EBPβ that allows dimerization through its C-terminal leucine zipper domain (Kim, Tang, Li, & Lane, 2007). Dimerization brings the adjacent basic regions into position to hold the C/EBP regulatory element of the gene in a scissor-like grip. All these actions of the dual phosphorylated C EBPβ facilitate its DNA-binding and transcriptional regulatory activities.

Our recent studies showed that phosphorylation also contributes to the stability of C/EBPβ. Both *ex vivo* and *in vitro* experiments indicated that phosphorylation on Thr^{188} by MAPK/cyclin A/cdk2 protected C/EBPβ from calpain-mediated proteolysis (Zhang et al., 2012).

Although the JNK pathway has not been directly implicated in adipocyte differentiation, addition of inhibitors of the MAPK p38 early in the differentiation process of 3T3-L1 preadipocytes decreases C/EBPβ phosphorylation (Engelman, Lisanti, & Scherer, 1998) and adipocyte formation compared to untreated cells. Other reports demonstrate that p38 phosphorylates C/EBP homologous protein 10 (CHOP10) on Ser^{78} and Ser^{81} that is dominant-negative regulator of C/EBPβ (Wang & Ron, 1996). A mutant of CHOP10 that cannot be phosphorylated ($CHOP^{Ser78,\ 81Ala}$) inhibits adipogenesis poorly as compared to wide-type CHOP10, suggesting that p38's MAPK activity blocks adipocyte differentiation.

2. PHOSPHOPROTEOMIC ANALYSIS

For investigation of the effect of site-specific phosphorylation on transcriptional activity in 3T3-L1 preadipocyte differentiation, the power of the online yin–yang multidimensional liquid chromatography tandem mass spectrometry (MDLC-MS/MS) system resulted in the large-scale identification of both the proteome and phosphoproteome in one analysis. When combined with stable isotope labeling by amino acids in cell culture (SILAC), this allowed simultaneous quantification of protein abundance and site-specific phosphorylation, thus giving direct insight into the effect of phosphorylation on transcriptional regulatory activity during adipogenesis (Wu et al., 2009).

2.1. Required materials

Stable isotope-containing amino acid ($^{13}C_6$, $^{15}N_2$) lysine (Cambridge Isotope Laboratories, Andover, MA)
Complete protease inhibitor cocktail tablets (Roche Diagnostics GmbH, Mannheim, Germany)
Urea, dithiothreitol (DTT), ammonium bicarbonate, and iodoacetamide (Bio-Rad Laboratories, CA)
Trypsin (Promega, Madison, WI)
Milli-Q system (Millipore, Bedford, MA)
Linear ion trap (LTQ)-Orbitrap hybrid mass spectrometer (Thermo Electron, San Jose, CA) equipped with a nanospray ion source operated in data-dependent mode in which acquisition was automatically switched between MS in the Orbitrap and MS/MS in the LTQ
D9785 deficient medium (Dulbecco's modified Eagle's medium/F-12 1:1 mixture deficient in L-lysine and three other amino acids) (Sigma-Aldrich, St Louis, MO)
Sodium orthovanadate (Sigma-Aldrich, St Louis, MO)
Sodium fluoride (Sigma-Aldrich, St Louis, MO)
Acetone (Sigma-Aldrich, St Louis, MO)
Ethanol (Sigma-Aldrich, St Louis, MO)
Acetic acid (Sigma-Aldrich, St Louis, MO)
NH_4HCO_3 (Sigma-Aldrich, St Louis, MO)
Dexamethasone (Sigma-Aldrich, St Louis, MO)
3-Isobutyl-1-methylxanthine (Sigma-Aldrich, St Louis, MO)
Insulin (Sigma-Aldrich, St Louis, MO)

2.2. Cell culture and differentiation induction

1. 3T3-L1 preadipocytes were cultured in D9785 medium supplemented with 10% dialyzed bovine serum containing either light or heavy ($^{13}C_6$, $^{15}N_2$) lysine, respectively, in an atmosphere of 10% CO_2 at 37 °C.

2. Complete replacement of light lysine or heavy lysine was ensured prior to the experiment. Adapt one population of 3T3-L1 preadipocytes to D9785 medium with either light or heavy ($^{13}C_6$, $^{15}N_2$) lysine at 37 °C, 10% CO_2 for at least five passages.
3. Two days postconfluence, 3T3-L1 fibroblasts cultured with light lysine were stimulated with 1 μg/ml insulin, 115 μg/ml 3-isobutyl-1-methylxanthine, and 0.39 μg/ml dexamethasone for 1 h at 37 °C.
4. The 3T3-L1 cells cultured with heavy lysine were left untreated as a control.

2.3. Sample preparation

1. Following stimulation, cells from the light and heavy lysine-containing cultures were lysed in buffer containing 8 *M* urea, 4% CHAPS, 40 m*M* Tris, 65 m*M* DTT, protease inhibitor mixture, 1 m*M* NaF, and 1 m*M* Na_3VO_4.
2. Cell lysates were ultrasonicated and then centrifuged at 25,000 × *g* at 4 °C for 1 h to remove the insoluble material.
3. After centrifugation, the protein concentration of each cell lysate was quantitated by the Bradford assay, and then equal amounts of light and heavy cell lysates were combined for the subsequent steps.

2.4. Protein precipitation and digestion

1. 250 μg each of heavy and light lysates were mixed briefly, 2 μl of 1 *M* DTT was added, and the mixture was incubated at 37 °C for 2.5 h.
2. 10 μl of 1 *M* iodoacetamide was added and incubated with the mixture for an additional 40 min at room temperature in darkness.
3. The protein mixtures were subjected to precipitation at 1:5 (v/v) with 50% acetone, 50% ethanol, and 0.1% acetic acid.
4. After precipitation for 20 h at −20 °C, the mixtures were centrifuged at 14,000 × *g* and washed with cold ethanol twice.
5. The precipitates were dissolved in 50 m*M* NH_4HCO_3 and incubated with trypsin (25:1) at 37 °C for 20 h.
6. The trypsinized peptide mixtures were collected and lyophilized for further analysis.

2.5. The online Yin–yang MDLC-MS/MS system

1. The peptide fractionation and MS identification were performed on the online yin–yang MDLC system coupled with a mass spectrometer

(Dai et al., 2007). Two subsystems were built. One subsystem involves a strong cation exchange (SCX) column (320 μm × 100 mm; Column Technology Inc.), two C_{18} trap columns (300 μm × 5 mm; Agilent Technologies), and an analytical C_{18} column (75 μm × 150 mm; Column Technology Inc.). For the other subsystem, a strong anion exchange (SAX) column (320 μm × 100 mm; Column Technology Inc.) was utilized in place of the SCX column in the first subsystem. The SCX column and SAX column subsystems were equilibrated by pH 2.5 and 8.5 buffers, respectively.

2. The peptide mixture was dissolved in pH 2.5 buffer solution (0.01 *M* NaH_2PO_4-H_3PO_4) and loaded by the Surveyor Autosampler (Thermo Electron, San Jose, CA) into the subsystem with the SCX column mounted firstly, and the flow-through peptide mixture was collected and lyophilized.
3. The flow-through peptide mixture from the SCX column was redissolved in 80 μl of pH 8.5 buffer (0.01 *M* NaH_2PO_4-Na_2HPO_4) and loaded into the SAX column subsystem. The pH continuous gradient elution was performed from pH 2.5 to 8.5 for the SCX subsystem and from pH 8.5 to 2.0 for the SAX subsystem, respectively.
4. Ten fractions for each subsystem were used for peptide separation from the SCX or SAX column to the C_{18} trap column followed by further analysis by reverse-phase high-performance liquid chromatography (HPLC).
5. The HPLC solvents for the reverse phase were 0.1% formic acid (v/v) aqueous (A) and 0.1% formic acid (v/v) in acetonitrile (B). The reverse-phase gradient was from 2% to 35% mobile phase B in 165 min at a 120 μl/min flow rate before the split and at 250 nl/min after the split.
6. The mass spectrometer was set so that one full MS scan was followed by MS/MS scans on the 10 most intense ions from the full MS spectrum with the following dynamic exclusion TM settings: repeat counts, 2; repeat duration, 30 s; and exclusion duration, 120 s. The resolving power of the LTQ-Orbitrap mass analyzer was set at 100,000 for the precursor ion scans ($m/\Delta m$ 50% at m/z 400).

2.6. Database searching, phosphorylation site localization, and quantitative analysis using census

1. All data files were created using Bioworks 3.2 with precursor mass tolerance of 500 ppm, and all acquired MS/MS spectra were automatically searched against the mouse International Protein Index protein sequence

database (version 3.35) containing 51,490 protein entries combined with real protein and reverse sequences of proteins by using the TurboSEQUEST searching program (University of Washington, licensed to Thermo Finnigan). Trypsin was designated as the protease with two missed cleavage sites allowed. Carbamidomethylation was searched as a fixed modification, whereas isotope-labeled lysine (+8.014199 Da), phosphorylation of serine/threonine/tyrosine residues (+79.96633 Da), and oxidized methionine (+15.99492 Da) were allowed as variable modifications.

2. For nonphosphopeptide identification and phosphorylation site localization, all accepted SEQUEST results had a ΔCn score of at least 0.1 regardless of charge state, because $\Delta Cn \geq 0.1$ is significant for discriminating the first candidate peptide from the second candidate peptide.
3. All output results were combined using a software application developed in-house called Build Summary (PMID:22217156), and peptides and phosphopeptides were calculated separately and filtered to assure a false discovery rate of $\leq 0.5\%$. The false discovery rate was calculated based on the following formula: $\%\ \text{fal} = n_{\text{rev}}/(n_{\text{rev}} + n_{\text{real}})$ where n_{rev} is the number of peptide hits matched to "reverse" protein and n_{real} is the number of peptide hits matched to "real" protein (Elias & Gygi, 2007).
4. Each phosphopeptide spectrum was manually checked that it conformed to the following criteria:
   - **a.** Its fragment ion peaks had a high signal-to-noise ratio.
   - **b.** It showed sequential members of the b- or y-ion series with phosphorylation site included.
   - **c.** It showed intense proline-directed fragment ions.
   - **d.** In addition, the phosphoric acid neutral loss peaks were checked to assist in phosphorylation site identification. When a scan was matched to phosphopeptides with a common sequence and the same number of phosphates but with different phosphorylation sites, the site was considered as ambiguous when ΔCn was <0.08. Moreover, all identified phospho- and nonphosphopeptides were filtered with a precursor ion mass tolerance of 10 ppm.
5. To eliminate redundancy, if the same peptide(s) was assigned to multiple proteins, then the multiple proteins were clustered into a "protein group." If all of the peptides in one protein group (group A) were covered by another protein group that has an additional peptide assigned to it (group B), then protein group A was removed.

6. A single protein with the highest sequence coverage was selected from each protein group for further analysis.
7. For quantitative analysis, only the lysine-containing peptides were subjected to the Census program (version 1.28) as quantification candidates to determine the $^{12}C_6$,$^{14}N_2$-peptide/$^{13}C_6$,$^{15}N_2$-peptide ratio (Park, Venable, Xu, & Yates, 2008). The results of quantification were filtered so that peptides with determinant scores (R^2) ≥ 0.5 were retained, and the correction factor (ln) was set at 0.0 when data were exported. Peptides with negative R^2 were removed, and singleton peptides were discarded. To determine the quantification of protein expression, we measured the SILAC ratio of all nonphosphorylated peptides except for those that had corresponding phosphopeptides with the same sequence because changes in phosphorylation state may alter the portion of the protein in the nonphosphorylated form. Outliers were eliminated from all peptides that were assigned to the same protein using the biweight algorithm (Goodall, 1983), whereas the weighted mean of the peptide ratio was determined as the protein expression ratio. For the phosphopeptide quantification, outlier elimination and weighted mean determination were also processed with the biweight algorithm for all peptide hits of a unique peptide.

3. WESTERN BLOT ANALYSIS

Western blotting is the most common method used for assessing the phosphorylation state of a protein. Following separation of the biological sample with sodium dodecyl sulfate-polyacrylamide gel electrophoresis (SDS–PAGE) and subsequent transfer to a polyvinylidene fluoride (PVDF) membrane or nitrocellulose membrane, a phospho-specific antibody can be used to identify the protein of interest.

3.1. Required materials

Bradford assay reagents (Bio-Rad Laboratories, CA)
PVDF membrane (pore size = 0.2 μ*M*) (Immobilon-P^{SQ} transfer membrane, Millipore Corporation, Billerica, MA)
Pierce enhanced chemiluminescence (ECL) substrate (Life Technologies Corporation, Grand Island, NY)
Digital imaging system (ImageQuant LAS 4000 mini, GE Healthcare, UK)

3.2. Western blotting

1. 3T3-L1 preadipocytes were grown in high-glucose DMEM supplemented with 10% new born bovine serum. Two days postconfluence, the cells were stimulated with 1 μg/ml insulin, 115 μg/ml 3-isobutyl-1-methylxanthine, and 0.39 μg/ml dexamethasone. The cells were harvested with lysis buffer (2% SDS and 50 m*M* Tris–HCl, pH 7.0).
2. The total protein concentrations were determined with Bradford assay. Each sample containing 30 μg total proteins was loaded into separate wells and separated by SDS–PAGE.
3. The proteins were transferred from the gel onto PVDF membranes.
4. The PVDF membranes were blocked in blocking buffer (150 m*M* NaCl; 5 m*M* EDTA; 25 m*M* Tris–HCl, pH 7.5; 0.05% Tween-20; and 5% nonfat dry milk) at room temperature for 1 h.
5. The PVDF membranes were incubated with the corresponding primary antibodies diluted 1:1000 in blocking buffer at 4 °C overnight.
6. The PVDF membranes were washed three times in TTBS buffer (150 m*M* NaCl; 5 m*M* EDTA; 25 m*M* Tris–HCl, pH 7.5; and 0.05% Tween-20) at room temperature for 5 min each wash.
7. The PVDF membranes were incubated with horseradish peroxidase (HRP)-conjugated secondary antibodies (Santa Cruz Biotechnology, Inc.) diluted 1:10000 in blocking buffer at room temperature for 1 h.
8. The PVDF membranes were washed three times in TTBS buffer at room temperature for 10 min each wash.
9. Bound phospho-specific antibodies were detected with ECL according to the manufacturer's instructions and photographed with a digital imaging system.

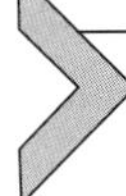

4. PHOSPHORYLATION ANALYSIS BY *IN VITRO* RADIOLABELING

4.1. Required materials

pGEX-6P (GE Healthcare, Piscataway, NJ)
Escherichia coli strain BL21(DE3) pLysS (Novagen, San Diego, CA)
Spectrophotometer (BioTek Instruments, Winooski, VT)
Triton X-100 (Sigma-Aldrich, St Louis, MO)
DNase I (Sigma-Aldrich, St Louis, MO)
RNase A/T1 (Sigma-Aldrich, St Louis, MO)

GSH Sepharose affinity resin (GE Healthcare, Piscataway, NJ)
PreScission Protease (GE Healthcare, Piscataway, NJ)
Bradford assay reagents (Bio-Rad Laboratories, CA)
MAPK (Calbiochem, San Diego, CA) or cdk2-cyclin A (Upstate Biotechnology, Lake Placid, NY) or GSK3β (Upstate Biotechnology, Lake Placid, NY)
All chemical reagent (Sigma-Aldrich, St Louis, MO)

4.2. Expression of GST fusion proteins

1. A cDNA encoding the protein of interest was cloned downstream of the glutathione S-transferase (GST) coding region in pGEX-6P.
2. The plasmid was transformed into *Escherichia coli* strain BL21 (DE3) pLysS. A single colony was propagated overnight in 3 ml of LB media containing ampicillin and chloramphenicol and was then diluted (1:100) into 500 ml of fresh LB media the next day and cultured until an A600 of 0.6–0.7 was reached.
3. Expression of the GST fusion proteins was induced by addition of 0.5 m*M* isopropyl β-D-thiogalactoside (IPTG) for 3 h.
4. The cells were harvested and resuspended in 25 ml of PBS (137 m*M* NaCl, 2.7 m*M* KCl, 10 m*M* Na_2HPO_4, and 2 m*M* KH_2PO_4) containing 1% Triton X-100. After lysis by one cycle of freeze-thawing, the cell suspension was treated with DNase I and RNase A/T1 on ice for 10 min in the presence of 0.5 *M* NaCl and 5 m*M* DTT to disrupt DNA–protein interactions.
5. After centrifugation at 12,000 × *g* for 10 min, 250–500 μl bead volume of GSH-Sepharose4B affinity resin (GE Healthcare, Piscataway, NJ) was added to the supernatant and mixed overnight at 4 °C.
6. The beads were washed three times with PBS, once with PreScission cleavage buffer (50 m*M* Tris–Cl, pH 7.5; 150 m*M* NaCl; 1 m*M* EDTA; and 1 m*M* DTT), and then treated with 80 units of PreScission Protease (GE Healthcare) diluted in 960 μl of buffer.
7. Following purification, proteins were resuspended in NETN buffer (20 m*M* Tris–HCl; 0.1 *M* NaCl; 1 m*M* EDTA; and 0.5%NP-40, pH 8.0), and their concentrations were determined by Bradford assay.

4.3. Protein kinase assays

1. Two micrograms of GST fusion proteins were incubated with activated kinase (MAPK, cdk2-cyclin A, or GSK3-β) or a combination of two

kinases in kinase buffer (50 m*M* Hepes, pH 7.0; 10 m*M* $MgCl_2$; 1 m*M* DTT) plus 20 μCi [γ-^{32}P] ATP (1Ci = 37 GBq) at 30 °C for 30 min.
2. The proteins were separated by SDS–PAGE.
3. Phosphorylation was detected with X-ray film by autoradiography.

5. ENZYME-LINKED IMMUNOSORBENT ASSAY

Enzyme-linked immunosorbent assay (ELISA) is a powerful method for measuring protein phosphorylation. ELISA is more quantitative than Western blotting and is particularly useful for studying kinase and phosphatase activity. The format for this microplate-based assay typically utilizes a capture antibody against the protein of interest and a detection antibody specific for the phosphorylation site to be analyzed.

5.1. Required materials

Polyvinylchloride (PVC) microtiter plate (R&D Systems, Minneapolis, MN)
Capture antibody for target protein
Site-specific phosphorylation antibody for target protein
HRP-conjugated secondary antibody (GE Health, Piscataway, NJ)
TMB substrate (Sigma-Aldrich, St Louis, MO)
HRP reaction STOP solution (2 *M* H_2SO_4) (Sigma-Aldrich, St Louis, MO)
Spectrophotometer (BioTek Instruments, Winooski, VT)

5.2. Sandwich ELISA

1. Obtain a PVC microtiter plate and rinse three times with PBS.
2. Immobilize the capture antibody on the bottom of each well by adding approximately 50 μl of antibody diluted to 20 μg/ml in PBS to each well. PVC will bind approximately 100 ng/well.
3. Incubate the plate overnight at 4 °C to allow complete binding.
4. Recover the antibody solution for reuse, if desired, by flicking the plate over a suitable container. Wash the wells twice with PBS. A 500 ml squirt bottle is convenient for this wash step.
5. Any remaining sites for protein binding on the microtiter plate must be saturated by incubating with blocking buffer. Fill the wells to the top with 3% BSA in PBS. Incubate for 2 h to overnight in a humid atmosphere at room temperature.

6. Discard the blocking buffer and wash the wells twice with PBS.
7. Add 50 μl of the samples containing the antigen (purified protein or cell lysate) to each well. Use titrations of each sample to facilitate accurate quantitation. All dilutions should be done in the blocking buffer (3% BSA in PBS). Incubate for at least 2 h at room temperature in a humid atmosphere.
8. Discard the samples and wash the plate four times with PBS.
9. Add the site-specific phosphorylation antibody. The amount added can be determined in preliminary experiments. For accurate quantitation, the second antibody should be used in excess. All dilutions should be done in the blocking buffer.
10. Incubate for 2 h or more at room temperature in a humid atmosphere.
11. Discard the antibody solution and wash each well four times with PBS.
12. Add 100 μl of HRP-conjugated secondary antibody diluted 1:10000 in blocking buffer to each well. Incubate the plate for 30 min at 37 °C in a humid atmosphere.
13. Add 100 μl of TMB substrate (0.05 mg/ml tetramethylbenzidine 0.03% H_2O_2; 0.1 *M* citric acid; 0.2 *M* Na_2HPO_4, pH 5.0) to each well and incubate for 30 min at room temperature in a humid atmosphere.
14. Add 100 μl of STOP solution (2 *M* H_2SO_4) to each well to stop the reaction. Shake gently for a few seconds.
15. Read absorbance at 450 nm within 30 min after adding STOP solution.

6. MUTATION OF PHOSPHORYLATION SITES

We use site-directed mutagenesis to assess the importance of phosphorylation events on the function of phosphoproteins. Generally, we mutate phosphorylated residues to alanines to mimic the unphosphorylated state and substitute glutamic acid or aspartic acid for the phosphorylated residues to mimic the phosphorylated state.

6.1. Required materials

TaKaRa site-directed mutagenesis kit (TaKaRa Bio Group, Dalian, China)

Thermal cycler (Eppendorf, Hamburg, Germany)

6.2. Site-directed mutagenesis

1. Using guidelines provided in the kit, design the forward primer to include the mutation.
2. Set up the following reactions as indicated in the succeeding text, each in a total volume 50 μl:

Pyrobest DNA polymerase(5 U/μl)	0.25 μl
10 × Pyrobest buffer II	5 μl
dNTP mixture(each 2.5 m*M*)	4 μl
Forward primer (20 μ*M*)	1 μl
Reverse primer (20 μ*M*)	1 μl
Template	0.01–1 ng
ddH_2O	up to 50 μl

3. Place tubes into thermal cycler and run the reaction for 30 cycles with the following cycle parameters:
   94 °C, 30 s
   55 °C, 30 s
   72 °C, 5 min
4. Separate the nucleic acids in the reaction by electrophoresis using a 1% agarose gel, and harvest the mutagenized product.
5. Set up blunting kination reaction as follows and incubate at 37 °C for 10 min, then incubate at 70 °C for 10 min.

DNA fragment	1 pmol
10 × Blunting kination buffer	2 μl
Blunting kination enzyme mix	1 μl
ddH_2O	up to 20 μl

6. After the blunting kination reaction, put 0.25 pmol (5 μl) and 5 μl ligation solution I together, mix gently, and incubate at 16 °C for 1 h.
7. Add all the ligation reaction into 100 μl competent cells and transform to create clones of cells carrying the mutated cDNA construct.

ACKNOWLEDGMENTS

This work was supported by the National Key Basic Research Project Grant 2011CB910201 and 2013CB530601, the State Key Program of National Natural Science Foundation 31030048, and the National Natural Science Foundation Grant 81270954 (for X. L.); the department is supported by Shanghai Leading Academic Discipline Project B110 and 985 Project 985III-YFX0302.

REFERENCES

Aouadi, M., Laurent, K., Prot, M., Le Marchand-Brustel, Y., Binétruy, B., & Bost, F. (2006). Inhibition of p38MAPK increases adipogenesis from embryonic to adult stages. *Diabetes*, *55*, 281–289.

Bost, F., Aouadi, M., Caron, L., & Binetruy, B. (2005). The role of MAPKs in adipocyte differentiation and obesity. *Biochimie*, *87*, 51–56.

Camp, H. S., Tafuri, S. R., & Leff, T. (1999). c-Jun N-terminal kinase phosphorylates peroxisome proliferator-activated receptor-gamma1and negatively regulates its transcriptional activity. *Endocrinology*, *140*, 392–397.

Cha, H. C., Oak, N. R., Kang, S., Tran, T. A., Kobayashi, S., Chiang, S. H., et al. (2008). Phosphorylation of CCAAT/Enhancer-binding protein α regulates GLUT4 expression and glucose transport in adipocytes. *The Journal of Biological Chemistry*, *283*, 18002–180011.

Choi, J. H., Banks, A. S., Estall, J. L., Kajimura, S., Bostrom, P., Laznik, D., et al. (2010). Anti-diabetic drugs inhibit obesity-linked phosphorylation of PPARγ by Cdk5. *Nature*, *466*, 451–457.

Cohen, P. (2002). The origins of protein phosphorylation. *Nature Cell Biology*, *4*, E127–E130.

Dai, J., Jin, W. H., Sheng, Q. H., Shieh, C. H., Wu, J. R., & Zeng, R. (2007). Protein phosphorylation and expression profiling by Yin-yang multidimensional liquid chromatography (Yin-yang MDLC) mass spectrometry. *Journal of Proteome Research*, *6*, 250–262.

Elias, J. E., & Gygi, S. P. (2007). Target-decoy search strategy for increased confidence in large-scale protein identifications by mass spectrometry. *Nature Methods*, *4*, 207–214.

Engelman, J. A., Lisanti, M. P., & Scherer, P. E. (1998). Specific inhibitors of p38 mitogenactivated protein kinase block 3T3–L1 adipogenesis. *The Journal of Biological Chemistry*, *273*, 32111–32120.

Goodall, C. (1983). M-estimators of location: An outline of the theory. In D. C. Hoaglin, F. Mosteller, J. W. Tukey, D. C. Hoaglin, F. Mosteller, & J. W. Tukey (Eds.), *Understanding robust and explanatory data analysis* (pp. 339–403). New York: John Wiley & Sons, 2000.

Hu, E., Kim, J. B., Sarraf, P., & Spiegelman, B. M. (1996). Inhibition of adipogenesis through MAP kinase-mediated phosphorylation of PPAR gamma. *Science*, *274*, 2100–2103.

Kim, J. W., Tang, Q. Q., Li, X., & Lane, M. D. (2007). DNA binding and transcriptional activation by C/EBP beta: Effect of phosphorylation and S-S bond-induced dimerization. *Proceedings of the National Academy of Sciences of the United States of America*, *104*, 1800–1804.

Li, P., Fan, W. Q., Xu, J., Lu, M., Yamamoto, H., Auwerx, J., et al. (2011). Adipocyte NCoR knockout decreases PPARg phosphorylation and enhances PPARg activity and insulin sensitivity. *Cell*, *147*, 815–826.

Li, X., Kim, J. W., Gronborg, M., Lane, M. D., & Tang, Q. Q. (2007). Role of cdk2 in the sequential phosphorylation/activation of C/EBPβ during adipocyte differentiation. *Proceedings of the National Academy of Sciences of the United States of America*, *104*, 11597–11602.

Mann, M., Ong, S. E., Gronborg, M., Steen, H., Jensen, O. N., & Pandey, A. (2002). Analysis of protein phosphorylation using mass spectrometry: deciphering the phosphoproteome. *Trends In Biotechnology, 20*, 261-268.

Moreno-Navarrete, J. M., Ortega, F. J., Ricart, W., & Fernandez-Real, J. M. (2009). Lactoferrin increases ^{172Thr}AMPK phosphorylation and insulin-induced ^{p473Ser}AKT while impairing adipocyte differentiation. *International Journal of Obesity, 33*, 991–1000.

Park, S. K., Venable, J. D., Xu, T., & Yates, J. R. (2008). A quantitative analysis software tool for mass spectrometry-based proteomics. *Nature Methods, 5*, 319–322.

Rosen, E. D., Walkey, C. J., Puigserver, P., & Spiegelman, B. M. (2000). Transcriptional regulation of adipogenesis. *Genes & Development, 14*, 1293–1307.

Ross, S. E., Radomska, H. S., Wu, B., Zhang, P., Winnay, J. N., Bajnok, L., et al. (2004). Phosphorylation of C/EBP alpha inhibits granulopoiesis. *Molecular and Cell Biology, 24*, 675–686.

Shao, D., Rangwala, S. M., Bailey, S. T., et al. (1998). Interdomain communication regulating ligand binding by PPAR-gamma. *Nature, 396*, 377–380.

Tang, Q. Q., Gronborg, M., Huang, H., Kim, J. W., Otto, T. C., Pandey, A., et al. (2005). Sequential phosphorylation of CCAAT enhancer-binding protein β by MAPK and glycogen synthase kinase 3β is required for adipogenesis. *Proceedings of the National Academy of Sciences of the United States of America, 102*, 9766–9771.

Tong, J., Zhu, M. J., Underwood, K. R., Hess, B. W., Ford, S. P., & Du, M. (2008). AMP-activated protein kinase and adipogenesis in sheep fetal skeletal muscle and 3T3-L1 cells. *Journal of Animal Science, 86*, 1296–1305.

Wang, X. Z., & Ron, D. (1996). Stress-induced phosphorylation and activation of the transcription factor CHOP (GADD153) by p38 MAP Kinase. *Science, 272*, 1347–1349.

Wu, L., Cai, X., Dong, H., Jing, W., Huang, Y., Yang, X., et al. (2010). Serum regulates adipogenesis of mesenchymal stem cells via MEK/ERK-dependent PPAR expression and phosphorylation. *Journal of Cellular and Molecular Medicine, 14*, 922–932.

Wu, Y. B., Dai, J., Yang, X. L., Li, S. J., Zhao, S. L., Sheng, Q. H., et al. (2009). Concurrent quantification of proteome and phosphoproteome to reveal system-wide association of protein phosphorylation and gene expression. *Molecular & Cellular Proteomics, 8*, 2809–2826.

Zhang, Y. Y., Li, S. F., Qian, S. W., Zhang, Y. Y., Liu, Y., Tang, Q. Q., et al. (2012). Phosphorylation prevents C/EBPβ from the calpain-dependent degradation. *Biochemical and Biophysical Research Communications, 419*, 550–555.

AUTHOR INDEX

Note: Page numbers followed by "*f*" indicate figures and "*t*" indicate tables.

C

D

I

M

N

W

X

SUBJECT INDEX

Note: Page numbers followed by "*f*" indicate figures and "*t*" indicate tables.

B

F

G

H

M

N

O

R

S

Y

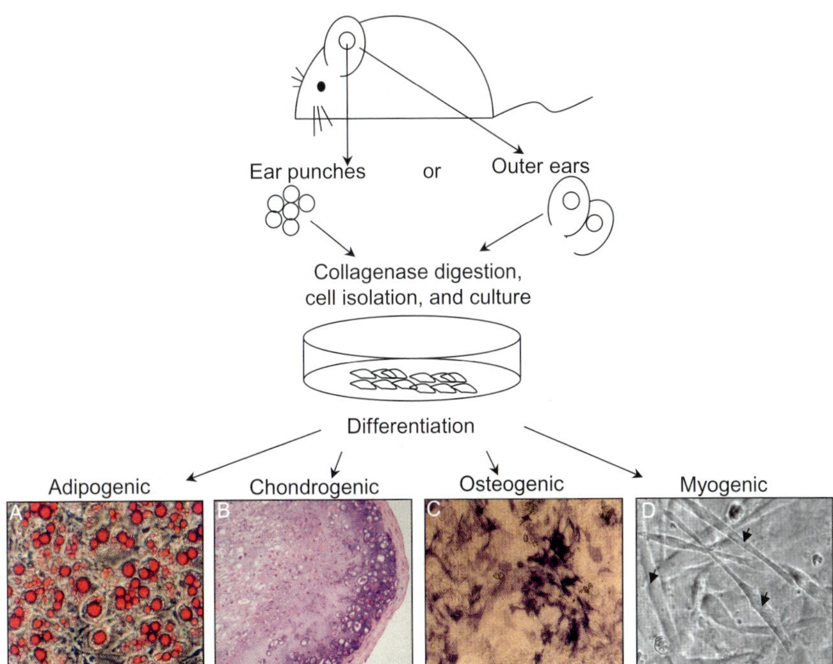

Barbara Gawronska-Kozak, Figure 1.1 EMSC isolation, culture, and adipogenic (A), chondrogenic (B), osteogenic (C), and myogenic (D) differentiation. (A) Confluent EMSC were stimulated with adipogenic differentiation medium. Adipogenesis is indicated by accumulation of lipid vacuoles that stain with Oil Red O; (B) EMSC in high-glucose micromass culture promoted the chondrogenic phenotype, hematoxylin, and eosin staining of paraffin-embedded pelleted cell sections; (C) EMSC differentiation along the osteoblast lineage was monitored with alkaline phosphatase staining; (D) phase-contrast photomicrograph of EMSC differentiated for 7 days in myogenic medium; arrows indicate spontaneously contracting myocytes.

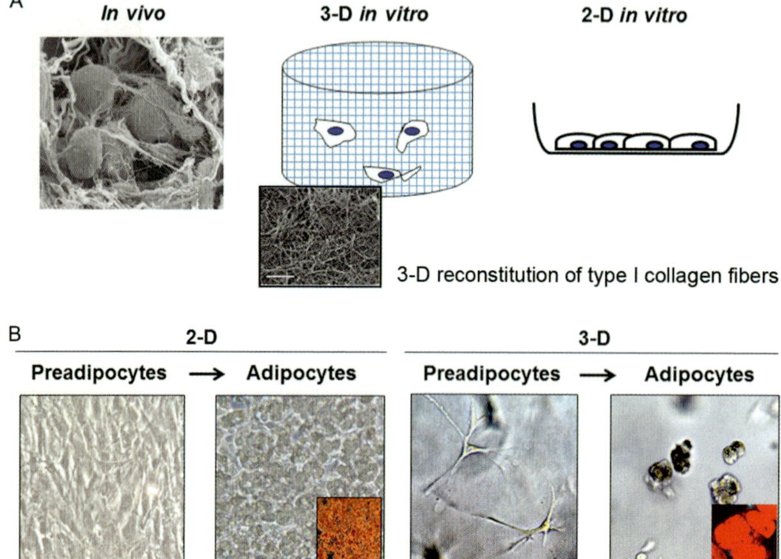

Tae-Hwa Chun and Mayumi Inoue, Figure 2.1 3-D *in vitro* adipogenesis. (A) 3-D but not 2-D *in vitro* adipogenesis recapitulates 3-D ECM environment found *in vivo*. *In vivo* adipocytes and preadipocytes are enwrapped by the network of collagen fibers (see scanning electron micrograph—SEM, left). By reconstituting rat tail-derived type I collagen as a gel (SEM, middle), we can recreate three-dimensional collagen-rich environment *in vitro*, which is as easily observable under a microscope as 2-D cultured cells. (B) Microscopic observation of 3-D adipogenesis. By adding adipogenic cocktail (insulin, IBMX, dexamethasone), vascular stromal cells (preadipocytes) isolated from white adipose tissue differentiate into adipocytes 2-D *in vitro*. Lipid accumulation is stained with Oil Red O (inset). Likewise, 3-D embedded preadipocytes differentiate into lipid-laden adipocytes 3-D *in vitro*. Lipid stained with Nile red (inset).

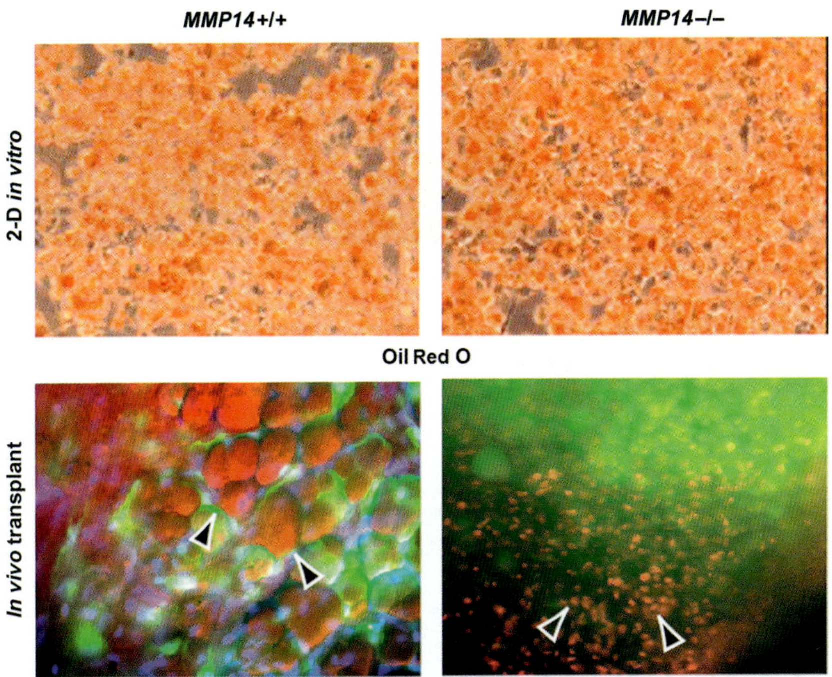

Tae-Hwa Chun and Mayumi Inoue, Figure 2.3 MMP14 as *in vivo*-specific cell-autonomous adipogenesis regulator. *MMP14* deficiency does not interfere with 2-D *in vitro* adipogenesis; however, the loss of *MMP14 in vivo* leads to aborted adipogenesis. Primary vascular stromal cells are differentiated 2-D *in vitro*, and lipids are stained with Oil Red O; GFP-labeled wild-type and *MMP14–/–* cells are transplanted subcutaneously into wild-type host, and lipid droplets are detected with Nile red. *Reprinted from Chun et al., 2006 with permission from Elsevier.*

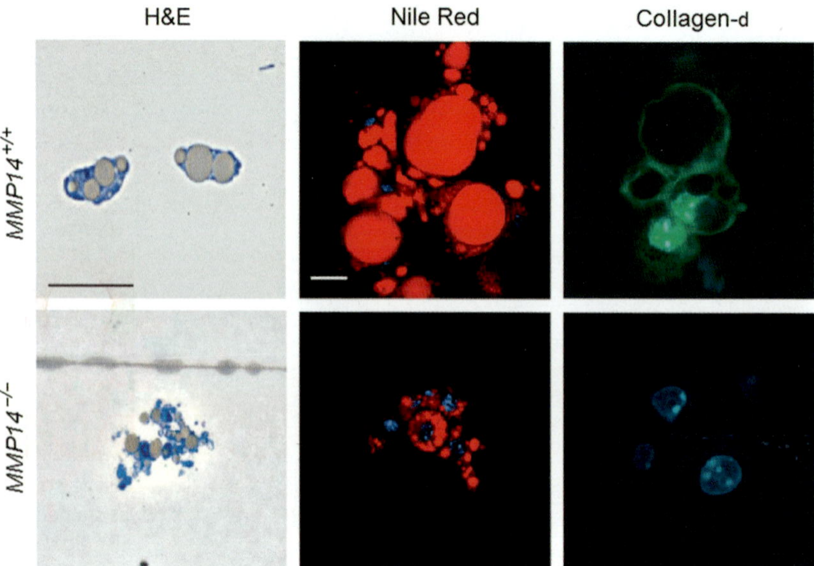

Tae-Hwa Chun and Mayumi Inoue, Figure 2.4 *MMP14*-dependent adipogenesis in 3-D collagen gels. Loss of *MMP14* impairs adipogenesis *in vitro* within 3-D collagen gels. MMP14 is essential for hypertrophic maturation of adipocytes, which is coupled with collagen degradation. Lipid droplets shown with Nile red (red), collagen degradation (green), and nucleus (blue). *Reprinted from Chun et al., 2006 with permission from Elsevier.*

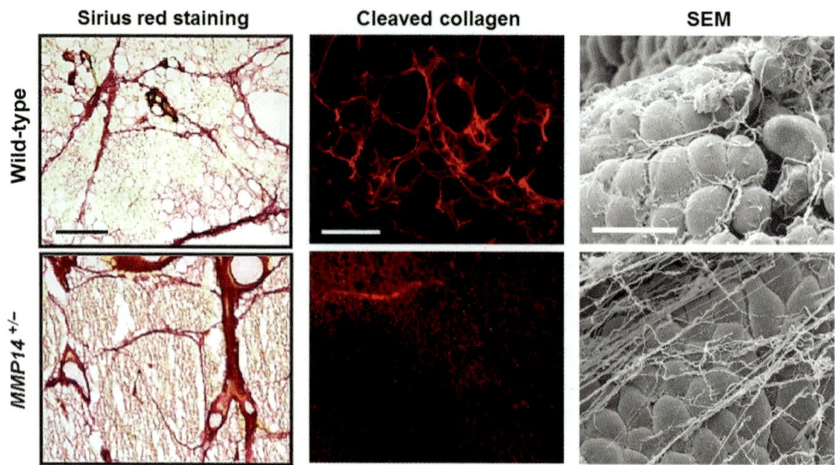

Tae-Hwa Chun and Mayumi Inoue, Figure 2.5 *MMP14*-dependent collagen turnover in high-fat diet-induced obesity. Cross-linked fibrillar collagen is detected with Sirius red staining. In wild-type mice, high-fat diet decreases the density of collagen fibers via MMP14-dependent collagen cleavage. Cleaved collagen products are detected with anticleaved collagen antibody. SEM confirms the disruption of collagen fibers upon HFD challenge. This diet-induced collagen turnover is markedly suppressed in MMP14 heterozygous mice. *Reprinted from Chun et al., 2010 with permission.*

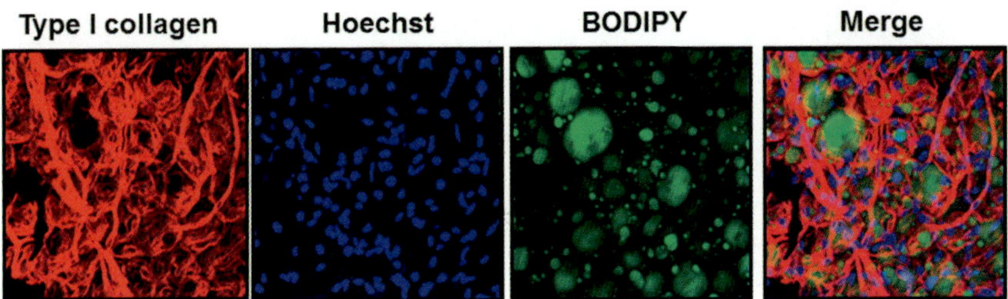

Tae-Hwa Chun and Mayumi Inoue, Figure 2.6 Whole-mount adipose tissue staining for type I collagen and lipids. A piece of mouse inguinal adipose tissue was stained with anti-type I collagen antibody (red), Hoechst 33342 (nuclei, blue), and BODIPY (lipid droplets, green). A merged image is shown on the far right. *Modified from Chun, 2012.*

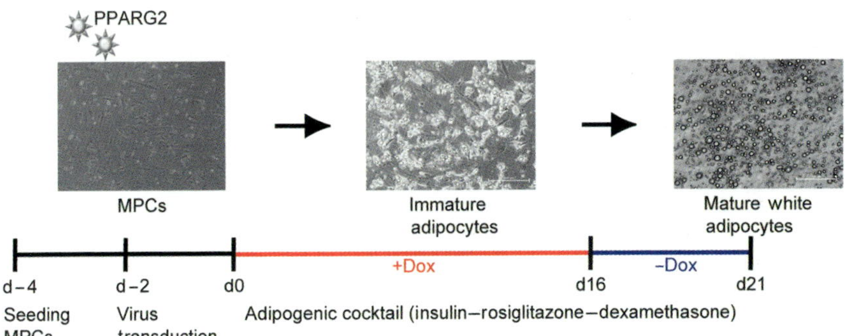

Youn-Kyoung Lee and Chad A. Cowan, Figure 3.2 Experimental scheme for the differentiation of hPSC into white adipocytes. MPCs were transduced with lentivirus constitutively expressing the lenti-rtTA M2 domain and PPARG2. Cells were cultured in adipogenic differentiation medium containing doxycycline for 16 days and maintained without doxycycline until 21 days. Bright-field images showing different stages during the differentiation of white adipocytes. Pictures were taken at day 10 for immature white adipocytes and at day 21 for mature white adipocytes, which exhibit a single large lipid droplet (×200 magnification).

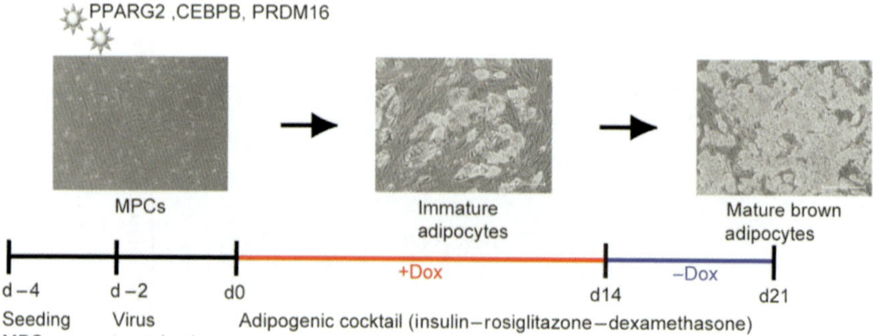

Youn-Kyoung Lee and Chad A. Cowan, Figure 3.3 Experimental scheme for differentiation of hPSC into brown adipocytes. MPCs were transduced with lentivirus constitutively expressing the lenti-rtTA M2 domain and PPARG2, CEBPB, and PRDM16. Cells were cultured in adipogenic differentiation medium containing doxycycline for 14 days and maintained without doxycycline until 21 days. Bright-field images showing different stages during the differentiation of brown adipocytes. Majority of mature brown adipocytes at day 21 exhibit multiple small lipid droplets (×200 magnification).

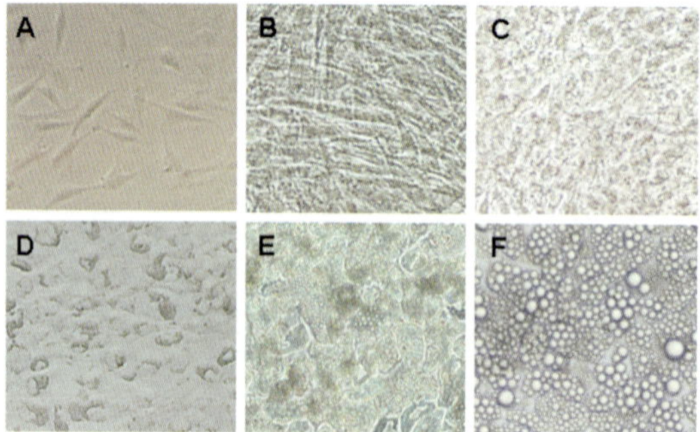

Mi-Jeong Lee and Susan K. Fried, Figure 4.1 Microscopic images of human adipose stromal cells at various stages. Microscopic images are taken at preconfluence (A), 2 days postconfluence (B), d3 (C), d7 (D), d14 (E), and d35 (F) of differentiation.

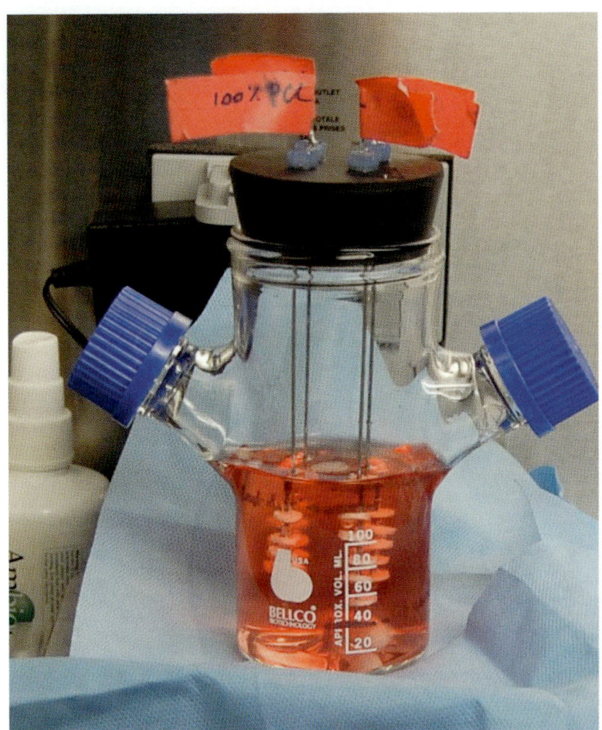

Ammar T. Qureshi *et al.*, Figure 5.1 Spinner flask assembly on magnetic hot plate/stirrer.

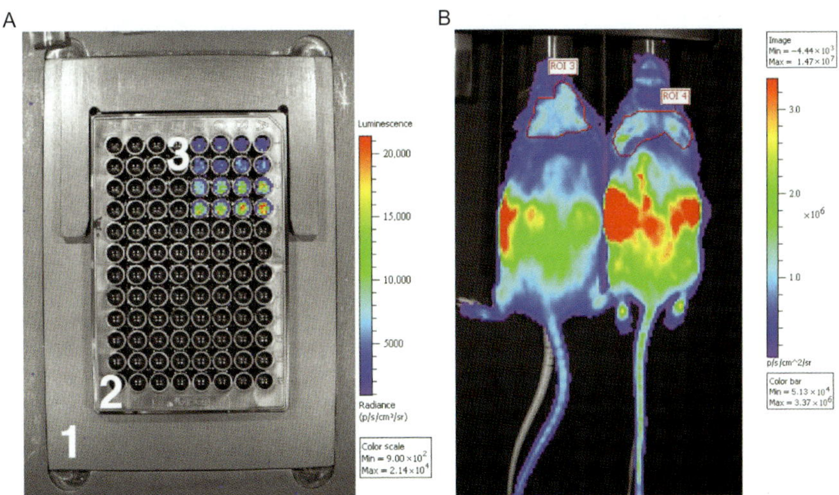

Elena Dubikovskaya *et al.*, Figure 7.4 Bioluminescent fatty acid uptake assays. (A) Using a plate adapter (1), a 96-well plate (2) containing mature 3T3-L1 adipocytes and undifferentiated fibroblasts (3) was placed into an IVIS Spectrum. Fatty acid uptake rates were monitored following the addition of 100 μl of 100 μ*M* BSA-bound 100 μ*M* FFA-SS-luc. (B) L2G85 mice injected with 100 μl of a 200 μ*M* solution intraperitoneally were imaged using autoexposure settings. Red regions of interest denote the BAT area in both animals.

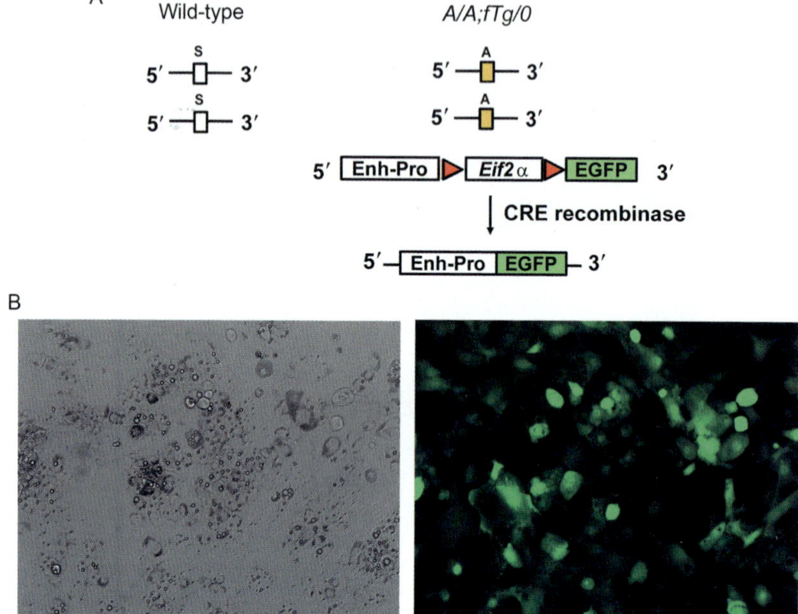

Jaeseok Han and Randal J. Kaufman, Figure 8.1 Deletion of *Eif2α* transgene by CRE expression. (A) Diagram illustrating the genotypes of wild type and *A/A;fTg/0*. *A/A;fTg/0* MEFs harbor two Ser51Ala mutant *Eif2α* alleles (*A/A*) as well as a floxed wild-type *Eif2α* transgene (*fTg*) driven by the CMV enhancer and chicken β-actin promoter. LoxP sequences (red arrowheads) allow excision of floxed wild-type *Eif2α* transgene and turn on EGFP expression. S indicates endogenous wild-type *Eif2α* allele and A indicates the Ser51Ala mutant allele of *Eif2α*. (B) Primary *A/A;fTg/0* MEFs are differentiated into mature adipocytes as described in the text. On day 10 after differentiation, adenoviruses expressing CRE recombinase gene under CMV promoter are infected at MOI 100. After deletion of the wild-type *Eif2α* transgene (*fTg*), EGFP expression is expressed. DIC (differential interference contrast, left panel) and GFP fluorescence (right panel) images are taken after 3 days of adenovirus infection. Magnification: 100×.

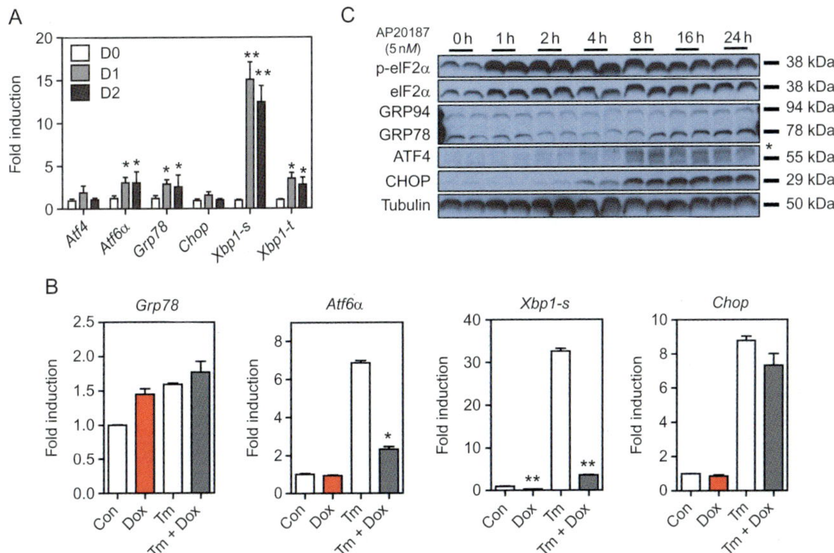

Jaeseok Han and Randal J. Kaufman, Figure 8.2 Inducible system for IRE1α and eIF2α phosphorylation. (A) Gene expression profile for 3T3-L1 cells expressing wild-type IRE1α under the control of a tetracycline-inducible promoter. Total RNAs were collected on day 0 (D0), day 1 (D1), and day 2 (D3) after doxycycline (5 μg/ml) treatment, followed by quantitative real-time PCR. Data are presented as means ± SEM of three independent experiments with triplicates. (B) Gene expression profile for stable 3T3-L1 cells expressing dominant-negative RNase mutant IRE1α (K907A) under the control of a tetracycline-inducible promoter in the absence or presence of Tm-induced ER stress. Data are presented as means ± SEM of three independent experiments with triplicates. (C) Protein expression profile in a stable cell line that expresses chimeric Fv2E-PERK. Cell lysates were collected at the indicated times after AP20187 treatment (5 n*M*) for Western blot analysis. * indicates nonspecific band. *From Han, Murthy, et al. (2013) with permission.*

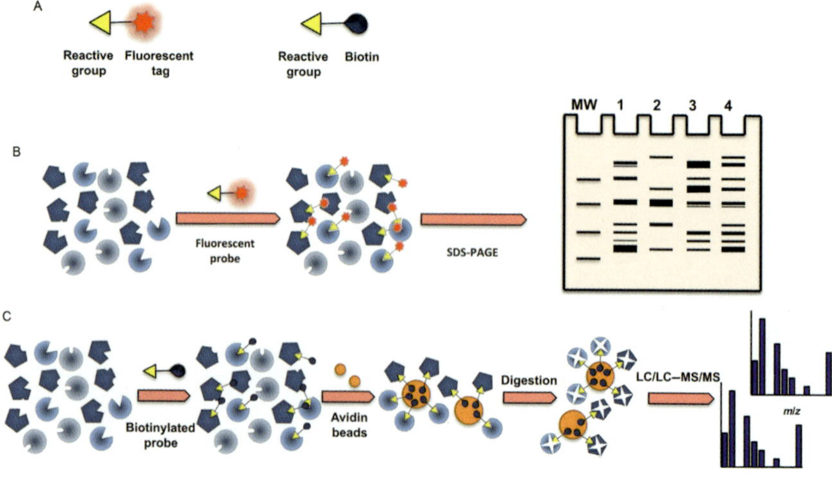

Andrea Galmozzi *et al.*, Figure 9.1 Basics of activity-based protein profiling (ABPP). (A) Schematic of ABPP probes showing their basic features: (1) a reactive head group that targets a specific enzyme class and (2) a reporter tag. (B) Gel-based ABPP. A fluorophosphonate-reactive group can be coupled to a fluorophore tag (e.g., rhodamine) to covalently label and detect active serine hydrolases by SDS-PAGE and in-gel fluorescence scanning. (C) ABPP–MudPIT. Biotin-conjugated ABPP probes can be used to label, enrich, and identify by mass spectrometry active serine hydrolases (SH).

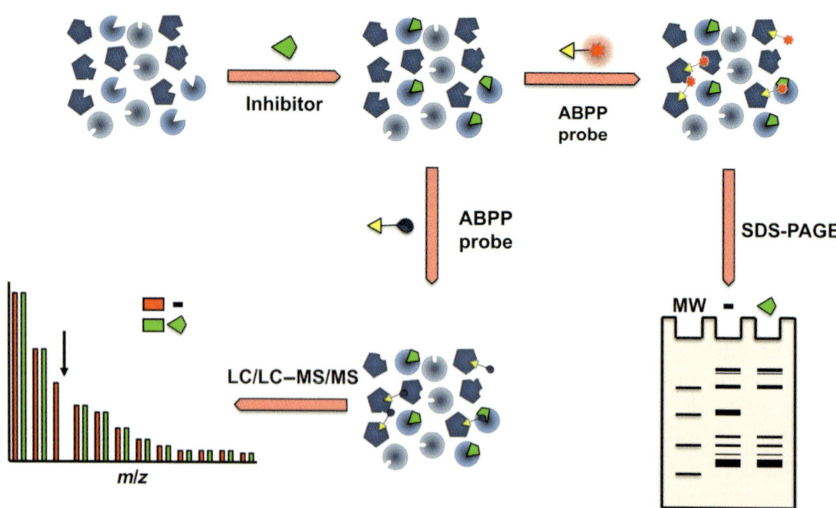

Andrea Galmozzi *et al.*, Figure 9.2 Competitive ABPP for enzyme and inhibitor discovery. In a competitive activity-based protein profiling experiment, cells, animals, or prepared proteomes are first treated with either an inhibitor or a vehicle and subsequently labeled with a broad ABPP probe. Gel-based ABPP or ABPP–MudPIT is then used to either visualize or identify active enzymes (i.e., labeled by the ABPP probe). An enzyme that is the target of an inhibitor will show reduced signals in the inhibitor-treated samples relative to vehicle controls.

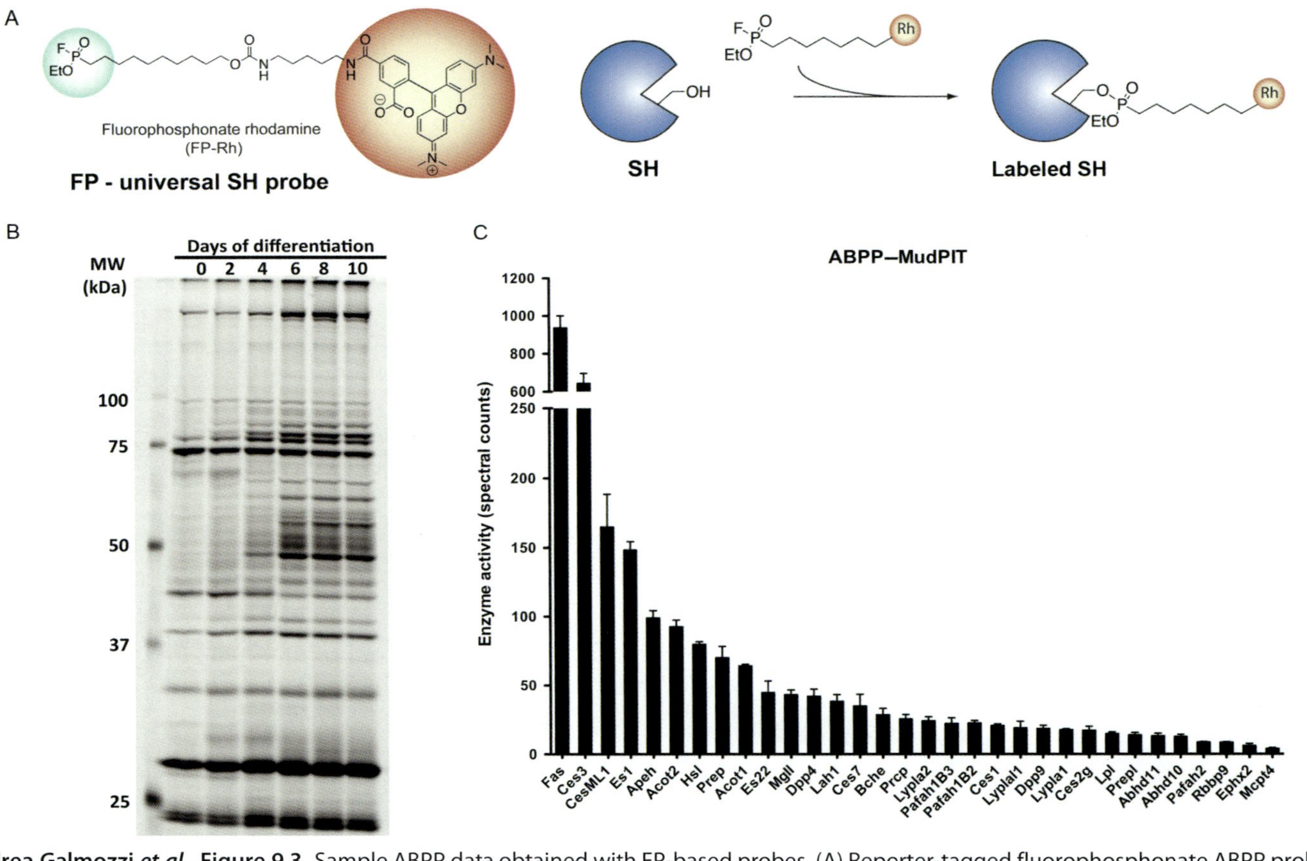

Andrea Galmozzi *et al.*, Figure 9.3 Sample ABPP data obtained with FP-based probes. (A) Reporter-tagged fluorophosphonate ABPP probes that react covalently only with the active form of serine hydrolases can be used to (B) profile the pattern of serine hydrolase activity by gel-based ABPP during differentiation of 10T1/2 adipocytes or (C) identify active serine hydrolases in mouse white adipose tissue using ABPP–MudPIT.

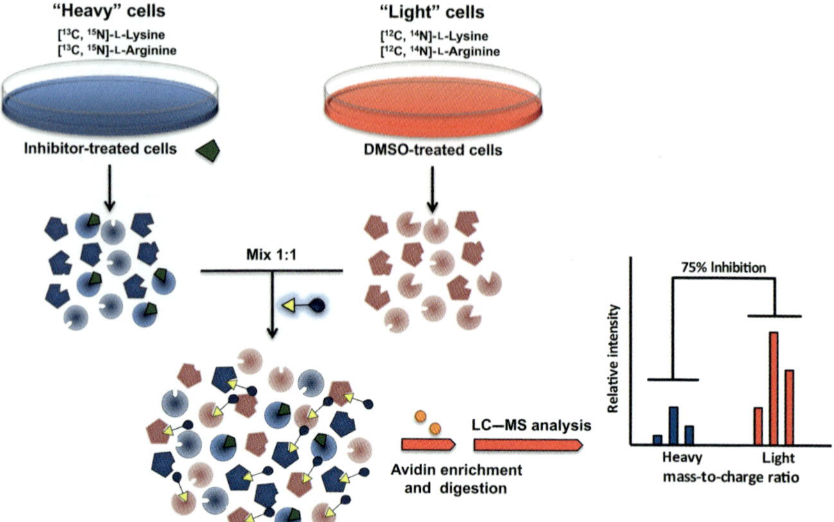

Andrea Galmozzi *et al.*, Figure 9.4 Competitive ABPP–SILAC to identify the target of enzyme inhibitors. Cells are grown in different media with isotopically stable amino acids. Proteomes prepared from compound-treated "heavy" and vehicle-treated "light" cells are labeled with a biotin-conjugated ABPP probe and mixed in a 1:1 ratio. Following avidin enrichment and on-bead digestion, samples are analyzed by mass spectrometry and representative peptides belonging to treated or control cells are identified by mass–charge shift.

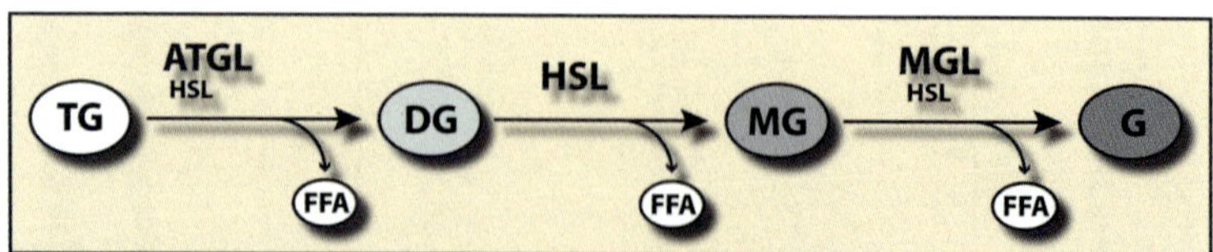

Martina Schweiger *et al.*, Figure 10.1 Coordinated breakdown of triglycerides in the course of lipolysis.

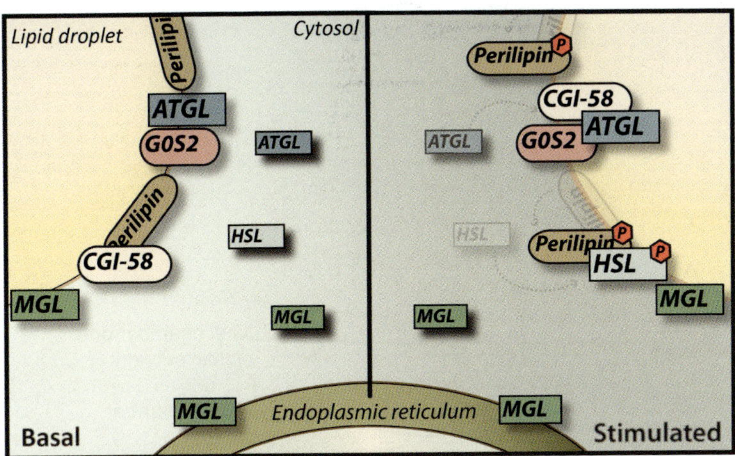

Martina Schweiger *et al.*, Figure 10.2 Intracellular localization of lipases and regulatory proteins in the basal/non-lipolytic-stimulated state and upon lipolytic stimulation. In the basal/non-lipolytic-stimulated state, CGI-58 is bound to perilipin-1 and unavailable for ATGL binding/activation. HSL resides in the cytosol. β-Adrenergic stimulation leads to the phosphorylation of HSL and perilipin-1, which causes HSL to translocate onto the LD. ATGL also translocates from the cytosol to the LD surface. In addition, phosphorylated perilipin-1 releases CGI-58, which is now available for ATGL binding and activation. MGL is localized on the LD, the cytosol, and the ER independent of the metabolic state of the cell.

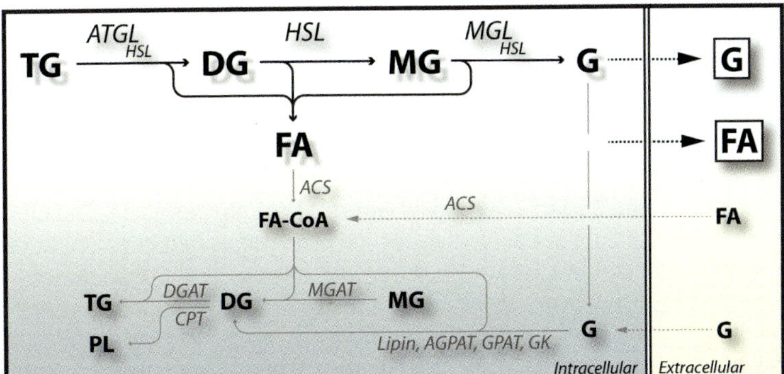

Martina Schweiger *et al.*, Figure 10.3 The fate of free fatty acids (FA) and glycerol (G) generated during intracellular lipolysis. FAs generated by the activity of ATGL, HSL, and MGL are either released into the circulation or activated by acyl-CoA synthetases (ACS) to FA-CoA. These FA-CoAs serve as building blocks for the synthesis of MG, DG, TG, and phospholipids (PL), by the activities of MGAT (monoacylglycerol-acyltransferase), DGAT (diacylglycerol-acyltransferase), or CPT (cholinephosphotransferase). Glycerol (G) is recycled by GK (glycerol kinase), GPAT (glycerolphosphate-acyltransferase), AGPAT (acylglycerolphosphate-acyltransferase), and lipin to DGs.

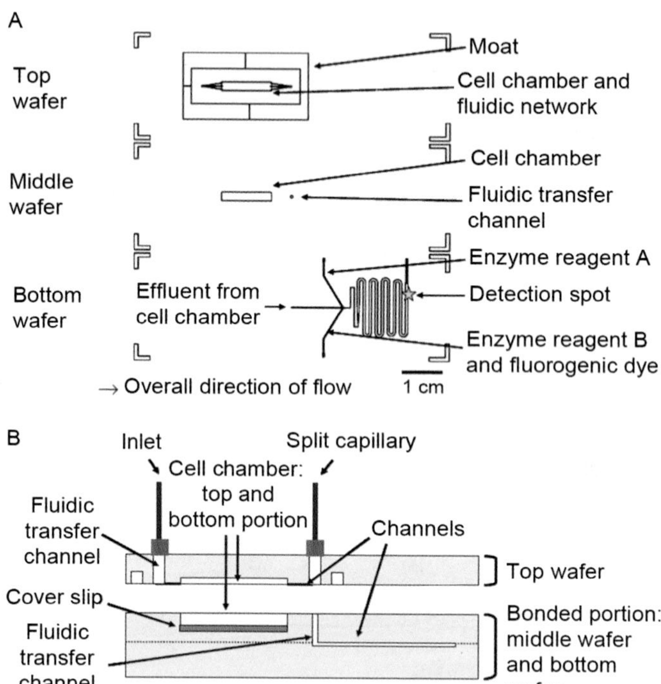

Colleen E. Dugan and Robert T. Kennedy, Figure 11.4 Chip design. (A) The multilayer device was comprised of three separately etched glass wafers that integrated a cell perfusion chamber and fluidic channels for online mixing of the fluorescence-based enzyme assay. (B) A side view of the cell chamber depicts the bonded and reversibly sealed portions of the device. *Adapted with permission from Clark et al. (2010). Copyright 2010 Springer.*

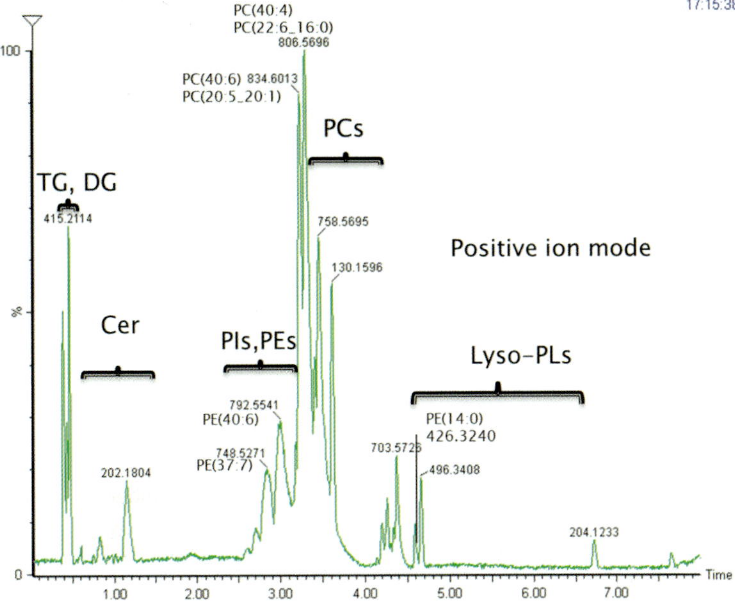

Lee D. Roberts *et al.*, Figure 12.3 Total ion chromatogram of the intact lipids detected in cardiac tissue using amide HILIC chromatography. Notice the different order of lipid classes separated compared with the reverse-phase method in Fig. 12.2.

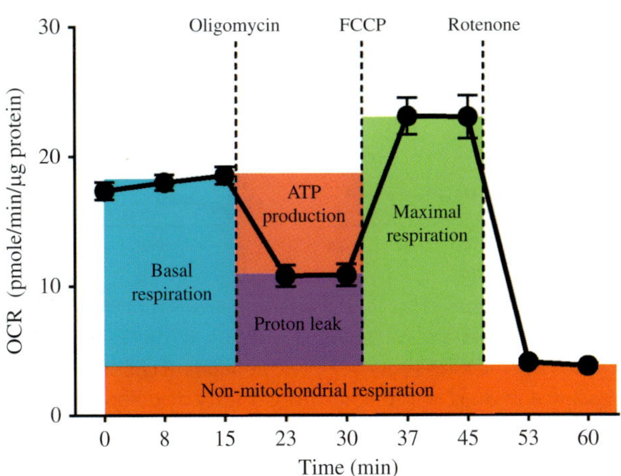

Anne Bugge *et al.*, Figure 13.3 Interpretation of an XF24 cell mitochondrial stress experiment investigating mitochondrial function in mouse primary inguinal adipocytes, represented as by Seahorse Bioscience (http://seahorsebio.com/products/consumables/kits/cell-mito-stress.php). Maximal respiratory capacity and the contribution of ATP producing and uncoupled respiration (proton leak) to the basal respiration are determined by observing how the OCR changes in response to drugs that modulate mitochondrial activity. Oligomycin inhibits the ATP synthase, FCCP disperses the proton gradient by permeabilizing the inner mitochondrial membrane, and rotenone abrogates the mitochondrial electron transport chain through the inhibition of complex I ($n=10$, SEM is denoted by vertical bars).

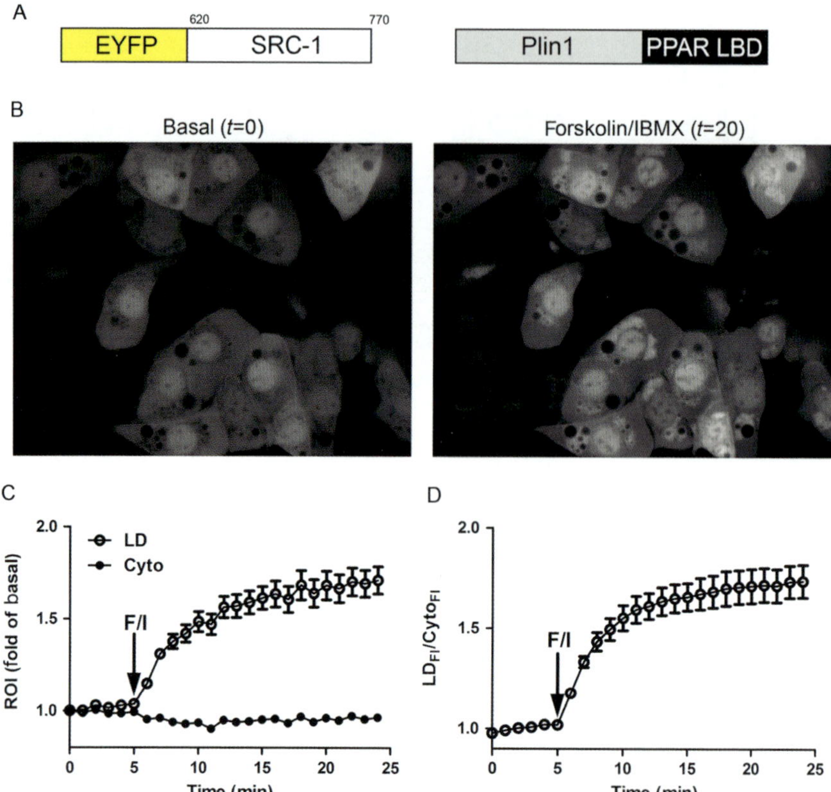

Emilio P. Mottillo *et al.*, Figure 15.2 Fluorescence sensor for imaging intracellular fatty acid production. (A) A schematic of the constructs used for the fluorescence fatty acid sensor (amino acids for SRC1 are shown). PLIN1, perilipin-1; LBD, ligand-binding domain. (B) EYFP fluorescence images of the lipid droplet fatty acid sensor stably expressed in BA cells. Basal image capture at $t=0$ and at $t=20$ min, 15 min after addition of forskolin/IMBX (F/I). (C) Quantification of the region of interest (ROI) from lipid droplet (LD) or cytoplasm (Cyto) and expressed as a fold of basal. (D) Data were further expressed as a ratio of the fluorescent intensity from the LD (LD_{FI}) over the fluorescence intensity from the cytoplasm ($Cyto_{FI}$).